ORGANIC CHEMISTRY SERIES
Series Editor: J E Baldwin, FRS

Volume 3

Total Synthesis of Natural Products: The 'Chiron' Approach

Related Pergamon Titles of Interest

Total Synthesis
of Natural Products:
The 'Chiron' Approach

STEPHEN HANESSIAN

Département de Chimie, Université de Montréal, Québec, Canada

PERGAMON PRESS

OXFORD · NEW YORK · TORONTO · SYDNEY · PARIS · FRANKFURT

MT

U.K.	Pergamon Press Ltd., Headington Hill Hall, Oxford OX3 0BW, England
U.S.A.	Pergamon Press Inc., Maxwell House, Fairview Park, Elmsford, New York 10523, U.S.A.
CANADA	Pergamon Press Canada Ltd., Suite 104, 150 Consumers Road, Willowdale, Ontario M2J 1P9, Canada
AUSTRALIA	Pergamon Press (Aust.) Pty. Ltd., P.O. Box 544, Potts Point, N.S.W. 2011, Australia
FRANCE	Pergamon Press SARL, 24 rue des Ecoles, 75240 Paris, Cedex 05, France
FEDERAL REPUBLIC OF GERMANY	Pergamon Press GmbH, Hammerweg 6, D-6242 Kronberg-Taunus, Federal Republic of Germany

First edition 1983

Reprinted 1984 (twice)

Library of Congress Cataloging in Publication Data

Hanessian, Stephen.
Total synthesis of natural products: the ''Chiron''
approach.
(Organic chemistry series ; v. 3)
Includes bibliographies.
1. Natural products. 2. Chemistry, Organic—Synthesis.
I. Title. II. Series.
QD415.H26   1983   547.7'0459   83-19307

British Library Cataloguing in Publication Data

Hanessian, Stephen
Total synthesis of natural products: the 'Chiron'
approach—(Organic chemistry series; v. 3)
1. Chemistry, Organic—Synthesis
I. Title   II. Series
547'.2   QD262
ISBN 0-08-029247-X (Hardcover)
ISBN 0-08-030715-9 (Flexicover)

In order to make this volume available as economically and as rapidly as possible the author's typescript has been reproduced in its original form. This method unfortunately has its typographical limitations but it is hoped that they in no way distract the reader.

Printed and bound in Great Britain by William Clowes Limited, Beccles and London

To the memory of my Father

If there is one way better than another,
it is the way of Nature

Aristotle

FOREWORD

The spectacular advances in synthetic organic chemistry of
the last decade, during which time the achievement of total
synthesis of functionally and stereochemically complex structures
has become almost routine, has also placed an added requirement for
the chemist, ie the synthesis of optically pure molecules. One
approach to this objective has been the employment of chiral
starting materials or chiral synthons, usually readily available
natural products, particularly of the carbohydrate class. In this
book Professor Stephen Hanessian, one of the most successful
exponents of this art, describes how chemists can easily utilise
this important procedure. His treatment of methodology, strategy
and the available starting materials themselves, will find wide use
among chemists who are involved in synthetic problems of all types
in the field of organic chemistry.

J E Baldwin, FRS

Oxford

ACKNOWLEDGMENTS

I am indebted to a large group of coworkers who proofread portions of the original manuscript as well as the references, namely, Dr. P.C. Tyler, Dr. T. Sugawara, Dr. A. Ugolini, D. Delorme, J. Kloss, B. Vanasse, G. Caron, D. Dubé, D. Desilets, S. Beaudoin, S. Léger, and A. Glamyan. The major portion of the manuscript was patiently typed by Ms. D. Coupal with occasional help from Mrs. C. Vincent-Major and Ms. C. Potvin. The extensive art work was superbly done by Ms. L. St-Onge. Final proof-reading and pasting of schemes were done by Lucine Hanessian, while sister Nora and brother Greg helped with alphabetizing the index and general cosmetics of the final product. To all those and many others, I extend my hearty thanks and appreciation.

In any undertaking of this magnitude, particularly in view of the large number of structural formulae, errors will be inevitable. I do hope, however, that serious as these may be, they will not detract from the original theme of the book or lessen its impact.

PREFACE

Natural product synthesis is hardly a novelty on the synthetic chemist's menu nowadays. Some may in fact consider it to be the appetizer, for under the guise of total synthesis, much innovation and creativity in methodology has come forth. For whatever the reason, natural product synthesis continues to inject spark and excitement into organic chemistry.The elements of design and strategy, though left very much to the explorer's choice, present boundless challenges that may be surpassed only by the exhilaration of the final conquest of the target molecule - maybe even a vintage bottle of champagne.

Several years ago, during a visit to a major east coast university in the US, where I discussed the concept of synthetic design with "chiral templates" derived from small naturally-occurring starting materials, I was urged by a respected colleague and an authority in natural product synthesis, to write on the subject. What started out as an article, mushroomed into a "how-to" monograph. My intention was to engage in a visual dialogue with molecules, no matter how complex, using a language common to both of us, namely, stereochemistry. This led to disconnections of strategic bonds with minimum perturbation of chiral centers to generate optically active or preferably enantiomerically pure synthons. We have christened these 'chirons', which in turn can be obtained from "chiral templates" such as amino acids, carbohydrates, hydroxy acids and terpenes _ in other words, Nature's gifts to the synthetic chemist. The main thrust of this book deals with strategy in synthetic design using 'chirons' derived from carbohydrates, although the concept can be expanded to other chiral starting materials as well. It was thought that the most effective way to indoctrinate the reader in the 'chiron' approach, was to make generous use of flow-charts and schemes. The inherent aversion to stereochemistry, particularly as it relates to carbohydrates, has been hopefully made palatable by a short discussion of this class of naturally-occurring building blocks in modern "organic" terms, emphasizing not their mutual stereochemical relationships, but their versatile carbon framework in terms of patterns, shapes and other topological features.

Sixteen chapters describe strategies used in the synthesis of three basic categories of target structures depending on how easily one can "see" a carbohydrate in its intricate molecular architecture. Each synthesis is presented in a chart made of two parts. First, a retro-synthetic analysis that uncovers 'chirons' and the appropriate starting carbohydrate, as seen through the eyes of the author (with some surprises!). Second, the synthesis proper, in which important transformations together with the reagents are shown. Thus, in addition to aspects of design and strategy, a large volume of synthetic transformations, consequently a compendium of reagents and conditions can be found, which may be helpful for personal reference or for pedagogic purposes. The discussion part analyzes the strategies in the schemes and highlights certain features of the syntheses. For each target, at least one major background reference is given, mostly from the primary literature (isolation, structure determination, etc.). Syntheses by other strategies are given at the end of every section, thus providing the reader with a primary source of references for a large collection of natural products. In all, some 400 references are cited, many with multiple citations, for over 100 syntheses, most of which are schematically analyzed in more than 150 flow-charts.

It was not the intention to provide a comprehensive account dealing with the subject in question. The examples chosen are therefore representative cases, although I personally feel that coverage has been good. In the event of any omissions or citations with less emphasis than deserved, I ask for your indulgence and tolerance. On the positive side, an effort was made to stay up-to-date until well into the final typing stage. References covering the first-half of 1983 were included whenever possible, but new schemes could not be drawn for these. I therefore hope that in spite of some shortcomings, the book will be not only a source of information, but also a stimulus for more creativity and invention in the art of natural product synthesis.

Stephen Hanessian
Montréal, July 12, 1983

CONTENTS

ABBREVIATIONS

Ac	acetyl
AIBN	azobisisobutyronitrile
Benz.	benzene
Bn	benzyl
Bu	butyl
Bz	benzoyl
Cbz	carbobenzyloxy
CSA	camphorsulfonic acid
m-CPBA	meta-chloroperbenzoic acid
DCC	N,N-dicyclohexylcarbodiimide
DEAD	diethylazodicarboxylate
DIBAL	diisobutylaluminum hydride
DMAP	4-dimethylaminopyridine
DME	dimethoxyethane
DMF	N,N-dimethylformamide
DMSO	dimethylsulfoxide
Et	ethyl
EDAC	ethyldiethylaminopropyl carbodiimide.HCl
HMPA	hexamethylphosphoric triamide
LAH	lithium aluminum hydride
LDA	lithium diisopropylamide
Me	methyl
MEM	methoxyethoxymethyl
Mol.	molecular
MOM	methoxymethyl
Ms	methanesulfonyl
NBS	N-bromosuccinimide
PCC	pyridinium chlorochromate
Ph	phenyl
TCC	trichloroethoxycarbonyl
TFA	trifluoroacetic acid
TFAA	trifluoroacetic anhydride
THF	tetrahydrofuran
TMEDA	tetramethylethylenediamine
TMS	trimethylsilyl
Tr	trityl
Ts	p-toluenesulfonyl

CHAPTER ONE

INTRODUCTION

The practice of organic synthesis,[1] particularly as it relates to natural products,[2] continues to flourish and expand in a relentless manner. New challenges of conquering increasingly complex synthetic targets seem to push the limitless horizons of achievement even further. A cursory look at the conceptual basis[1,3,4] of many completed syntheses, will partly explain why modern synthetic organic chemistry has been referred to as a form of "art" and "architecture", matching the skill and creative beauty traditionally associated with these activities. In a conceptual sense, organic synthesis, particularly in the field of natural products, is very much a form of art, and the synthetic chemist can indeed have the attributes of an artist and a molecular architect as well as being a logician, strategist and a creative explorer.[4] The combination of these traits, put in the context of the microcosmic yet attainable realm of atoms and molecules, has produced our present day masterpieces in organic synthesis. The synthetic organic chemist has marvelled and envied Nature's tools in producing the most awesome array of molecular structures with such efficacy, and stereochemical precision, in a time span that defies human reference terms. On the other hand, if there is an area where the chemist can boast as qualitatively "coming close" to a natural biosynthetic process, it is in the modest attempts to control regiochemistry and stereochemistry in a given transformation. Unfortunately, when a given reaction is not as regio- or stereoselective as desired, the outcome cannot be concealed since a written account is given. Criticism usually ensues. Nature however, can disguise such a flaw by a variety of processes. Therein lies a

1

"weakness" on which the chemist can capitalize by producing the other
isomer, incomparably slower on a time scale, but perhaps surer in
outcome.

Traditionally, optically pure targets have been obtained by
techniques of resolution at some stage in the synthetic scheme.
Occasionally, optically active starting materials have also been used
whenever available, which has obviated, totally or partially, the need
to resolve "en route" to the target, or once there. With the major
advances in instrumental techniques, along with the vast number of
synthetic reagents and other operational and technical amenities at our
disposal, it appears that the time is opportune to devise syntheses that
utilize readily available and operationally versatile optically-active
starting materials. Some of the basic considerations associated with
this type of strategy are summarized below.

TARGET	CHIRAL PRECURSOR
1. Functional group interrelations	1. Functional groups present
2. Carbon framework (patterns & shapes)	2. Flexibility of carbon framework
3. Stereochemistry	3. Stereochemical versatility
4. Topology	4. Availability and cost

The purpose of this monograph is to illustrate and discuss how carbohydrates can be utilized as starting materials in the synthesis of
natural products of varied structural complexity, and of organic
compounds containing predisposed centers of asymmetry. While it must be
appreciated that no single class of organic compounds, including carbohydrates, will become general, all-purpose building blocks in organic
synthesis, we hope to demonstrate that this little appreciated class of
naturally-occurring organic molecules can at times, come surprisingly
close.

Within the past few years several reviews have appeared dealing
with the general subject of using chiral starting materials in organic
synthesis. The use of readily available optically active hydroxy acids
such as (+) or (-)-tartaric acid among others have been duly

illustrated,[5] as well as the possibility of microbiological and
enzymatic transformations leading to optically active, synthetically
useful building blocks such as amino acids, hydroxy acids, etc.[6] The
use of carbohydrates as chiral building blocks has also been reviewed in
general,[7-9] and an account of research activities representing progress
in our laboratory covering a period up to 1978 has been given.[10] A
recent short review on synthesis with chiral precursors covers the
general area by a classification according to type of starting
material.[11] There also exist monographs and reviews on asymmetric
synthesis[12] in which the formation of enantiomerically enriched products
in reactions involving additions to carbonyl compounds, and other
compounds containing sp^2 carbon atoms, using a chiral reagent or in a
chiral environment is discussed. Superbly documented reviews discussing
recent strategies in stereocontrol in acyclic systems and aldol
methodology are available.[13]

In preparing this article, an effort was made to bring a large
collection of contributions in an area under one theme. Because of our
long-standing involvement in related research activities, the tendency
to use a personalized approach could not be avoided. It is the
intention therefore to discuss the general strategy of organic synthesis
with carbohydrates, based on elements of *design*, *discovery*, and *execution*.
A brief overview of basic concepts will give the reader a general
background to appreciate and hopefully adopt the carbohydrate route
whenever applicable. It will not elude the reader, that the conceptual
basis of this approach to the synthesis of enantiomerically pure organic
compounds has much broader applications. Although specific examples are
not schematically illustrated, it is evident that the strategy can be
extended to other chiral starting materials such as amino acids, hydroxy
acids and terpenes.

REFERENCES

1. See for example, R.B. Woodward in, Perspectives in Organic
   Chemistry, A.R. Todd, ed., Interscience, p. 155, 1956. R.E.
   Ireland, Organic Synthesis, Prentice Hall, Inc., Englewood Cliffs,
   New Jersey, 1969; I. Fleming, Selected Organic Syntheses, A
   Guidebook for Organic Chemists, John Wiley & Sons, Inc., New York,
   1972; S. Turner, The Design of Organic Synthesis, Elsevier
   Scientific Publishing Co., Amsterdam, 1976; S. Warren, Designing

Organic Syntheses, A Programmed Introduction to the Synthon
Approach, John Wiley & Sons, Inc., New York, 1978.

2. See for example, Natural Products Chemistry, vol. 1,2, K. Nakanishi,
    T.Goto, S. Ito, S. Natori and S. Nozoe, eds. Academic Press Inc. New
    York, N.Y., 1974; The Total Synthesis of Natural Products, Vol. 1-4,
    J. ApSimon, ed., Wiley-Interscience, New York, N.Y., 1973-81; T.K.
    Devon and A.I. Scott, Handbook of Naturally Occurring Compounds,
    Vol.1, Academic Press, Inc., New York, N.Y., 1975.

3. For a background in the logistics of synthetic design, see, J.B.
    Hendrickson, E.Braun-Keller and G.A. Toczko, Tetrahedron, 37, 359
    (1981); see also J.B. Hendrickson, Top. Curr. Chem., 62, 49 (1976).
    J.S. Bindra and R. Bindra, Creativity in Organic Synthesis, vol. 1,
    Academic Press Inc., New York, N.Y., 1975.

4. E.J. Corey, Pure Appl. Chem., 14, 30 (1967).

5. D. Seebach, in Modern Synthetic Methods, R. Scheffold, ed., Otto
    Salle Verlag, Frankfurt am Main, Germany, 1980, p. 91.

6. A. Fischli, in Modern Synthetic Methods, R. Scheffold ed., Otto
    Salle Verlag, Frankfurt am Main, Germany, 1980 p. 269; see also J.B.
    Jones in, Applications of Biochemical Systems in Organic Chemistry,
    Part 1, J.B. Jones, C.J. Sih and D. Perlman, eds. Wiley-Inter-
    science, New York, 1976, p. 107; see also, L.R. Smith and H.J.
    Williams, J. Chem. Ed., 696 (1979).

7. B. Fraser-Reid and R.C. Anderson, Fortschr. Chem. Org. Naturst. 39,
    1 (1980); B. Fraser-Reid, Acc. Chem. Res., 8, 192 (1975).

8. T.D. Inch, Advan. Carbohydr. Chem. Biochem., 27, 191 (1972).

9. A. Vasella, in Modern Synthetic Methods, R. Scheffold, ed., Otto
    Salle Verlag, Frankfurt am Main, Germany, 1980, p. 173.

10. S. Hanessian, Acc. Chem. Res., 12, 159 (1979).

11. W.A. Szabo and H.T. Lee, Aldrichimica Acta, 13, 13 (1980); see also
    D. Seebach and H.-O. Kalinowski, Nachr. Chem. Techn. Lab., 24, 415
    (1976).

12. See for example J.D. Morrison and H.S. Mosher, Asymmetric Organic
    Reactions, Prentice Hall, Englewood Cliffs, New Jersey, 1971; see
    also H.B. Kagan and J.C., Fiaud, Topics in Stereochemistry, 10, 175
    (1978); D.Valentine, Jr., and J.W. Scott, Synthesis, 329 (1978);
    W. Klyne and J. Buckingham, Atlas of Stereochemistry, vol. 1, 2,
    Chapman and Hall, London, 1974, 1978; Y. Izumi and A. Tai, Stereo-
    differentiating Reactions, Academic Press Inc., New York, N.Y.,
    1977, etc., D. Seebach and V. Prelog, Angew. Chem. Int. Ed. Engl.,

21, 654 (1982); K. Drauz, A. Kleeman and J. Martens, Angew. Chem.
Int. Ed. Engl., 21, 584 (1982).

13. For some excellent recent reviews, see C.H. Heathcock, in
Comprehensive Carbanion Chemistry, Vol.II, T. Durst and
E. Buncel, eds., Elsevier, Amsterdam, 1981; D.A. Evans, J.V. Nelson
and T.R. Taber, Topics in Stereochemistry., 13, 1(1982); R.W.
Hoffmann, Angew. Chem., Int. Ed. Engl. 21, 555(1982); S. Masamune
and W. Choy, Aldrichimica Acta, 15, 47(1982); T. Mukaiyama, Org.
React., 28, 203 (1982); C.H. Heathcock, Science, 214, 395(1981).

PART ONE

BASIC CONCEPTS

Chiral Precursors

CHAPTER TWO

CARBOHYDRATES, AMINO ACIDS, HYDROXY ACIDS AND TERPENES

Over the years, carbohydrates have been recognized as naturally-occurring organic compounds endowed with a wealth of stereochemical attributes.[1] The thrust of research in this area has been inspired by biochemical events and the advent of antibiotics has fostered an accelerated effort in synthesis and chemical modification of component sugar units.[2,3] While the inherent traits have not changed over the years, attitudes have, and quite dramatically so. The organic chemist has thus discovered carbohydrates once again, hopefully to maintain a long-lasting and rewarding rapport in generations to come. The reasons are evident, for in the carbohydrates we find a large number of attributes that fulfill many requirements sought by the organic chemist in the quest of conquering enantiomerically pure synthetic targets. Briefly, carbohydrates are a relatively cheap and replenishable source of chiral carbon compounds, available in a variety of cyclic and acyclic forms, chain lengths, and oxidation or reduction states. They are endowed with a plethora of functional, stereochemical and conformational features which lend themselves to chemical exploitation. These very features also ensure a measure of regio- and stereocontrol in bond-forming reactions that are not easily matched by other classes of organic compounds. Some of these points as well as aspects of carbohydrate structures which are not shared by other chiral precursors are summarized below. With carbohydrates as starting materials, one has the option of using a cyclic or acyclic carbon atom framework consisting of 3-7 carbon atoms, to modify either extremity, and to create or destroy asymmetric centers at will by chemical manipulation of existing

CARBOHYDRATES

CARBON FRAMEWORK	ASYMMETRIC CENTERS
acyclic	1-5 (or 6)
cyclic	(includes anomeric center)
combination	
3-7 carbon atoms	SEQUENTIAL FUNCTIONALITY
	α-hydroxy aldehyde, etc.
SENSE OF CHIRALITY	α-amino aldehyde, etc.
	polyol, amino alcohol, etc.
2^n permutations	
generally D	

IN ADDITION:

a. variable chain length   c. topology and conformational
                                       bias
b. stereochemical duality   d. stereoelectronic effects

functionality by chain-extension, chain-branching or by cleavage. This
provides the synthetic chemist with room to manoeuver chemically and
stereochemically with obvious operational and practical advantages.
Chiral, readily available starting materials such as terpenes,[4] amino
acids,[5] hydroxy acids,[6] have their unique place in the practice of the
art of synthetic design, particularly when one or two asymmetric centers
are sought. Microbiological transformations[7] can also be very useful in
this regard. The salient features of these chiral building blocks as
well as selected examples of their utility in natural product synthesis
are summarized below.

AMINO ACIDS

CARBON FRAMEWORK	ASYMMETRIC CENTERS
generally acyclic	generally 1 or 2
(except proline, etc)	
3-6 carbon atoms	
SENSE OF CHIRALITY	SEQUENTIAL FUNCTIONALITY
generally L (natural)	α-amino acid
	α-amino β-substituted acid
	etc.

Some examples of the utilization of amino acids in natural product synthesis

L-cysteine   BIOTIN[8]   CEPHALOSPORIN[9]

(R)-glutamic acid   (S)-SULCATOL[10]

HYDROXY ACIDS

CARBON FRAMEWORK

acyclic
3-4 carbon atoms

SENSE OF CHIRALITY

R or S combinations

ASYMMETRIC CENTERS

1 or 2

SEQUENTIAL FUNCTIONALITY

α-hydroxy acid
α,β-dihydroxy acid, etc.

(S)-lactic acid   NONACTIN[11]

(S,S)-tartaric acid   DISPARLURE[12]

Some examples of the utilization of hydroxy acids in natural product synthesis

CHAPTER TWO

TERPENES

CARBON FRAMEWORK	ASYMMETRIC CENTERS
acyclic cyclic	generally 1 or 2

SENSE OF CHIRALITY	SEQUENTIAL FUNCTIONALITY
R or S	enone α-substituted ketone

Picrotoxinin [13] ⟹ (−)-Carvone

Steroid synthesis [14] ⟹ (−)-Camphor

(+)-Citronellol

Cytochalasin B [15] ⟹ L-phenylalanine   (S)-malic acid

Some examples of the utilization of terpenes in natural product synthesis

In dealing with targets containing multiple centers of chirality and functionality, the carbohydrate-derived 'precursor' presents definite stereochemical and operational advantages, but by no means constitutes the ideal approach. The question is often asked concerning the wisdom and practicality in terms of cost and effort, of using a carbohydrate to generate a chiral unit or a target as a whole, in which only one or two asymmetric centers are located. Obviously, an unbiased judgement should prevail in comparing several approaches and considering various parameters. If the carbohydrate route involves a cheap starting material, and reasonably efficient but few steps, then it could be quite

competitive. Destroying three asymmetric centers in a six-carbon sugar
derivative in order to preserve one center, may extend the approach to
its limits. Another practical problem often arises with the unavoidable
steps involving temporary protection and eventual deprotection of
functional groups when working with carbohydrates. This can be
tolerated when crystalline derivatives are involved as is often the
case, or when separation by chromatography is possible. In dealing with
targets containing carbon chains exceeding four atoms however, the
carbohydrate-derived precursor may present some advantages, particularly
if a D- form is required. As in any other strategy involving organic
synthesis, combining practicality with innovation seems to present the
best compromise. Access to some natural products from a carbohydrate on
the one hand and another chiral precursor on the other is illustrated
below. Note that the carbohydrate precursor provides a greater segment
of the carbon skeleton of the target molecule, compared to other
precursors.

(R,R)-Tartaric acid [16]   Anisomycin   D-glucose [17]

(S,S)-Tartaric acid [18]   6-epi-LTA4 intermediate   2-deoxy-D-erythro-pentose [19]

(S)-Malic acid [20]   PGF2α   D-Glycero-D-gulo-heptose [21]

L-Cysteine [8]   Biotin   D-Glucose [22]

Access to natural products from amino acids hydroxy acids and carbohydrates

A. Stereochemistry, stereoelectronic effects and conformation in
    carbohydrates.

    The simplest definition of a carbohydrate is that it is a poly-
hydroxy aldehyde or ketone, capable of existing in cyclic or acyclic
forms. In this regard, (<u>R</u>)-(+) or (<u>S</u>)-(-)-glyceraldehyde can be regarded as the simplest acyclic carbohydrate. In more evolved systems, a
carbohydrate is an assembly of carbon atoms with predisposed senses of
chirality, each of which can be manipulated or used as a stereochemical
reference in an absolute sense. An understanding of the factors that
control such processes and the knowledge of structural and stereochemical requirements and their consequences are basic tenets for utilizing
carbohydrates in organic synthesis. Taking D-glucose as an example, it
can be seen that the acyclic polyhydroxy aldehyde form is in equilibrium
with two cyclic forms (Scheme 1). The six-membered pyranose ring is

Scheme 1

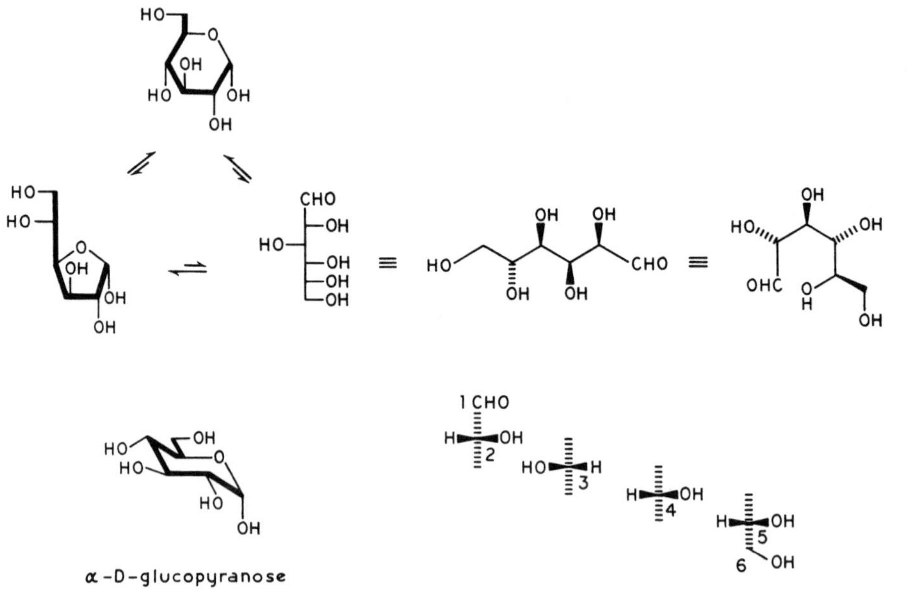

α-D-glucopyranose

favored over the five-membered and the seven-membered forms, as would be
expected on conformational grounds. In the $^4C_1$ conformation, α-D-glucopyranose has all its substituents equatorial except for the anomeric
position. This form of D-glucose can be conveniently manipulated as a
six-membered cyclic derivative, much the same way as a cyclohexane

derivative, with allowance made for the stereoelectronic consequences of replacing a ring carbon by an oxygen atom.[23] Under the constraints of ring-forming reactions, D-glucose can be used as a 6- or 5-membered structural entity. For example, methyl α-D-glucopyranoside readily forms a 4,6-0-benzylidene acetal derivative of the *trans*-decalin type, which is a crystalline and an extremely versatile starting material[24] (Scheme 2). In the presence of acetone and an acid catalyst, D-glucose

Scheme 2

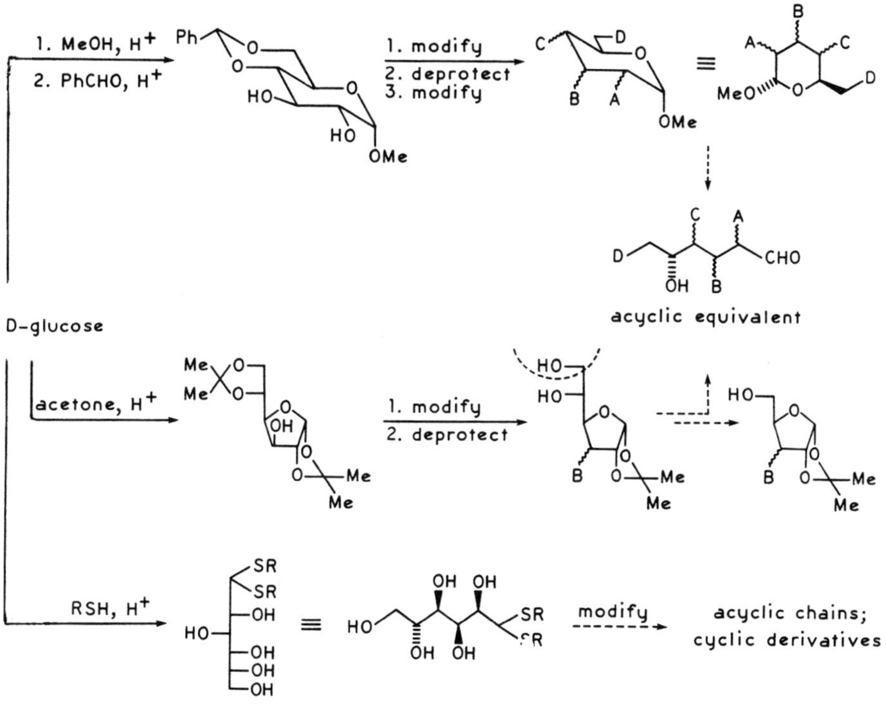

D-glucose

acyclic equivalent

acyclic chains; cyclic derivatives

gives the highly crystalline 1,2:5,6-di-0-isopropylidene-α-D-gluco-furanose,[25] another cornerstone in organic synthesis from carbohydrates. Finally, D-glucose can be "trapped" as the kinetic acyclic derivative by treatment with a thiol and an acid to give the

corresponding dialkyl dithioacetal.[26]   Now we can consider a six-carbon
acyclic chain for synthetic purposes. Of course, this option is in
principle, always available with modified pyranose and furanose forms,
hence the advantage of achieving stereocontrol in acyclic systems
indirectly by prior manipulation of an acyclic or cyclic derivative.
Thus, depending on the structural requirements of the desired precursor,
one can make the proper choice of a starting material derived from a
given sugar. In general, the same rules for obtaining different ring
forms or acyclic derivatives apply to other sugars as well but the
individual stereochemical and conformational characteristics have to be
appreciated. For example, ketones will usually lead to dioxolane rings,
but aldehydes can give dioxolane or dioxane rings (compare D-glucose and
D-galactose). As in cyclohexane derivatives, non-bonded interactions
are in general destabilizing in a pyranose derivative.   The preference
for an electronegative substituent at the anomeric carbon atom in a
cyclic pyranose derivative to adopt an antiparallel orientation with
respect to one of the lone pairs on the ring oxygen is termed the
anomeric effect.[23]   As a consequence of an axially disposed group at the
anomeric position of a D-hexopyranose in a $^4C_1$ conformation for example,
a good deal of regio- and stereochemical events occurring in reactions
involving the remaining ring carbon atoms can be anticipated[27] and
consequently incorporated in the synthetic blueprint (Scheme 3). These
notions are experienced in reductions of unsaturated systems, in
nucleophilic displacement reactions, and, among others, in addition
reactions to carbonyl functions. Another feature of predictable
reactivity is concerned with the opening of epoxides with nucleophiles
in conformationally biased systems. In general, these systems obey the
Fürst-Plattner rule and afford the diaxial products in preponderance.

These relatively simple notions concerning carbohydrate structure
and stereochemistry, and the logical "rules of thumb" concerning reactivity of some common intermediates will be helpful in following the
initial steps in schemes utilizing D-glucose and other sugars in organic
synthesis.

Scheme 3

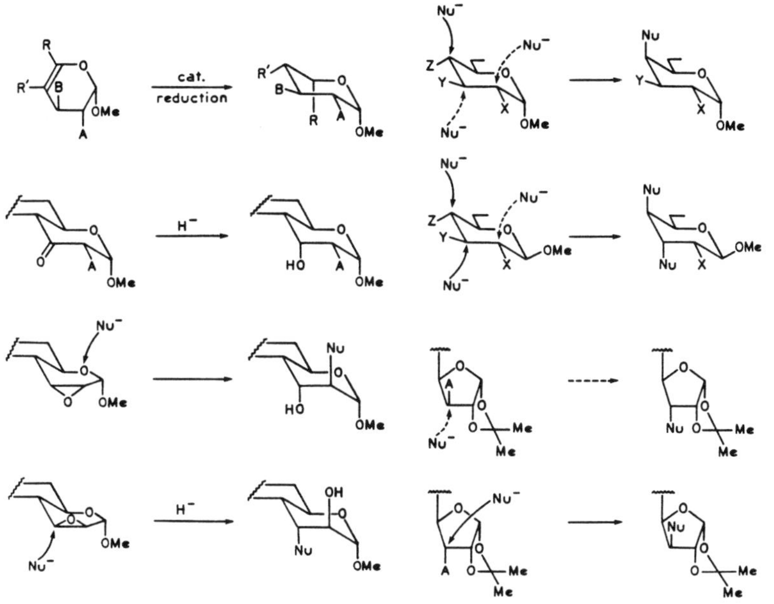

B. Symmetry relationships

Because the carbohydrate molecule contains different oxidation
states at either extremity (aldehyde and hydroxymethyl, aldehyde and
carboxyl, etc,), it lends itself to stereochemical duality. For example,
if the aldehyde and hydroxymethyl groups in a D-glucose derivative were
interchanged, a derivative of L-gulose would result. Thus, a rare sugar
and a different relative disposition of functional groups is unveiled
(Scheme 4). In general, "L-sugars" except for L-rhamnose, L-fucose and
L-arabinose, are not readily available from commercial sources. Moving
over to acyclic derivatives we find that reduction of D-glucose gives
the alditol, D-glucitol (or L-gulitol). The hydroxymethyl groups in
this compound are diastereotopic since oxidation at either end leads to

Scheme 4

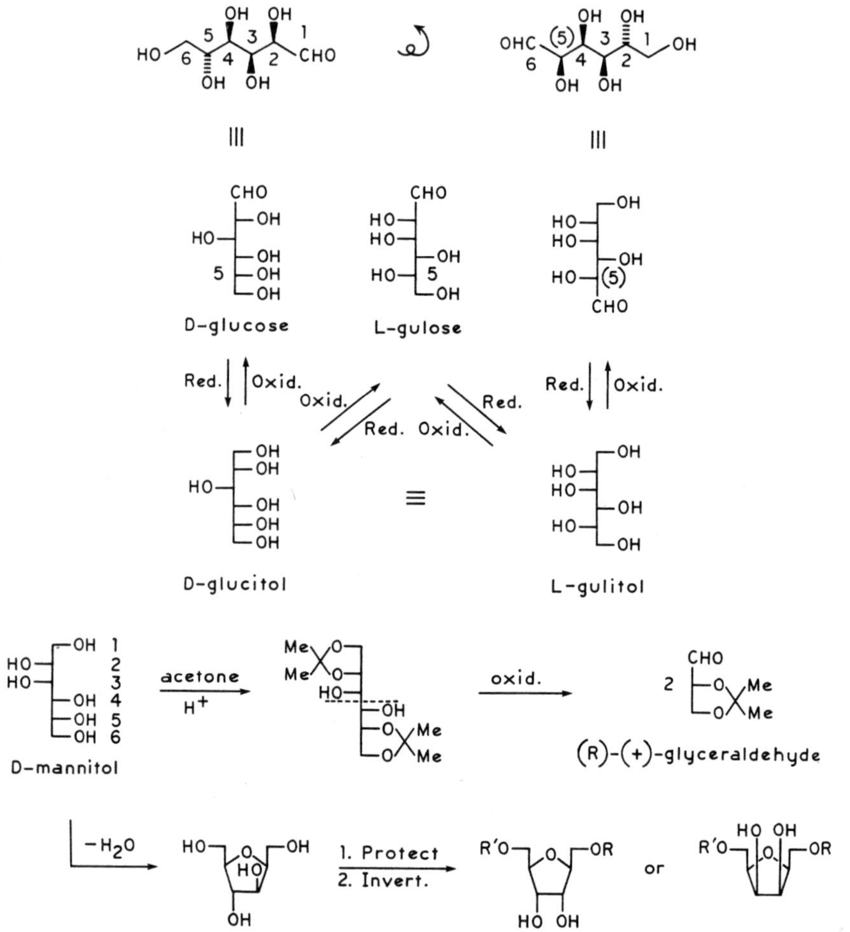

III

III

D-glucose   L-gulose

OH

CHO

Red. | Oxid.   Oxid. / Red. Oxid. | Red.   Red. | Oxid.

D-glucitol   ≡   L-gulitol

OH 1
HO   2
HO   3
OH 4
OH 5
OH 6

D-mannitol

$\xrightarrow[H^+]{acetone}$

$\xrightarrow{oxid.}$

2

(R)-(+)-glyceraldehyde

$\xrightarrow{-H_2O}$

$\xrightarrow[2.\ Invert.]{1.\ Protect}$ or

diastereoisomeric sugars. D-Mannitol on the other hand, is endowed with
symmetry which offers unique synthetic possibilities. For example, it
is a precursor to (R)-(+)-glyceraldehyde as well as to chiral 2,4-
disubstituted tetrahydrofurans (2,5-anhydro-D-alditols). Configurational inversion can provide symmetrical, and in this case optically in-
active derivatives. However, prior distinction between the two hydroxymethyl groups by preferential substitution affords asymmetric derivatives useful in the synthesis of targets containing tetrahydrofuran
units of the type illustrated or their deoxygenated analogs such as can
be found in certain ionophores.

With these basic notions, it will be possible to discuss the
elements of *design*, *discovery*, and *execution* embodied in the strategy of
natural products synthesis in particular, and organic chemistry in
general based on the concept of "chiral templates" derived from carbohydrates.

REFERENCES

1. For pertinent reading, see for example, J.F. Stoddart, Stereo-
    chemistry of the Carbohydrates, Wiley-Interscience, New York, N.Y.,
    1971; Rodd's Chemistry of Carbon Compounds, vol. 1, part 1F, S.
    Coffey, ed., Elsevier, Amsterdam, 1967; MTP Internationl Review of
    Science, vol. 7, Carbohydrates, G.O. Aspinall, ed., University Park
    Press, Baltimore, Md., 1978, 1976; The Carbohydrates, W.A. Pigman
    and D. Horton, eds, Academic Press Inc., New York, N.Y. 1972;
    Advan. Carbohydrate Chem. Biochem., Academic Press Inc., New York,
    N.Y., Methods in Carbohydr. Chem., R.L. Whistler and J.M.
    BeMiller, eds. Academic Press Inc., New York, N.Y.; etc; A.F.
    Bochkov and G.E. Zaikov, Chemistry of the 0-glycosidic bond: for-
    mation and cleavage, Pergamon Press, Oxford, U.K., 1979.

2. For antibiotics containing sugars, see S. Hanessian and T.H.
    Haskell in The Carbohydrates, vol. II A, 139 (1970); S. Umezawa in,
    MTP International Science Review, Vol. 7, Carbohydrates, Organic
    Series Two, 149 (1976). R.J. Suhadolnik, Nucleoside Antibiotics,
    Wiley-Interscience, 1970; see also, F. Johnson, in The Total Syn-
    thesis of Natural Products, Vol.1, J. ApSimon, ed., Wiley-Inter-
    science, N.Y., 1973, p.33.

3. For some recent reviews on aminoglycoside antibiotics, see J. Reden
    and W. Dürckheimer, Topics in Current Chem., 83, 105(1979); Drug
    Action and Drug Resistance in Bacteria - Aminoglycoside Antibio-
    tics, S. Mitsuhashi, ed. University Park Press, 1975; D.A. Cox, K.
    Richardson and B.C. Ross, Topics in Antibiotic Chem., 1, 1(1977);
    S. Umezawa, Advan. Carbohydr. Chem. Biochem., 30, 183(1974).

4. See for example, W.A. Szabo and H.T. Lee, Aldrichimica Acta, 13,
    13(1980) and references cited therein; see also, K. Mori, in The
    Total Synthesis of Natural Products, J. ApSimon, ed., Vol.4, Wiley-
    Interscience, New York, N.Y. 1981, p.1.

5. A. Fischli, in Modern Synthetic Methods , R. Scheffold, ed., Otto
    Salle Verlag, Frankfurt am Main, Germany, 1980, p.269, see also

K. Drauz, A. Kleeman and J. Martens, Angew. Chem., Int. Ed. Engl., 21, 584 (1982); L.R. Smith and H.J. Williams, J. Chem. Ed., 56, 696 (1979).

6. D. Seebach, in Modern Synthetic Methods, R. Scheffold, ed., Otto Salle Verlag, Frankfurt am Main, Germany, 1980, p.91; for an excellent review on the synthesis of insect pheromones in which hydroxy acids are used as starting materials among other strategies, see also K. Mori, in The Total Synthesis of Natural Products, J. ApSimon, ed., Wiley-Interscience, New York, N.Y., Vol.4, 1981, 1.

7. See for example ref.5; see also J.B. Jones in, Applications of Biochemical Systems in Organic Chemistry, Part 1, J.B. Jones, C.J. Sih and D. Perlman, eds., Wiley-Interscience, New York, N.Y. 1976, p.107; see also J.B. Jones, in Asymmetric Synthesis, J.D. Morrison, ed., Academic Press, in press.

8. P.N. Confalone, G. Pizzolato, E.G. Baggiolini, D. Lollar and M.R. Uskoković, J. Am. Chem. Soc., 99, 7020(1977).

9. R.B. Woodward, Science, 153, 487(1966); R.B. Woodward, K. Heusler, J. Gosteli, P. Naegeli, W. Oppolzer, R. Ramage, R. Ranganathan and H. Vorbrüggen, J. Am. Chem. Soc., 88, 852(1966).

10. K. Mori, Tetrahedron, 31, 3011(1975).

11. U. Schmidt, J. Gombos, E. Haslinger and H. Zak, Chem. Ber., 109, 2628(1976).

12. K. Mori, T. Takigawa and M. Matsui, Tetrahedron Lett., 3953(1976).

13. E.J. Corey and H.L. Pearce, J. Am. Chem. Soc., 101, 5841(1979).

14. R.V. Stevens and F.C.A. Gaeta, J. Am. Chem. Soc., 99, 6105(1977).

15. G. Stork, Yoshiaki Nakahara, Yuko Nakahara, and W.J. Greenlee, J. Am. Chem. Soc., 100, 7775(1978).

16. C.M. Wong, J. Buccini and J. TeRaa, Can. J. Chem., 46, 3091(1968).

17. J.P.H. Verheyden, A.C. Richardson, R.S. Bhatt, B.D. Grant, W.L. Fitch and J.G. Moffatt, Pure Appl. Chem., 50, 1363(1978).

18. N. Cohen, B.L. Banner and R.L. Lopresti, Tetrahedron Lett., 21, 4163(1980).

19. J. Rokach, R. Zamboni, C.-K. Lau and Y. Guindon, Tetrahedron Lett., 22, 2759(1981).

20. K.G. Paul, F. Johnson and D. Favara, J. Am. Chem. Soc., 98, 1285(1976).

21. G. Stork, T. Takahashi, I. Kawamoto and T. Suzuki, J. Am. Chem. Soc., 100, 8272(1978).

22. T. Ogawa, T. Kawana and M. Matsui, Carbohydr. Res., 57, C31(1977).

23. For a discussion of the anomeric effect, see R.U. Lemieux and S. Koto, Tetrahedron, 30, 1933(1974) and references cited therein; see also, The Anomeric Effect – Origin and Consequences, W.A. Szarek and D. Horton, eds., ACS Symposium Series, No.87, American Chemical Society, Washington, D.C. 1979; See also ref.1.

24. N.K. Richtmyer, Methods Carbohydr. Chem., 1, 107(1962), and references cited therein.

25. O. Th. Schmidt, Methods Carbohydr. Chem., 2, 318(1963), and references cited therein.

26. M.L. Wolfrom and A. Thompson, Methods Carbohydr. Chem., 2, 427(1963) and references cited therein.

27. A.C. Richardson, Carbohydr. Res., 10, 395 (1969).

PART TWO

DESIGN

CHAPTER THREE

THE CONCEPT OF "CHIRAL TEMPLATES" AND 'CHIRONS' — A VISUAL DIALOGUE WITH THE TARGET

As previously discussed, one has the flexibility of manipulating functional groups independent of each other in virtually all four classes of chiral starting materials. The situation is much more evident with carbohydrates since, in addition to functional and stereochemical features one can also manipulate various forms and chain lengths. It should, in principle, be possible to utilize them in the construction of polyfunctional acyclic, carbocyclic or heterocyclic compounds containing one or more centers of chirality. A strategy that provides a practical approach to such targets is based on the concept of "chiral templates"[1]* and the generation of 'chirons'** which represent enantiomerically pure synthons. In essence, this involves scrutinizing the molecular architecture of a given target to locate elements of symmetry***, chirality and function-

---

*The term "chiral template" is used in a descriptive context, hence the quotation marks. Current usage of the term 'chiral' in a generalized manner such as, 'chiral synthesis', 'chiral approach', etc. is misleading and incorrect. Chirality is related to the geometric and topological properties of objects, molecules and the like. For a recent review, see V. Prelog and G. Helmchen, Angew. Chem., Int. Ed. Engl., 21, 567 (1982).

**According to Greek mythology, Chiron was a wise and learned centaur who tutored Achilles, Jason, Hercules and Asclepius in music, morals and medicine. We thank Dr. G. Cassinelli (Farmitalia-Carlo Erba, Milan) for this historical insight.

***The term 'symmetry' is used in a figurative sense to reflect upon the ensemble of stereochemical, functional, and topological features present in the carbon framework of the target, as they can be related to those found in, or derivable from, a chiral precursor. When we talk of 'carbohydrate-type symmetry' in a target molecule for example, we focus attention to that segment of the carbon framework which can be related to, or derived from a carbohydrate with corresponding (but not necessarily identical) inter-relationships of chiral and functionalized centers.

ality, decoding such information, and transposing it onto the carbon
framework of suitable synthetic precursors ('chirons'), which in turn
are obtained from chiral starting materials (chiral templates) by systematic functionalization. The term "chiral template" derives its origin[1]
from the following consideration: it is a 'template' in the sense that
the carbon framework is a replica of a segment of the target, and it is
'chiral' because of its inherent natural chirality which can be related
to the original stereochemical code of the target.

    Mention should be made of the widely used term 'synthon', intro-
duced by Corey[2] when he first announced his innovative strategies for
the construction of complex molecules by considering a retrosynthetic
analysis. Unfortunately the term 'synthon' (presently also referred to
as 'retron') is sometimes loosely used to mean an intermediate. This
generalization deviates from the original designation. In the original
version,[2] the term was introduced to describe an idealized fragment,
usually a carbonium ion, a carbanion or a synthetic equivalent, produced
by bond disconnection during a retrosynthetic analysis. It may also involve some simple starting materials or reagents. In the synthon-based
approach to organic synthesis, the strategy is based on bond-
disconnection according to logical, but not necessarily the most
strategic sites which take into account stereochemical features. Thus,
by breaking a bond β- to a carbonyl group in a β-keto alcohol we
generate two synthons (ions) which can be considered as synthetic
equivalents of an aldehyde and an enolate. The reverse process, namely
the synthesis of the segment in question can therefore be contemplated
via an aldol condensation as dictated by the initial bond-disconnection
process. With the obvious need to synthesize enantiomerically pure (or
enriched) molecules, the synthon approach to this segment must make
additional provisions for stereochemical control. This can be achieved
by ensuring a measure of stereoselection in the aldol condensation.
Alternatively, we may choose to use optically active, chiral starting
materials to construct the segment in question.

    The 'chiron' strategy differs in concept in that strategic bond
disconnections in the target molecule proceed by locating segments
containing a number of chiral centers. 'Chirons' are then retro-
synthetically generated by <u>minimum perturbation of chiral centers</u> in the
target.

    The operational aspects of this strategy is schematically illus-
trated below by considering representatives of the four classes of

chiral building blocks. We will deal primarily with carbohydrates as starting materials, and refer to the others as the discussion warrants it. The combination of natural chirality, conformational bias, and inherent topology in a carbohydrate, combine to provide an ideal carbon framework or platform for achieving a high level of regio and stereocontrol, hence "chiral and functional overlap" with the 'chiron' and

Synthetic design with "Chiral templates" and the generation of 'chirons'

ultimately with the target. Arbitrarily, and whenever applicable, the 'chiron' contains the highest level of functionalization and stereochemical overlap with the target, yet it maintains the basic carbohydrate structure (such as a cyclic or acyclic derivative). Beyond that stage, a 'chiron' may no longer resemble a carbohydrate and in a visual sense, it could lose formal structural ties with the precursor from which it was derived.

The format of presentation in this book is therefore intended to show how to locate possible carbohydrate-like symmetry in a given molecular target, how to effectively disconnect certain precursors and even-

tually to a carbohydrate progenitor, and finally, how to utilize the
latter for the reverse process, that is, the construction of critical
intermediates ('chirons') and eventually the target structure itself.
With this type of presentation, the constitutional structure and stereochemistry of the starting carbohydrate or its derivative is considered
to be of secondary character and a matter of operational convenience.
In order to indoctrinate the reader to such a philosophy of "looking at
molecules" and engaging in a visual dialogue, it will be helpful to
establish some general guidelines. The concept of retrosynthetic or
antithetic analysis[2] of a target structure is pictorially and operationally useful in this approach and it will be systematically used. The
figure shown below illustrates how the 'chiron' is derived from the
target by retrosynthetic analysis, and produced from the chiral
precursor(s) ("templates") via synthetic operations.

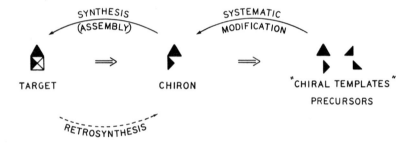

Schematic representation of the 'chiron' approach

A. Types of targets derived from carbohydrate precursors by synthesis.

In considering the types of target structures that can be constructed totally or in part using carbohydrate precursors, there emerge
three general categories.

1. Apparent carbohydrate-type symmetry. The "carbo-
    hydrate portion" in such targets is apparent and
    can be singled out. This category includes a
    large number of compounds that comprise chiral
    tetrahydrofuran and tetrahydropyran structures,
    such as the polyether antibiotics, certain
    terpenes, metabolites, toxins, C-nucleosides,
    cascade products derived from arachidonic acid,
    pheromones and insect poisons. Also included are

monocyclic and bicyclic lactones and a number of
simple acyclic structures. It is evident that in
a   synthetic strategy aimed at such targets, a
suitable cyclic or acyclic carbohydrate derivative
can be utilized to achieve the desired level of
stereochemical convergence with the target. The
terms 'acyclic' and 'cyclic' transfer of chirality
have been used by Fraser-Reid.[3] A partial list of
such targets has been reported in a previous
review.[1]

2.   Partially hidden carbohydrate-type symmetry. The
"carbohydrate portion" in such targets is only
partially evident and may constitute an extension
of the total structure. Certain amino acids,
terpenes, various cyclic and acyclic compounds or
structures containing acyclic chiral appendages
among others, may be considered to belong to this
category. Such molecules may be constructed by
chain-extension, or ring formation using one or
more carbon atoms of an appropriate carbohydrate
precursor after the desired level of "chiral and
functional overlap" has been achieved.

3.   Hidden carbohydrate-type symmetry. The "carbohydrate portion" in such targets is not evident
and must be located first. It may constitute a
segment containing one or more chiral centers in a
more complex structural framework bearing no
formal resemblance to a carbohydrate. Examples in
this category are more challenging in terms of
design, discovery, and execution. It is here,
when least expected based on past prejudices and
lack of operational directives, that the utilization of carbohydrates as chiral building blocks
emerges as a viable and practical strategy. The
concept of "chiral templates" and the generation
of 'chirons' by retrosynthetic analysis add a new
dimension to natural products synthesis based on

this approach. Alkaloids, macrolides,
prostaglandins,and heterocycles, to mention a few,
are likely targets as already illustrated in a
partial list.[1] The term 'transcription of
chirality' has been used by Fraser-Reid.[3]

Carbohydrate-type symmetry in the three categories is illustrated
below.

APPARENT

PARTIALLY HIDDEN

HIDDEN

REFERENCES

1. S. Hanessian, Acc. Chem. Res., 12, 159(1979).

2. E.J. Corey, Pure Appl. Chem., 14, 30(1969).

3. B. Fraser-Reid and R.C. Anderson, Fortschr. Chem. Org. Natust. 39,
    1(1980); see also, B.J. Fitzsimmons and B. Fraser-Reid, J. Am.
    Chem. Soc., 101, 6123(1979).

PART THREE

DISCOVERY

CHAPTER FOUR

WHERE IN THE CARBON SKELETON IS THE CHIRAL PRECURSOR?

A. Carbohydrates

In this section an effort will be made to establish some simple guidelines in order to locate carbohydrate-type symmetry in a given target. For targets that contain a tetrahydrofuran, tetrahydropyran or lactone units, and for relatively simple acyclic molecules, the 'chiron' as well as starting carbohydrate template derivative may be evident and will not be discussed. In other cases in which cyclic or acyclic units in the target are not directly convergent with a carbohydrate-type structure (hidden or partially hidden carbohydrate-type symmetry), the connection may be a bit remote, until certain parameters are considered.

1. Frameworks, perspectives, shapes – evident and disguised.

We have been accustomed to visualizing the carbohydrate molecule in terms of acyclic or cyclic five and six-membered ring structures. The natural substitution pattern consisting of hydroxyl groups, and an occasional amino or carboxyl function, has precluded their consideration as potential useful intermediates in organic synthesis. The central synthetic issues in carbohydrate chemistry have involved mostly peripheral transformations such as the introduction of amino, deoxy and occasionally carbon branching as dictated for example, by the necessities of antibiotic modification. The carbohydrate framework can, however, be used as a flexible platform for a variety of C-C and other bond-forming reactions since hydroxyl groups can be systematically

replaced by other functionalities with stereocontrol. With such an
added dimension, one can look at the carbohydrate structure as an
assembly of functional groups and substituents on a carbon backbone
which can be used as a template, and made to adopt the topology   and
"shape" of many targets and segments derived therefrom. A number of
"shapes" not normally associated with a carbohydrate structure, but
nevertheless inherent in its flexible acyclic or cyclic modifications
are illustrated below. No other group of chiral compounds can offer the
synthetic chemist the combined advantages of variable chain length,
functionality and stereochemistry provided by carbohydrates.

*Patterns and shapes derived from the carbon framework of various
target molecules - convergence with the framework of carbohydrate
starting materials (templates).*

2. Locating the "carbohydrate portion" - Bond disconnections -
Strategic bonds.

    For a given target in which the carbohydrate-type symmetry is
hidden or partially hidden, there may be one or more ways to disconnect
bonds and derive suitable 'chirons'. While there are no general rules
and the choice remains entirely arbitrary, the following guidelines may
facilitate the task. Since the 'chiron' is derived from retrosynthetic
reasoning, it is important that the reverse process be a practical and
feasible one.

        a. The "hydroxy-aldehyde or ketone connection" - (ex. the
"rule of five").

        In a large number of syntheses based on carbohydrate precursors,
the 'chiron' is derived from cyclic derivatives. Hence, the relationship of a hetero atom substituent particularly the ring oxygen atom, to
the anomeric carbon (formally an aldehyde or ketone) is important in
locating the "carbohydrate portion" in the target. A simple protocol
consists in establishing the spatial relationships between carbon atoms
bearing an oxygen (or other hetero atom) and a given carbon atom located
four or five atoms away along the chain or ring. This sets up a series
of at least four or five consecutive atoms, in which an sp^2 carbon, for
example, may correspond to the anomeric carbon and the hetero atom
containing carbon to C-4 or C-5 in a furanose or pyranose derivative
respectively. The possible functional overlap of the carbon framework
of a suitable carbohydrate with a segment of the target structure
regardless of configuration is thus found, and substituents present "in
between" can then be introduced by systematic manipulation of the
hydroxyl groups including eventually the ring oxygen. Of course, this
protocol does not preclude the utilization of carbohydrates as
percursors to targets or segments of targets which are devoid of hetero
atoms. In such cases, the carbohydrate chain must be extensively deoxygenated and predisposed centers replaced by hetero atoms, carbon
substituents, etc. based on the notions already developed. The
disposition of a ring oxygen and the anomeric carbon in a hexopyranose
derivative for example is five atoms (or bonds) away, and a "rule of
five" can be arbitrarily advanced. Such a "rule" is applicable to a
large number of recorded syntheses, even if, in retrospect, the
connection was not evident at the inception of the original plan. It
may indeed help in the planning of future syntheses from carbohydrates.
An illustration of how this "rule" can be used in locating potential
carbohydrate-type symmetry in maytansine (see page 256) by performing
bond disconnections that generate six possible six-carbon segments is
shown below. Each of these segments comprises a formal "C-5 oxygen
atom" with a predetermined sense of chirality (arrow) and an "anomeric
carbon" (indicated as (1)) of a potential hexose precursor. For
synthetic purposes, the choice of a particular segment, the generation
of a 'chiron' and its synthesis from an appropriate carbohydrate, is a
matter of operational convenience. Rather, consideration should be
given to the nature, the location, and the stereochemistry of

substitution along each chain and the prospects of achieving maximum "chiral and functional overlap" by the most efficient route from a given carbohydrate (D- or L-?).

Cyclic sugar derivatives can be functionalized at predetermined sites based on considerations of steric control, conformational bias, stereocontrol, and stereoelectronic control. Modified derivatives are then converted into other cyclic or acyclic derivatives which could coincide with the target structure or a portion of it. Further manipulation may be necessary to assemble the entire molecule.

MAYTANSINE

R = COCHN

"CHIRAL TEMPLATE"

C-1 → C-6 segment

C-6 → C-11 segment

C-3 → C-8 segment

C-9 → C-14 segment

C-6 → C-11 segment

C-3 → C-8 segment

Location of hidden carbohydrate-type symmetry in maytansine - application of the "rule of five". Note two reference points in the target: a) sp, sp², etc. carbon atom (overlap with anomeric carbon atom of carbohydrate); b) a hetero atom (O,N,S, etc.) located five bonds away from sp, sp² carbon atom (overlap with ring oxygen).

3. Functionalization and control of stereochemistry by manipulation of hydroxyl groups.

a. **Steric control** – An example of steric control is encountered in the synthesis of an amino acid derived from the antibiotic pyridomycin (Scheme 5, see page 148) which can be categorized as containing partially hidden carbohydrate-type symmetry. Inspection of the target structure indicates the presence of four different substituents on an acyclic five-carbon chain with a carboxyl terminus. The stereochemistry of the hydroxyl-containing carbon atom can be considered as the primary reference point with regard to the nature and type of

Scheme 5

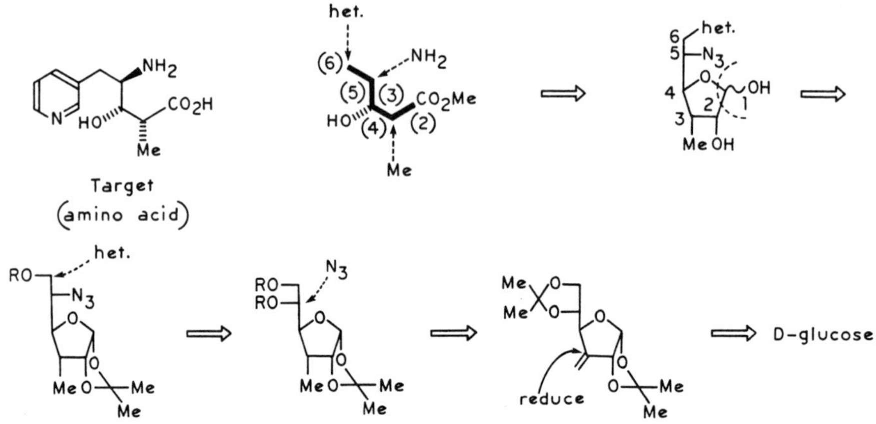

precursor sugar to be used. It has the same sense of chirality as that
at C-4 in a D-glucose derivative and can be "protected" during
manipulations by involving it in ring formation as in a furanose
derivative. Its β-relationship to the carboxyl group allows us to
create the latter functionality by oxidative cleavage of a diol, thus
sacrificing C-1 of the original D-glucose. The introduction of a
potential amino group as well as the heterocycle can be easily performed
by manipulating the C-5 and C-6 side-chain. Note that the amino group
must be introduced with retention of configuration, hence the necessity
for a double inversion process at C-5 starting with a D-glucose
configuration. The C-methyl group can be introduced on a suitable keto
derivative via Wittig methodology and the desired α-orientation can be
achieved based on preferential hydrogenation (steric control) of an
exocyclic methylene intermediate. Applying a logical sequence of
retrosynthetic steps, one arrives at 1,2:5,6-di-O-isopropylidene-α-D-
glucofuranose, hence D-glucose as a practical building block, since
regio- as well as stereochemical control can be ensured based on well
known chemical transformations.

        Scheme 6 shows another example of steric control in the intro-
duction of the acetic acid side chain in a thromboxane B_2 chemical precursor (see page 79). Hydrogenation of the olefinic intermediate is
virtually assured to occur from the β- (top) side to provide the desired
orientation of the acetate side-chain.

Scheme 6

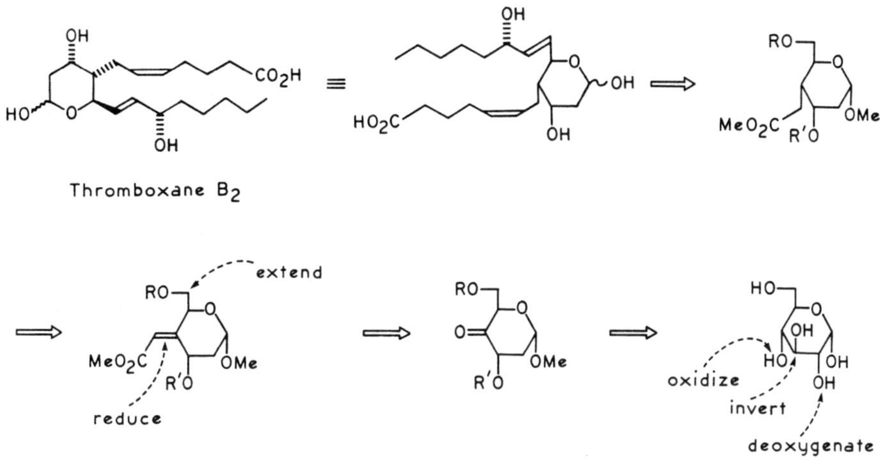

Thromboxane B$_2$

b.   Steric control, conformational bias and anomeric stereoselection - In a related approach, but using in addition conformational bias
and anomeric stereoselection, one can construct acyclic segments of
macrolide antibiotics as well as other natural products that are derived
from the propionate or polyketide biosynthetic route. The hidden carbohydrate-type symmetry in erythronolide A for example, can be related to
two six-carbon segments (Scheme 7)(see page 239) corresponding to
'chirons' A and B.

In each acyclic segment we find two hydroxyl groups which
correspond to C-3 and C-5 of a hexose. Chiron A has two C-methyl groups
at C-2 and C-4 and belongs to the L-series. Chiron B has one C-methyl
group at C-2, a branch point at C-4 and a D-configuration. The two
'chirons' share the same substitution pattern and sense of chirality at
C-2 and C-3 and they each harbor the C-5 oxygen atom in the ring,
protected until needed. In addition, B contains a C-methyl group and a
tertiary hydroxyl group at C-4, hence it can be used as an intermediate
to produce A, provided that the configuration at C-5 is inverted and the
hydroxyl group at C-4 is removed. The sequence shown in Scheme 7
illustrates how this analysis can be capitalized upon. Starting with

Scheme 7

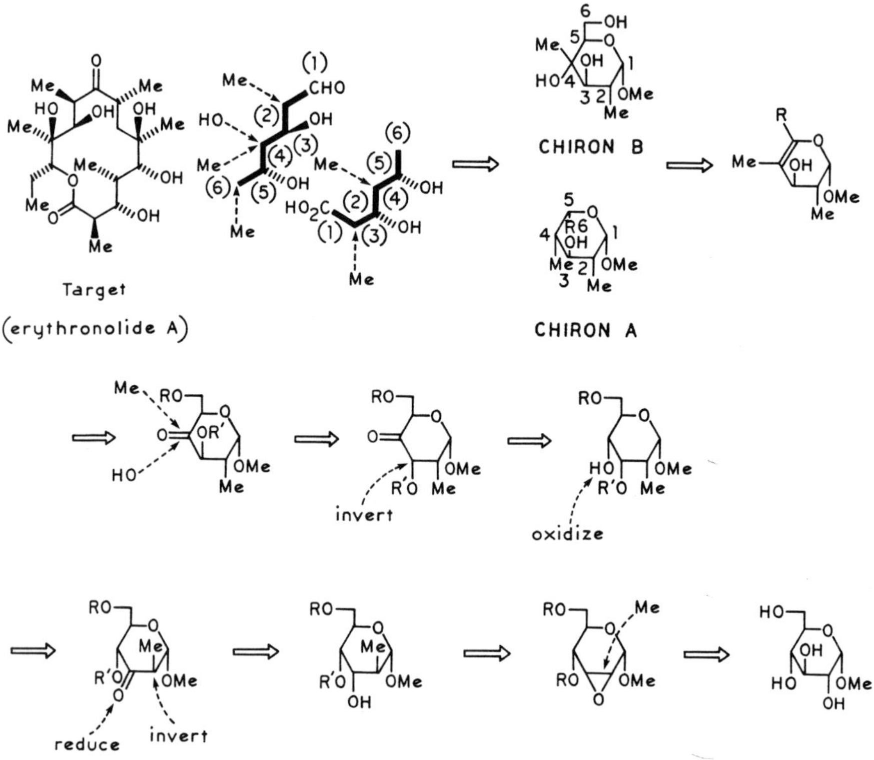

Target

(erythronolide A)

CHIRON B

CHIRON A

methyl α-D-glucopyranoside, one begins a systematic introduction of functional groups based on aspects of conformational bias and anomeric stereoselection provided by the axial orientation of the anomeric substituent (see Scheme 3). The 4-keto derivative serves as a pivotal intermediate for both 'chirons' A and B, which in turn, are converted to acyclic equivalents, chain-extended, and eventually linked to provide the carbon skeleton of the target (see page 245).

c. Transfer of chirality - A second unique strategy for controlling stereochemistry involves the utilization of existing chirality in a cyclic or acyclic carbohydrate structure to create another center of chirality, usually in β-position, via a Claisen-type or similar rearrangement. In this manner a useful methodology can be applied for the introduction of one or more carbon substituents in predetermined manner. Thus, in Stork's $PGF_{2\alpha}$ synthesis from D-glucose (see page 193), the focal point in terms of synthetic design is the convergence of a carbohydrate-derived segment containing two hydroxyl groups and a double bond, with C-11 — C-16 of the target (Scheme 8). Two centrally located hydroxyl groups in the original alditol intermediate are used to create a *trans*-double bond and one unprotected hydroxyl group is the anchor point for a stereospecific Claisen rearrangement to introduce the C-8 — C-12 bond in the target. Note again how preferential protection provides an ideal derivative which voluntarily succumbs to chemical manipulation. Ireland's synthesis of lasalocid A is another example involving transfer of chirality (see page 101).

Scheme 8

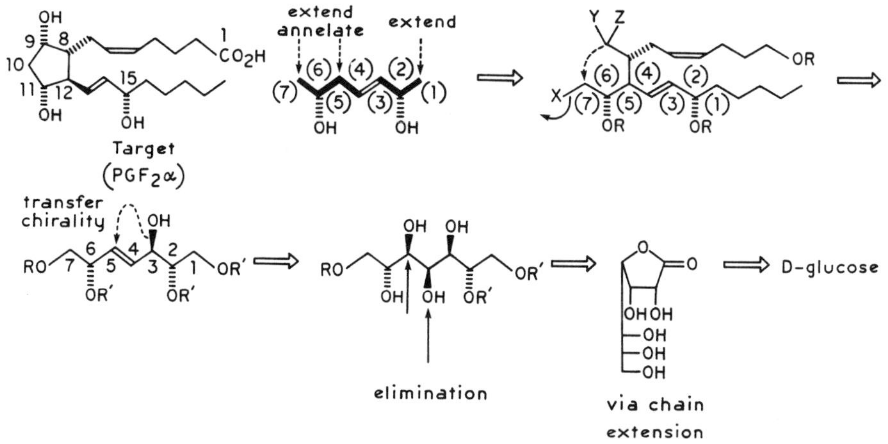

d. The "scaffold process" — A unique form of manipulating the carbohydrate structure is based on constructing a segment of the target "around" or "on top" a rigid entity, such as a ring, then once the desired level of functionality and chirality is reached, the original carbohydrate portion of the molecule is unfolded to unveil the target or a segment thereof. This type of strategy can be related to a "scaffold process" (Scheme 9). The synthesis of anisomycin for example (see page 221) is representative of this strategy. In this case the critical pyrrolidine ring is formed by intramolecular cyclization from a primary amine with concomitant inversion of configuration at a ring carbon atom. The original carbohydrate portion then becomes an appendage or an extension of the new heterocyclic ring. Note that in anisomycin, the hydroxyl group corresponds to that at C-5 in a D-glucofuranose

structure, and the acetoxy group is derived from the ring oxygen. The
original stereochemistry at C-5 must be inverted but that at C-4 is
carried over to the target. The synthesis of chrysanthemic acid can be
thought to proceed by a "scaffold process" also (see page 204).

Scheme 9

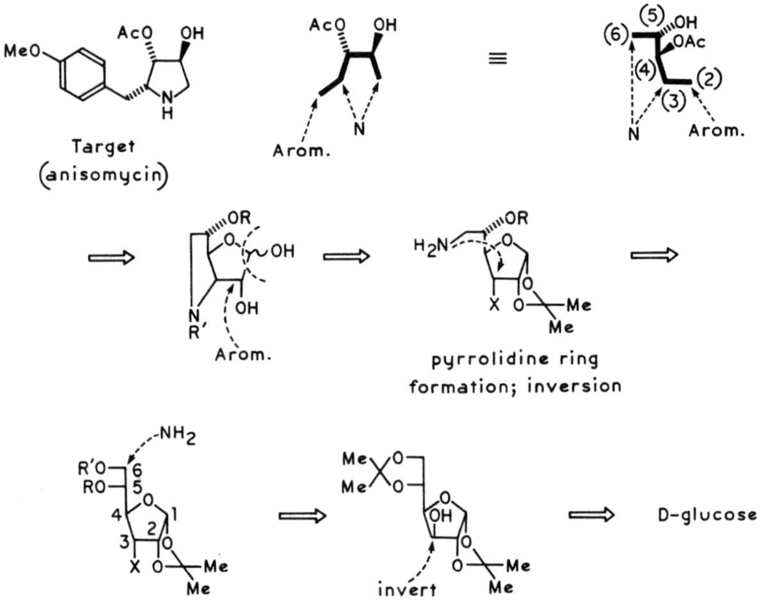

Target
(anisomycin)

Arom.

Arom.

Arom.

pyrrolidine ring
formation; inversion

D-glucose

invert

e. "Extended" and "fused" structures – In some targets a segment
of the carbohydrate portion may be utilized to construct an integral
part of the structure. Thus, in the synthesis of the imidazole ring in
L-erythro-β-hydroxy histidine a component of bleomycin (see page 170)
from 2-amino-2-deoxy-D-glucose, two carbon atoms of the carbohydrate
were "extended" into the heterocyclic structure (Scheme 10). The
remaining appendage comprising C-1-C-3 of the original amino sugar is
present with its intact functional groups and stereochemistry, hence the
partially-hidden symmetry. There are a number of other heterocyclic
structures which can be built from segments of carbohydrates (see page
228).

Scheme 10

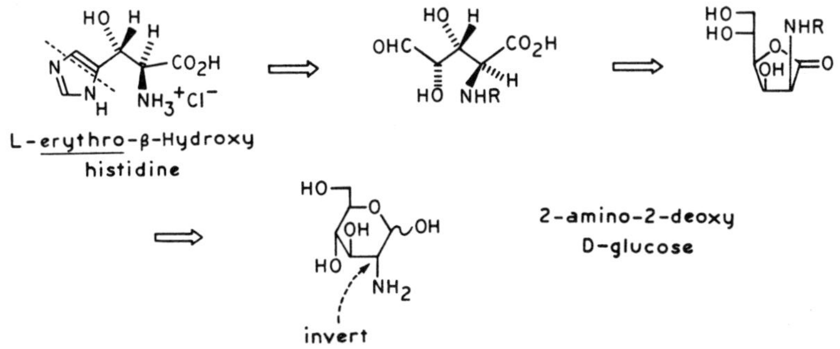

L-erythro-β-Hydroxy
histidine

2-amino-2-deoxy
D-glucose

invert

A different situation arises in some targets in which the carbohydrate symmetry is of the hidden type. Consider for example thienamycin (Scheme 11) which can be related to L-aspartic acid or L-threonine
as chiral synthetic precursors (see page 231). However, dissection of
the molecule in the manner indicated in Scheme 11 reveals a hidden
carbohydrate template which is "fused" to the remainder of the structure. Thus, C-1-C-9 can provide a highly functionalized six-carbon
segment containing the required three asymmetric centers. The known
acyclic amino acid or the corresonding lactone can be easily related to
a pyranoid carbohydrate precursor. Note how the "rule of five" applies
nicely in this case. Frontalin (see page 214) is another example of a
target in which a carbohydrate-derived segment is "fused" to the general
structure and becomes an integral part of it.

B. Discovery of amino acids, hydroxy acids and terpenes in target
structures.

It would be appropriate to generalize the concepts put forward for
discovering carbohydrate-type symmetry to other chiral precursors as
well. Aspects of amino acid, hydroxy acid and terpene structures were
tabulated in terms that would aid in locating them in the carbon framework of a given target (see pages 8-10). The sequential α-amino acid
functionality in amino acids for example is easy to detect in a target,
particularly if a carboxyl group is present. In other instances a β- or
γ-substituent such as in L-cysteine, L-threonine, and L-aspartic acid,
may provide a useful additional functional complement, particularly if
the sense of chirality in the precursor and the corresponding segment in

Scheme 11

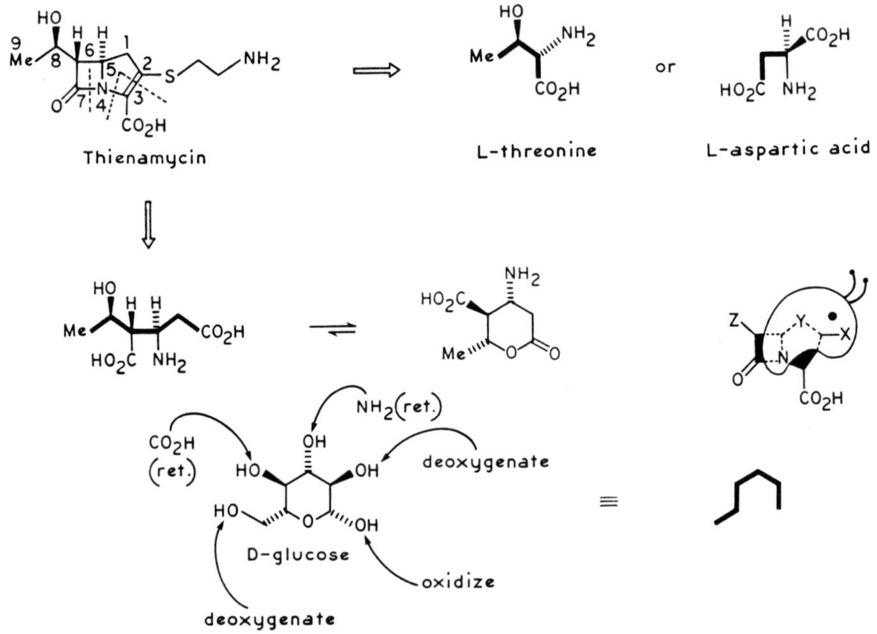

Thienamycin ⟹ L-threonine   or   L-aspartic acid

the target happen to match. Bifunctional amino acids such as cysteine
or aspartic acid may be extended at either end in such a way that the
extremities become an integral part of the target framework or ring
system. The examples on page 9 show excellent use of L-cysteine as a
chiral precursor to biotin and cephalosporin (hidden amino acid
symmetry). Scheme 11 shows how the carbon framework of L-threonine or
L-aspartic acid can overlap with a portion of the thienamycin
structure (compare D-glucose!). Amino acids are also useful in the
synthesis of α-hydroxy acids by deamination with retention of
chirality. The synthesis of sulcatol from L-glutamic acid (see page 9)
illustrates the exploitation of partially hidden amino acid symmetry.
Among the limitations associated with amino acid precursors are the
number of original chiral centers and functional groups, the choice of
chain length and the scarcity of most unnatural amino acids.[1] Moreover,
conformational bias cannot play a critical role, as in the case of
carbohydrates.

        The optically active isomers of hydroxy and dihydroxy acids such
as malic acid and tartaric acid share the same advantages and disadvan-
tages as the amino acids, although they too provide ideal substrates

when one or two (vicinal) asymmetric centers are sought. The tartaric acids have been extensively used in total synthesis and some examples can be found in the examples shown on pages 9,11 as for disparlure, anisomycin, etc. where the two asymmetric centers and all four carbon atoms are utilized. One inherent feature in tartaric acid is the presence of identical functionality at either end, hence a potential problem in distinguishing one hydroxyl group from the other. However, the symmetry encountered in these molecules allows chemical manipulation to provide one and the same product.[2]

Terpenes can provide carbon frameworks that are otherwise difficult to construct. The efficient and ingeneous use of (-)-carvone and (-)-camphor in the syntheses of picrotoxinin and a steroid nucleus respectively are classical examples of the discovery of terpene-like frameworks in these targets (see page 10). Unfortunately, these useful rigid and flexible templates provide few asymmetric centers and their utility must be exploited in generating chiral cycloalkanes or acyclic segments with a predetermined substitution pattern. The enone function in carvone is useful for further chain-branching to create a trisubstituted ring.

A particularly elegant demonstration of the utilization of optically active precursors in natural product synthesis is seen in Stork's synthesis of cytochalasin B, where all but a carbohydrate type precursor were astutely used in the elaboration of this intricate molecule (see page 10). Pertinent references for the utilization of amino acids, hydroxy acids and terpenes in natural product synthesis can be found in chapter two as well as in discussions of specific targets.

REFERENCES

1. There are however, specific processes for the production of non-natural (R)-amino acids; see for example, D. Dinelli, W. Marconi, F. Cecere, G. Galli, F. Morisi in, Enzyme Engineering, E.K. Pye and H.H. Weetall, eds., vol.3, Plenum, New York, N.Y., 1978, p. 477.

2. See for example, K.C. Nicolaou, D.P. Papahatjis, D.A. Claremon and R.E. Dole III, J. Am. Chem. Soc., 103, 6967 (1981); see also, D. Seebach and D. Wasmuth, Helv. Chim. Acta, 63, 197 (1980).

PART FOUR

EXECUTION

Molecules Containing Apparent Carbohydrate-type Symmetry

CHAPTER FIVE

ACYCLIC MOLECULES

     a. <u>(S)-Propanediol</u> – Perhaps one of the earliest recorded examples of the utilization of a chiral precursor derived from a carbohydrate was the synthesis of (<u>S</u>)-(+)-propanediol <u>12.4</u> from a derivative of (<u>R</u>)-(+)-glyceraldehyde <u>12.2</u>[1] (Scheme 12), which in turn is obtained from oxidative cleavage of the readily available 1,2:5,6-di-0-isopropylidene-D-mannitol <u>12.1</u>. Note that the $C2$ symmetry of D-mannitol leads to the formation of two molecules of (<u>R</u>)-(+)-glyceraldehyde.

Scheme 12

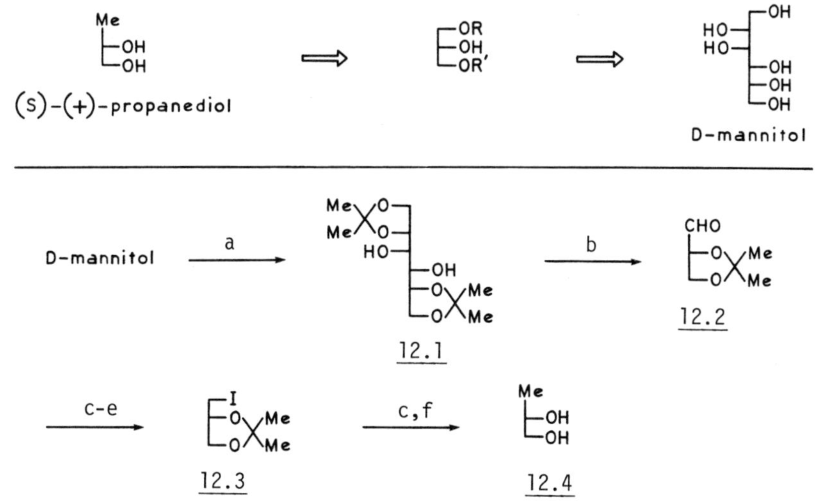

a. acetone, H^+; b. Pb(OAc)$_4$, benz.; c. Ra-Ni; d. TsCl, pyr.;
e. NaI, acetone; f. aq. H_2SO_4, EtOH

41

b. (R)-and (S)-Epichlorohydrin - Chiral three-carbon
oxiranes can be conveniently prepared from (R)-glyceraldehyde by
individual manipulation of both extremities (Scheme 13)[2]. Thus by
tosylation and deprotection of the (R)-glycerol derivative 13.1, one
obtains a monosubstituted asymmetric derivative 13.2, which can be
transformed to the chloride 13.3 hence to (S)-epichlorohydrin 13.4.
Alternatively, by first introducing the epoxide function then mesylation
and routine manipulation, the antipodal (R)-epichlorohydrin 13.7 is
obtained. Chiral heteroaryloxymethyl oxiranes have been prepared using
the same general strategy.[3]

Scheme 13

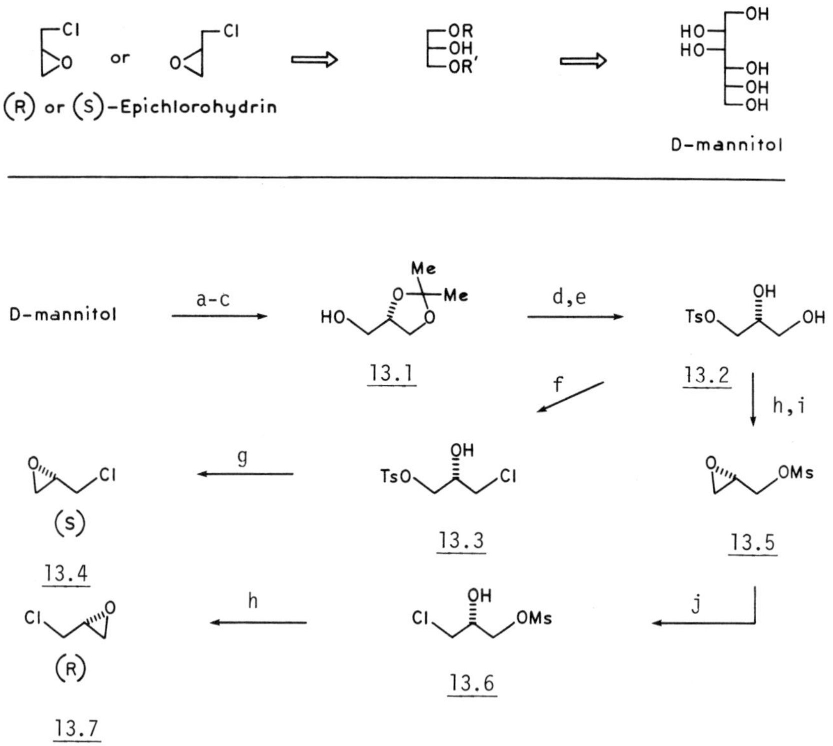

(R) or (S)-Epichlorohydrin

D-mannitol

13.1   13.2

13.3   13.5

13.4   13.6

13.7

a. acetone, H^+; b. Pb(OAc)$_4$; c. NaBH$_4$; d. TsCl, pyr.; e. aq. HCl;
f. Ph$_3$P, CCl$_4$; g. ethyleneglycol, Na; h. NaOMe, MeOH; i. MsCl, pyr.;
j. conc. HCl

c. O-Methyl-(S)-Mandelic acid - The synthesis of optically pure
O-methyl-(S)-(+)-mandelic acid 14.6 was achieved by Bonner[4] starting
with D-arabinose, which is depicted in Scheme 14 in its pyranose form.

Scheme 14

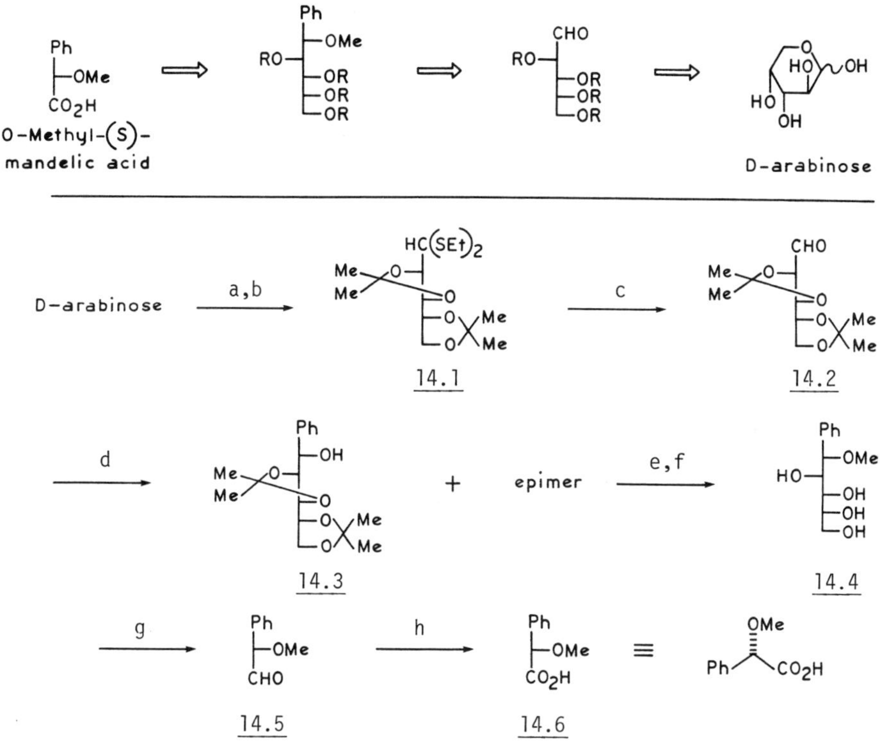

O-Methyl-(S)-
mandelic acid

D-arabinose

14.1   14.2

14.3   +   epimer   14.4

14.5   14.6

a. EtSH,H$^+$; b. acetone, H$^+$; c. HgCl$_2$, HgO, CdCO$_3$, aq. acetone; d. PhMgBr,
ether; e. MeI, Ag$_2$O; f. aq. AcOH; g. NaIO$_4$; h. Ag$_2$O, aq. acetone.

The objective here was to use an aldehydo form such as 14.2, readily
available from the dithioacetal derivative 14.1, in order to introduce
the phenyl group by a Grignard reaction. This was in fact a
stereoselective process giving the (S)-epimer as the preponderant
product which could be separated and processed to the corresponding
mandelic acid derivative. Note that in this sequence all the asymmetric
centers in the starting carbohydrate are destroyed, but the reward
resides in ultimately obtaining an optically pure acid. In a later
work, a high degree of stereoselectivity was observed in the reaction of
Grignard reagents of the type RMgX (R=Me, cyclohexyl) with the aldulose

derivative resulting from oxidation of 14.3.[5] Such derivatives comprise
"optically pure" benzylic carbon atoms and can be used as stereochemical
references for establishing the absolute stereochemistry at tertiary
centers in natural products. Moreover, it can be seen that even as
early as 1951, Bonner's work had demonstrated the operation of a high
degree of stereocontrol in C-C bond forming reactions with acyclic
systems derived from carbohydrates. In fact, the reaction of Grignard
reagents with carbohydrate derivatives goes back to 1906[6] and work up to
1956 has been reviewed.[7] Subsequent contributions[8,9] have been
sporadic, but demonstrate early examples of the phenomenon of "chelation
control"[10], and the operation of Cram's rules[11] in certain cases.

    d. (-)-Timolol - The potent β-adrenergic blocking agent Timolol,
3-(3-tert-butylamino-2-hydroxypropoxy)-4-N-morpholino-1,2,5-thiadiazole
15.2 has been prepared from the (R)-glyceraldehyde derivative 12.2 by
sequential reductive amination, deprotection-protection techniques[12]
(Scheme 15). It is of interest that only the levorotatory form so

    Scheme 15

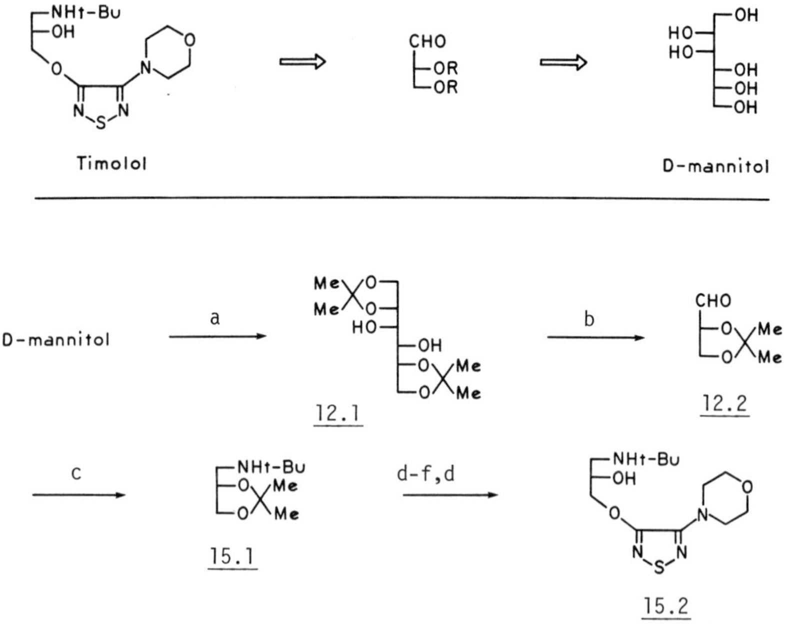

a. acetone, H^+; b. $Pb(OAc)_4$, THF ; c. t-$BuNH_2$, Pd/C, H_2, MeOH;
d. H_3O^+; e. PhCHO, $150°$; f. 3-chloro-4 (N-morpholino)-1,2,5-
thiadiazole, t-BuOK, t-BuOH;

produced is responsible for the β-adrenergic function. This synthesis
also confirms the absolute configuration of Timolol.

  e. L-(+)-Alanine - Another classic demonstration of the use of
carbohydrates for establishing absolute configuration of natural
products is seen in the synthesis of N-acetyl-L-(+)-alanine 16.3 from
2-amino-2-deoxy-D-glucose[13] (Scheme 16). The sense of chirality at C-2

Scheme 16

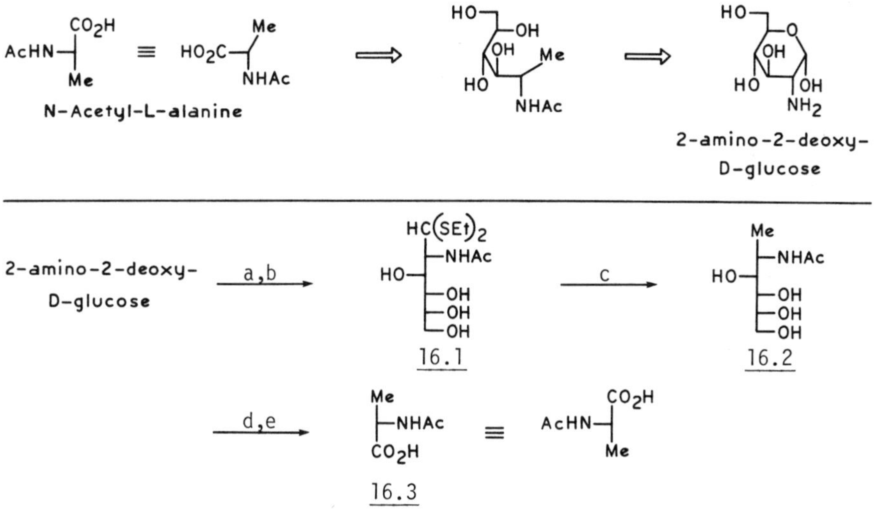

a. Ac_2O,MeOH, base; b. EtSH,H$^+$; c. Ra-Ni; d. Pb(OAc)$_4$, AcOH; e. aq. Br$_2$

in the amino acid was related to that in the amino sugar, which, as the
acyclic diethyl dithioacetal derivative 16.1, offered the possibility of
reductive desulfurization and oxidative cleavage. Three asymmetric
centers are destroyed in the process, but the stereochemical correlation
is the pivotal issue in such a scheme. This type of degradation has in
fact been used in other instances as for example in the establishment of
the partial stereochemistry of amino sugars in some aminoglycoside
antibiotics.[14,15] Access to other amino acids such as L-serine[16] and
L-threonine[17] from 2-amino-2-deoxy-D-glucose has been investigated. In
a different context, mention should be made of the synthesis of a chiral
ethanol-d from D-glucose.[18]

f. (-)-4-Amino-3-hydroxybutyric acid (GABOB) - The antiepileptic
and hypotensive drug 4-amino-3-hydroxybutyric acid 17.5 has been
prepared from L-ascorbic acid (Vitamin C) in an interesting
demonstration of its use as a chiral precursor[19] (Scheme 17). Thus,

Scheme 17

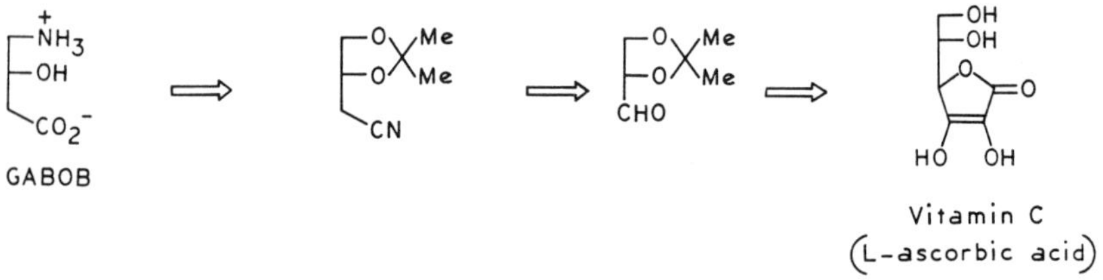

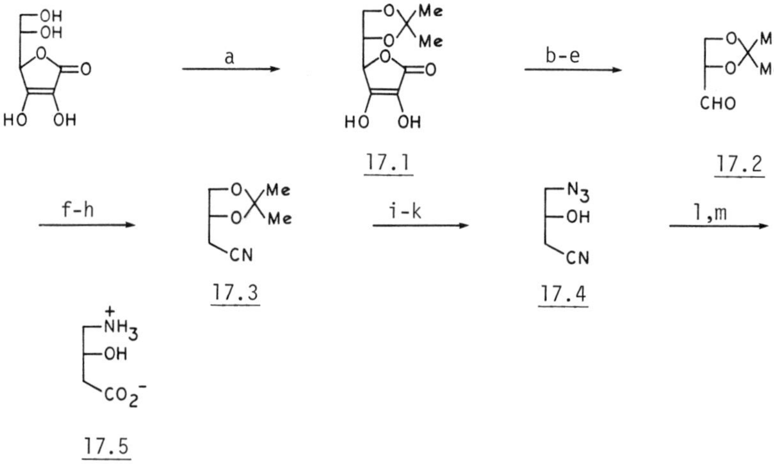

a. Acetone, H^+; b. NaBH$_4$, EtOH; c. NaOH; d. HCl; e. Pb(OAc)$_4$, EtOAc;
f. NaBH$_4$, EtOH; g. TsCl, Et$_3$N; h. KCN, DMSO, NaI; i. HCl; j. MsCl, pyr.;
k. KN$_3$, 18-crown-6, MeCN; l. Pd/C, H$_2$; m. H$^+$.

the only asymmetric center in the target is related to the C-5 hydroxyl
group in Vitamin C which is converted to the (S)-glyceraldehyde
derivative 17.2. A combination of substitution and cleavage reactions
was then used to arrive at the product. Note that the readily available
Vitamin C is a useful precursor to (S)-glyceraldehyde,which, unlike the
(R)-isomer is not as readily accessible.

g. Negamycin - The hydroxy amino acid (δ-hydroxy-β-lysine) in
negamycin[20] has been related to a 3,6-diamino-2,3,4,6-tetradeoxy-L-

threo-aldonic acid, and the corresponding lactone 18.6 has been synthesized from D-galacturonic acid by Umezawa and coworkers[21]   (Scheme
18). Inspection of the stereochemical features and the substitution

Scheme 18

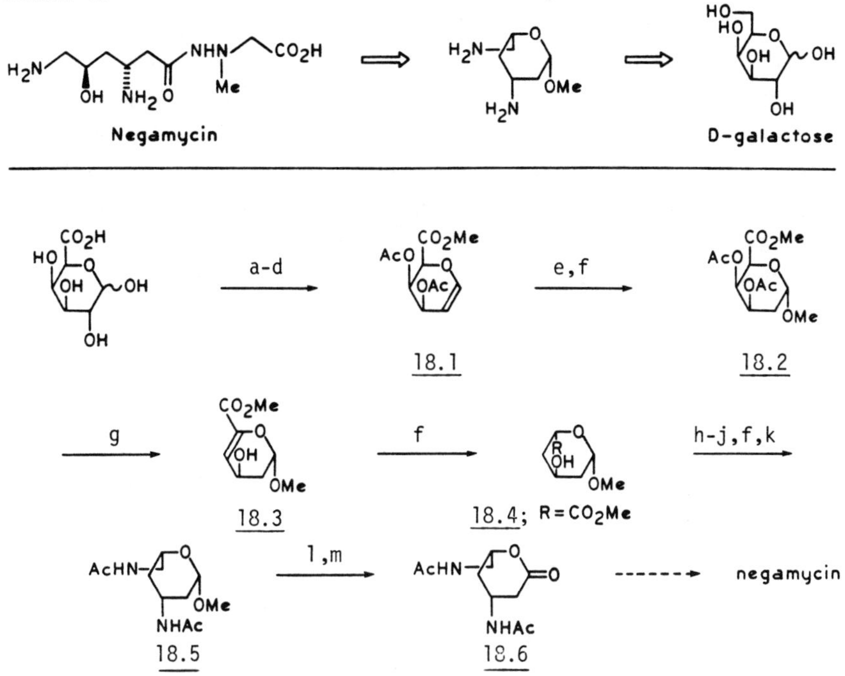

Negamycin   D-galactose

a. MeOH, H^+; b. Ac_2O, H^+; c. HBr, AcOH; d. Zn, AcOH; e. I_2, AgOAc, MeOH;
f. Pd/C, H_2; g. base; h. LAH; i. MsCl, pyr.; j. NaN_3; k. Ac_2O, pyr.;
1. H_3O^+; m. aq. Br_2.

pattern present in the target 18.6 reveals two features that must be
incorporated in the synthetic scheme. First, an L-sugar must be
produced, and second, extensive deoxygenation must be performed.
Deoxygenation at C-2 in hexopyranose derivatives can be achieved via the
glycal procedure among others,[22] and amination at C-3 can, in principle,
be accomplished by a nucleophilic displacement reaction. The major
problem therefore resides in the obtention of an intermediate with the
desired sense of chirality at C-5. The starting carbohydrate has the
D-configuration which is opposite to that at C-5 in the target; however,
the use of a uronic acid allows easy access to the L- configuration (or

5-(R)- in the target) by sequential β-elimination and catalytic
hydrogenation. The stereochemical outcome of such a process is
predictable based on literature precedents[23] and the α-orientation of
the anomeric carbon (anomeric stereoselection) (see Scheme 3). This
process also nicely introduces the 4-deoxy function. An intermediate
related to 18.5 has been obtained from methyl α-L-arabinofuranoside[24] by
sequential periodate oxidation, nitromethane cyclization, double
elimination and introduction of the remaining amino group at C-6. The
transformation of the lactone intermediate 18.6 as well as its antipode
(obtained from 3-amino-3- deoxy-D-glucose) into the natural negamycin
and its antipode respectively has been previously reported in connection
with degradative work on the antibiotic. A synthesis of (+)-negamycin
and (+)-epinegamycin from a derivative of dl,3-amino-5-oxo-6-chloro-
hexanoic acid has been recently reported.[25]

REFERENCES

1. E. Baer and H.O.L. Fisher, J. Am. Chem. Soc., 70, 609 (1948).

2. J.J. Baldwin, A.W. Raab, K. Mensler, B.H. Arison and D.E. McClure,
    J.Org. Chem., 43, 4876 (1978).

3. D.E. McClure, E.L. Engelhardt, K. Mensler, S. King, W.S. Saari,
    J.R. Huff and J.J. Baldwin, J. Org. Chem., 44, 1826 (1979).

4. W.A. Bonner, J. Am. Chem. Soc., 73, 3126 (1951).

5. T.D. Inch, R.V. Ley and P. Rich, J. Chem. Soc., C 1693 (1968).

6. C. Paal and F. Hornstein, Ber., 39, 1361, 2823 (1906).

7. W.A. Bonner, Advan. Carbohydr. Chem., 6, 251 (1956).

8. M.L. Wolfrom and S. Hanessian, J.Org. Chem., 27, 1800 (1962).

9. T.D. Inch, Carbohydr. Res., 5, 45 (1967).

10. W.C. Still and J.H. McDonald, III, Tetrahedron Lett., 21, 1031
    (1980) and references cited therein.

11. D.J. Cram and K.R. Kopecky, J. Am. Chem. Soc., 81, 2748 (1959).

12. L.M. Weinstock, D.M. Mulvey and R. Tull, J. Org. Chem., 41, 3121
    (1976).

13. M.L. Wolfrom, R.U. Lemieux and S.M. Olin, J. Am. Chem. Soc., 71,
    2870 (1949).

14. See for example, S. Hanessian and T.H. Haskell in The Carbohydrates,
    vol.IIA, 139(1970) and references cited therein.

15. See, S. Umezawa in, MTP International Science Review,
     G.O. Aspinall, ed., Vol.7, Carbohydrates, Organic Series Two,
     (1976), p.149, University Park Press, Baltimore, Md.

16. K. Turumi and S. Yamada, Tohoku J. Exptl. Med., 62, 329(1955).

17. S. Hanessian and S.P. Sahoo, unpublished results.

18. R.U. Lemieux and J. Howard, Can. J. Chem., 41, 308(1963).

19. M.E. Jung and T.J. Shaw, J. Am. Chem. Soc., 102, 6304 (1980).

20. S. Kondo, S. Shibahara, S. Takahashi, K. Maeda, H. Umezawa, and M.
     Ohno, J. Am. Chem. Soc,. 93, 6305 (1971).

21. S. Shibahara, S. Kondo, K. Maeda, H. Umezawa and M. Ohno, J. Am.
     Chem. Soc., 94, 4353 (1972).

22. See for example, S. Hanessian, Advan. Carbohydr. Chem., 21, 143
     (1966); R.J. Ferrier, Advan. Carbohydr. Chem., 20, 67 (1965); 24,
     199 (1969), and references therein.

23. H.W.H. Schmidt and H. Neukom, Carbohyd. Res., 10, 361 (1969); see
     also J. Kiss, Advan. Carbohydr.Chem. Biochem., 29, 230 (1974).

25. W. Streicher and H. Reinshagen, Carbohydr. Res., 83, 383 (1980);
     see also W. Streicher, H. Reinshagen and F. Turnowsky,
     J. Antibiotics, 31, 725 (1978).

25. G. Pasquet, D. Boucherot and N.R. Pilgrim, Tetrahedron Lett., 21,
     931 (1980).

CHAPTER SIX

MOLECULES CONTAINING A TETRAHYDROFURAN-TYPE RING

a. (+)-Muscarine – One of the earlier syntheses of a
natural product containing a tetrahydrofuran ring from a carbohydrate
precursor goes back to Hardegger and Lohse[1] who described the synthesis
of (+)-muscarine,[2] a toxic principle isolated from mushrooms and pos-
sessing cholinomimetic activity, from 2-amino-2-deoxy-L-gluconic acid
19.2 (Scheme 19). This starting material contains the six-carbon backbone of the target in addition to the required sense of chirality for
C-2 and C-3 (muscarine numbering; C-4, C-5 of the sugar acid). The key
reaction in the synthesis is a nitrous acid deamination with concomitant
tetrahydrofuran ring formation and retention of configuration at the
amine-containing carbon atom. Although the required acid is not commercially available, it can be prepared from the readily available
L-arabinose via the cyanohydrin procedure and isolated as the N-benzyl-
amino derivative 19.1. The cyclic acid 19.3 was converted to the amide
derivative 19.4, which upon reduction with lithium aluminium hydride
gave normuscarine 19.5 and finally, muscarine 19.6 after methylation.
The reduction step is of interest since it appears that only the primary
and one of the secondary tosyloxy functions were replaced by hydrogen.
The regioselective reduction at C-4 of the normuscarine precursor may
have occurred via direct hydride attack at C-4 (C-O cleavage) and hydrolysis of the C-3 tosyloxy ester (-O-SO$_2$ cleavage), but more likely it
proceeded via initial ester cleavage at C-3, epoxide formation and
regioselective hydride attack at C-4 to give the observed product.
Thus, 2-amino-2-deoxy sugars can be useful precursors to chiral tetrahydrofuran derivatives via the well-known deamination reaction,[3] and the

Scheme 19

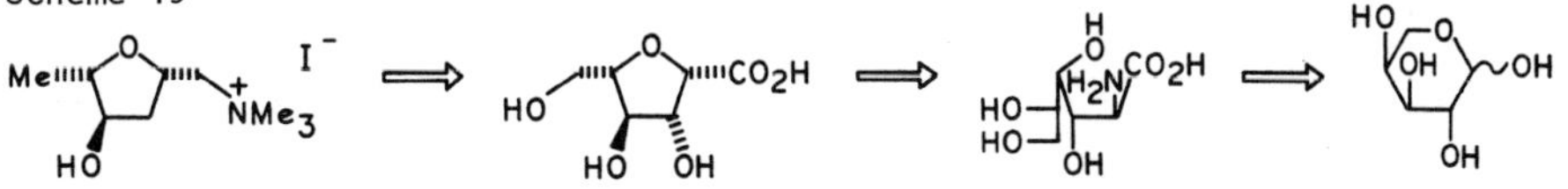

(+)-Muscarine

L-arabinose

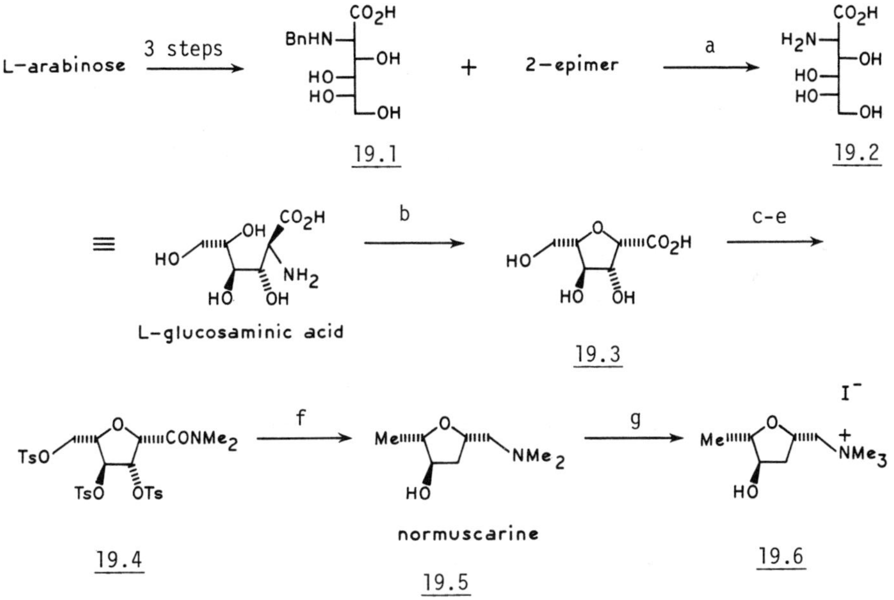

19.1 19.2

L-glucosaminic acid

19.3

19.4 19.5 19.6

normuscarine

a. Pd/C, H_2; b. $NaNO_2$, HCl; c. CH_2N_2, MeOH; d. Me_2NH, MeOH; e. TsCl, pyr.;
f. LAH,THF; g. MeI

stereochemistry of the newly created asymmetric center (formerly bearing
the amine function) can be predicted depending on the nature of neighboring carbon atom. As expected, amino acids give products with retention of configuration, but amino aldehydes, amino alcohols, etc. may
lead to inversion.[3]

In another approach[4] to the muscarine skeleton, the tetrahydrofuran ring was derived from D-mannitol based on literature precedents[5]
(Scheme 20). Intramolecular dehydration of D-mannitol is known to give
2,5-anhydro-D-glucitol 20.1 which contains the desired ring form and
side-chain substitution pattern to allow further elaboration into muscarine. The key transformations require the regioselective generation of
the C-3 hydroxyl group from the diol system present in 20.1. This was

Scheme 20

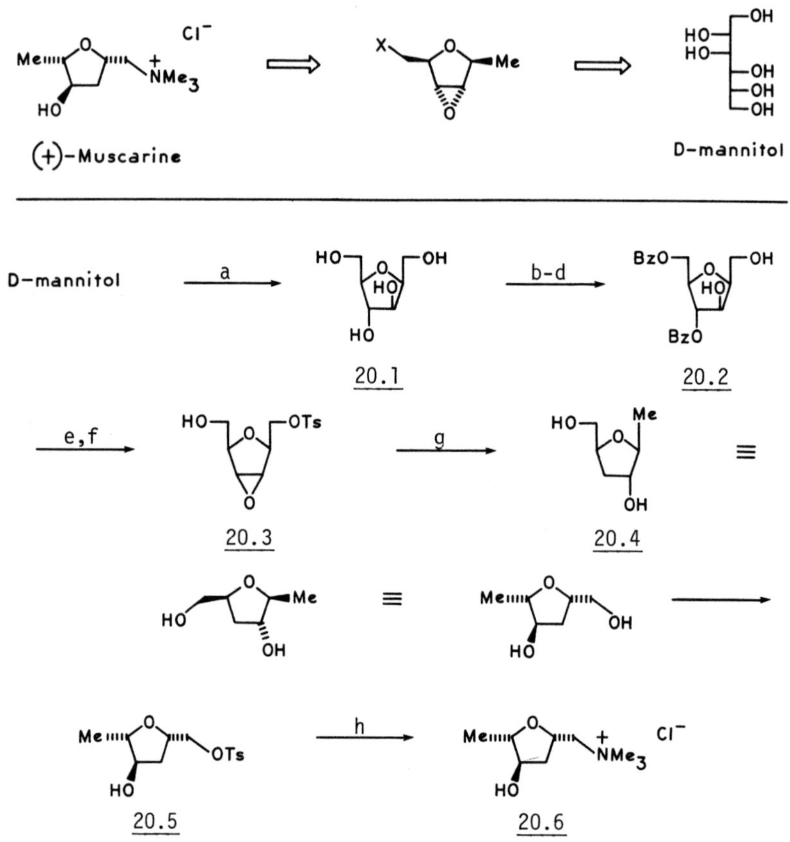

a. ref. 5; b. acetone, H^+; c. BzCl, pyr.; d. H_3O^+; e. TsCl, pyr.;
f. NaOMe; g. Red-al, THF; h. Me_3N, MeOH; then HCl.

achieved by sequential protection, deprotection and generation of intermediate 20.3, which, in one step, was reduced with sodium bis-(2-metho-
xyethoxy)-aluminum hydride (Red-al) to 20.4, an intermediate with the
desired oxidation state and sense of chirality at the ring carbon
atoms. It should be noted that the highly regioselective ring opening
(12:1) of the epoxide was assured by the nature of the hydride reagent
(compare LAH, 3:1) and the anchoring effect of the hydroxymethyl group.
A similar strategy had been previously used to create the tetrahydrofuran ring system in an oxidation product[6,7] derived from arachidonic

acid (see Scheme 26) and demonstrates not only the consequences of the *C2* symmetry present in D-mannitol but the possibility to manipulate the various hydroxyl groups in 20.1 with a high degree of stereo and regio-control. Other syntheses of optically active and racemic muscarines are also available[8].

b. (-)-Epiallomuscarine — The biological activity of musca-rine has fostered studies aimed at the synthesis of optically pure ana-logues such as epiallomuscarine[9] 21.6, where the stereochemistry at C-3 and C-5 is inverted (Scheme 21). The chiral tetrahydrofuran unit was

Scheme 21

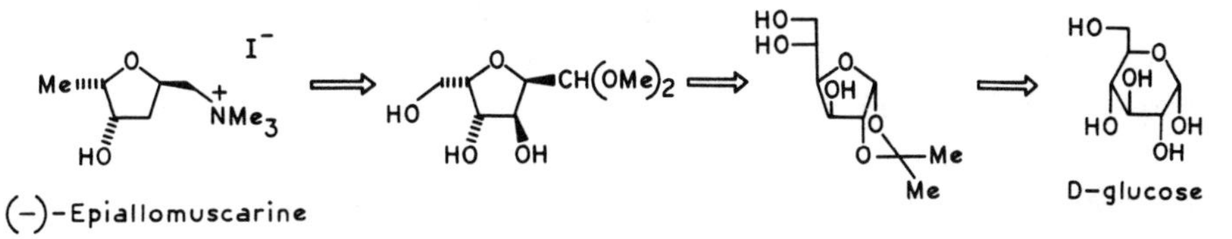

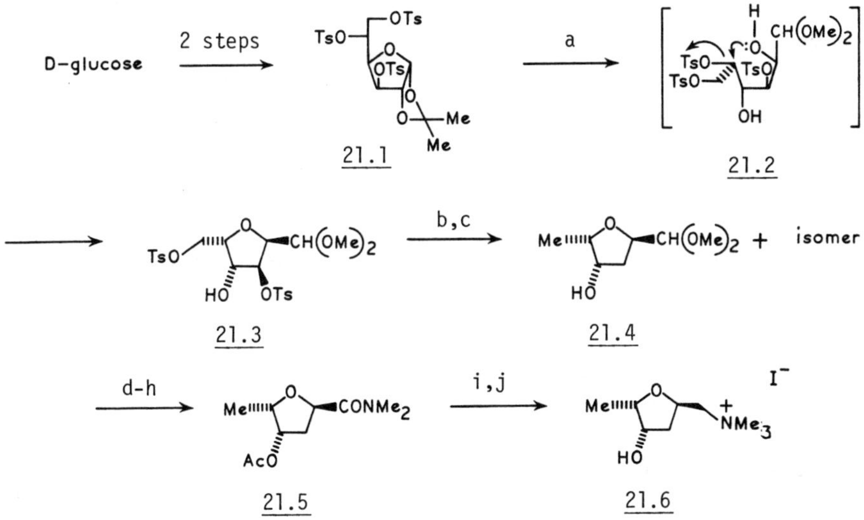

(-)-Epiallomuscarine

a. MeOH, HCl, reflux; b. NaH, THF; c. LAH, THF; d. Ac$_2$O, pyr., DMAP; e. 80% aq. AcOH; f. CrO$_3$; g. (COCl)$_2$, toluene; h. Me$_2$NH, toluene; i. LAH, THF; j. MeI

generated from a derivative of D-glucose based on the method reported by Ogawa and coworkers.[10] This consists in generating a partially tosy-lated dimethylacetal derivative 21.2 (obtained in 64% overall from D-glucose), which cannot be isolated but undergoes acid catalyzed intra-

molecular cyclization with inversion of configuration at C-5, to give
the tetrahydrofuran derivative 21.3 (a 2,5-anhydro-L-idose derivative).
Epoxide formation and hydride opening gave a 3:2 mixture of regioisomers
which were separated and the desired one 21.4, further transformed to
the amide 21.5 and ultimately to the target 21.6. Except for the formation of regioisomers in the reductive opening of the epoxide ring derived from 21.3, the method is useful for the generation of 2,5-disub-
stituted tetrahydrofurans and has been used as an approach to the synthesis of C-nucleosides.[11]

    c. (+)-Furanomycin - The structure of furanomycin 22.6 [12,13] has
been recently confirmed by a total synthesis which permitted the
correction of the original stereochemical assignment at C-5[14] (Scheme
22). The synthesis consisted in generating the functionalized
tetrahydrofuran derivative 21.3 from D-glucose[10], which was induced to
undergo a series of deoxygenations and the incorporation of an amino
acid unit. The presence of the tosyloxy esters ensured the desired
deoxygenation reaction via the epoxide 22.1[15] and the diselenide
derivative 22.2 was obtained in quantitative yield. The regiospecificity
can be ascribed to a preferred steric approach at C-4 of the bis-anhydro
aldose derivative 22.1. After introduction of the unsaturation, the
sensitive aldehyde obtained from 22.4 was subjected to the Ugi "four
component condensation reaction"[16,17] to give a 1:1 mixture of epimeric
amino acid derivatives. The desired isomer 22.5 was then converted to
(+)-furanomycin and found to be identical with a sample of the
authentic natural product. The overall yield from D-glucose was 6%.
Thus, the carbon framework of the target as well as two of three
asymmetric centers were derived from D-glucose. Except for the
formation of two stereoisomers in the Ugi reaction the synthesis can be
considered to be stereospecific. A previous synthesis of furanomycin in
0.02% overall yield is also available.[13]

    d. 2(S),4(S)-4-Amino-2-(hydroxymethyl)tetrahydrofuran-4-carboxylic
acid - The unusual amino acid 23.8 has been isolated from the acid hydrolyzate of urine from diabetics and also from similar treatment of
mixtures of hexoses and urea.[18] The absolute configuration has been
established as 2(S), 4(S) by a synthesis from D-ribose[19] (Scheme 23).

Scheme 22

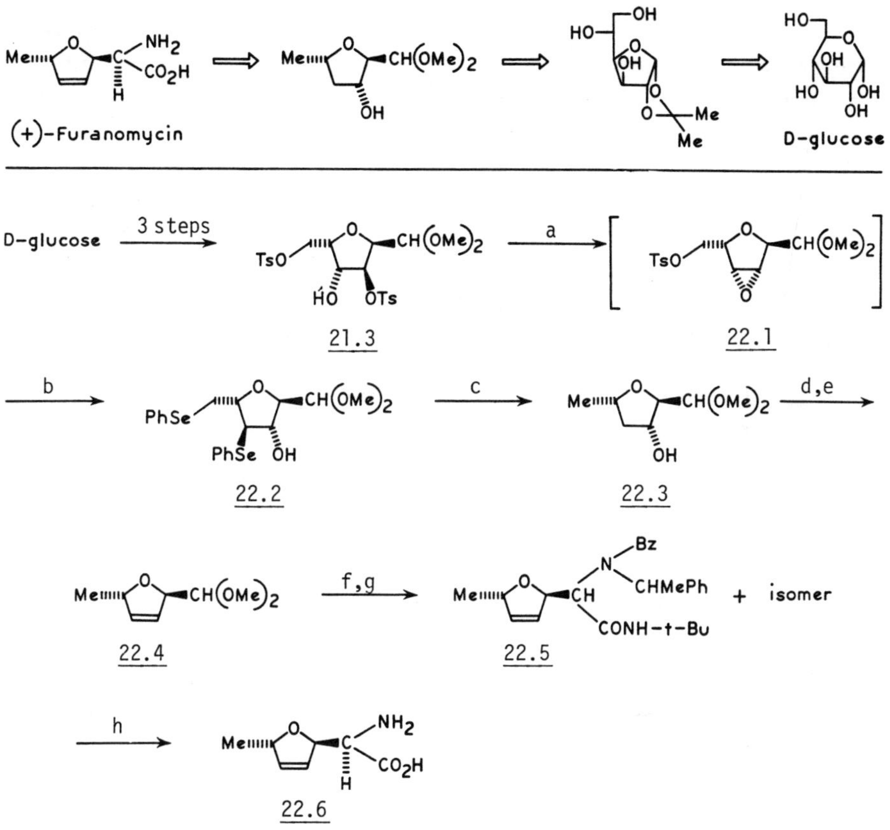

a.   NaH, THF; b. NaSePh, DMF; c. Ra-Ni, THF; d. TsCl, pyr.; e. NaOMe, MeOH;
f.   TsOH, THF; g. (+)-α-methylbenzylamine, PhCO$_2$H, t-BuNCO, MeOH; h. aq. HCl

Clearly the advantages of a carbohydrate-derived approach, resides in
the ability to generate the tetrahydrofuran ring system with a hydroxymethyl group of desired 2(S) chirality. A practical solution relies on
the selection of a D-pentofuranose derivative, which allows the introduction of the amino acid unit at C-2 of the sugar (C-4 of the target).
The choice of D-ribose also provided access to the diastereomeric 2(S),
4(R) derivative. The readily available methyl D-ribofuranoside derivative 23.1 was converted to the acyclic keto-diol derivative 23.4 via an

Scheme 23

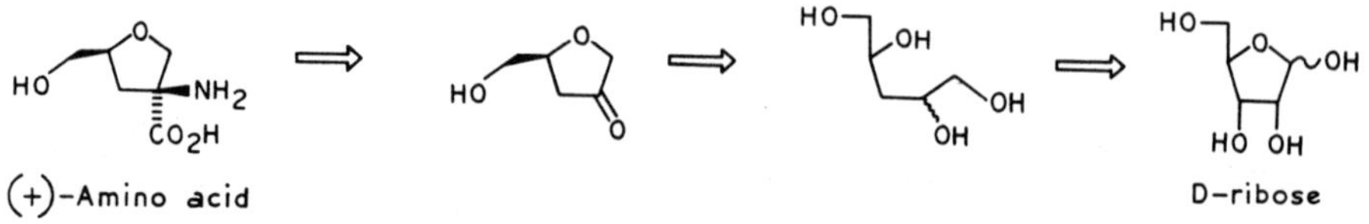

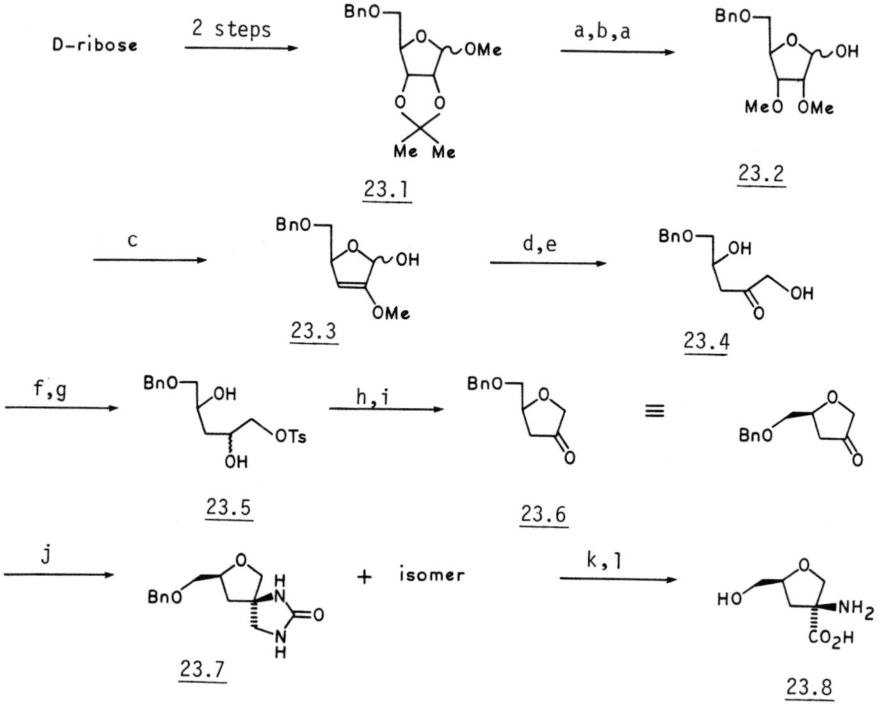

(+)-Amino acid   D-ribose

D-ribose $\xrightarrow{\text{2 steps}}$ 23.1 $\xrightarrow{\text{a,b,a}}$ 23.2

23.1

23.2

$\xrightarrow{c}$ 23.3 $\xrightarrow{d,e}$ 23.4

23.3

23.4

$\xrightarrow{f,g}$ 23.5 $\xrightarrow{h,i}$ 23.6 $\equiv$

23.5

23.6

$\xrightarrow{j}$ 23.7   + isomer $\xrightarrow{k,l}$ 23.8

23.7

23.8

a. H_3O^+; b. MeI, NaH, DMF; c. aq. Ca(OH)$_2$, 50°; d. NaBH$_4$; e. CG-50(H$^+$);
f. NaBH$_4$; g. TsCl, pyr., -15°; h. pyr., Et$_3$N (few drops); i. (CF$_3$CO)$_2$O,
DMSO; j. KCN, (NH$_4$)$_2$CO$_3$, MeOH, CO$_2$, 50 atm.; k. aq. Ba(OH)$_2$, 100°;
l. Pd/C, H$_2$

interesting β-elimination reaction. Tetrahydrofuran ring formation from
23.5 was found to be spontaneous and accelerated in the presence of a
few drops of triethylamine. With the ketone 23.6 in hand, it was
possible to form an epimeric 1:3.5 mixture of hydantoin 23.7 and its
isomer which were separated and individually converted into the desired
amino acid derivative 23.8 and its 4(R) epimer. The synthesis
illustrates the utility of a β-elimination reaction to indirectly deoxy-

genate at C-3 of an aldose, and the conversion of a furanose derivative
into a tetrahydrofuran (formally a 1,4-anhydroalditol) by intramolecular
cyclization of an alditol derivative. Unfortunately the stereoselection
in the hydantoin formation was not in favor of the natural isomer.

e. "Oxybiotin"- The oxygenated analog of biotin, in which oxygen
replaces sulfur was synthesized by Ohrui, Kuzuhara and Emoto[20] and named
"oxybiotin" 24.6. An examination of the target structure reveals a
tetrahydrofuran ring containing three contiguous substituents, including
a side-chain (Scheme 24). At first glance, this leads to a
hexofuranose-type precursor in which the C-5 chain can be elaborated
into the desired chain length and level of deoxygenation. The amino
functions in the ring have to be introduced individually and the
anomeric hydroxyl group must be replaced by hydrogen by a direct or
indirect process. The retrosynthetic analysis in Scheme 24 shows that
the side-chain was derived from C-1 and not C-5, via a process that
requires chain-shortening by one carbon (C-6). Furanose derivative 24.2
was prepared in high yield from D-glucose, albeit via a route that
involved two inversions of configuration at C-3. Note that nucleophilic
displacement reactions at C-3 in the D-allofuranose series (the
reduction product of 24.1) are facile, unlike the analogous reactions in
the D-glucose configuration (Scheme 3). Intermediate 24.1 therefore
provides easy access to an azido derivative with the desired stereochemistry. Transformation 24.3 → 24.4 involves intramolecular attack of
the C-2 hydroxyl group in an intermediate dimethyl acetal derivative
onto the terminal primary carbon atom containing the tosyloxy group.
The remaining amino group was introduced by a nucleophilic displacement
reaction and the side-chain was extended by conventional Wittig methodology. In terms of a stereochemical "balance sheet", only the
original C-2 carbon atom of D-glucose was preserved and "transferred" to the product, the stereochemistry at C-5 was sacrificed and
those at C-3 and C-4 were manipulated to incorporate the amino groups.

f. Methyl 8(R), 11(R), 15-trihydroxy-9(S),12-eicosa-(5Z),(13E)-
dienoate - An oxidation product of arachidonic acid - The synthesis of
the title compound 25.7 (Scheme 25) reported by Just and coworkers,[21]
confirmed the gross structure of a naturally occurring oxidation product
of arachidonic acid.[22] The starting carbohydrate was D-glucose whose
six-carbon framework was to become a 2,5-disubstituted tetrahydrofuran

Scheme 24

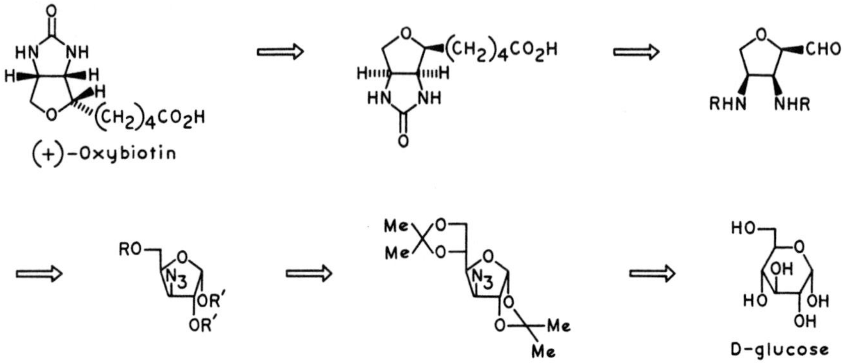

(+)-Oxybiotin

D-glucose

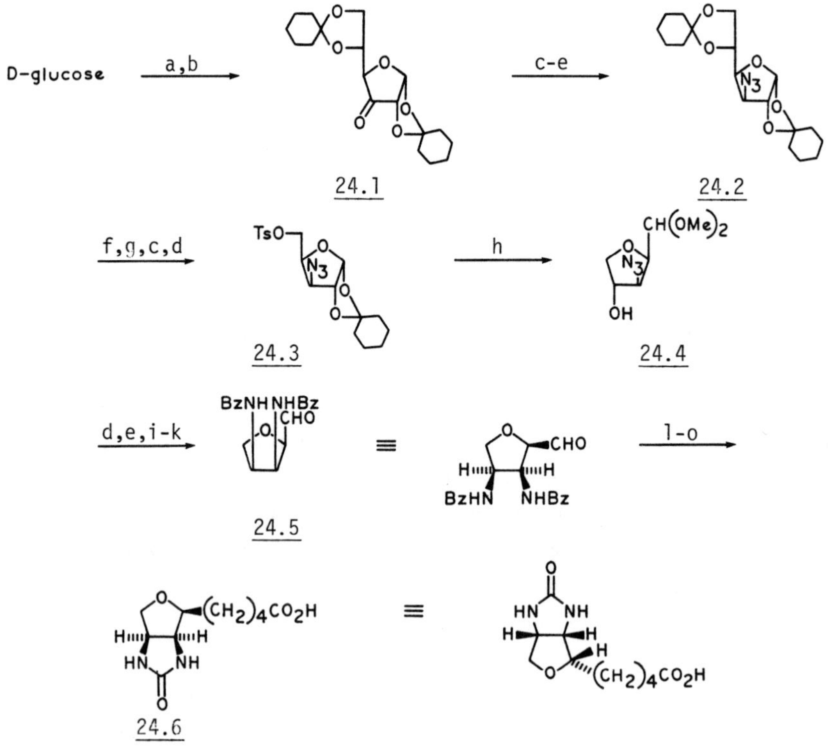

D-glucose $\xrightarrow{a,b}$ $\xrightarrow{c-e}$

24.1   24.2

$\xrightarrow{f,g,c,d}$ $\xrightarrow{h}$

24.3   24.4

$\xrightarrow{d,e,i-k}$ $\equiv$ $\xrightarrow{1-o}$

24.5

$\equiv$

24.6

a. cyclohexanone, H^+; b. RuO_4, CCl_4; c. $NaBH_4$, MeOH; d. TsCl, pyr.;
e. NaN_3, DMF; f. AcOH; g. $NaIO_4$; h. MeOH, HCl; i. Zn, MeOH;
j. BzCl, pyr.; k. H_3O^+; l. $Ph_3P=CH(CH_2)_2CO_2Et$; m. PtO_2, H_2;
n. $Ba(OH)_2$; o. $COCl_2$

Scheme 25

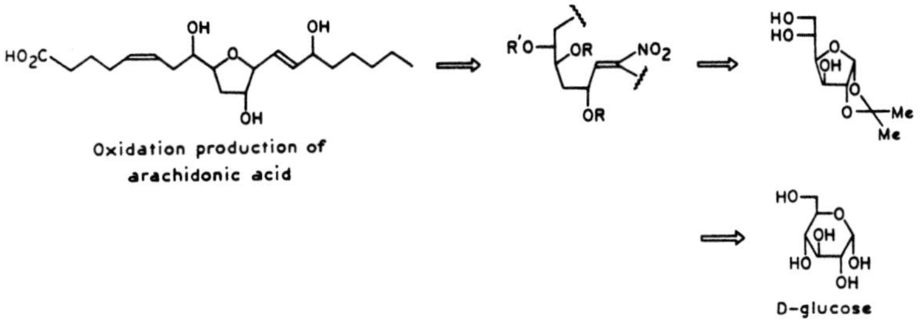

Oxidation production of
arachidonic acid

D-glucose

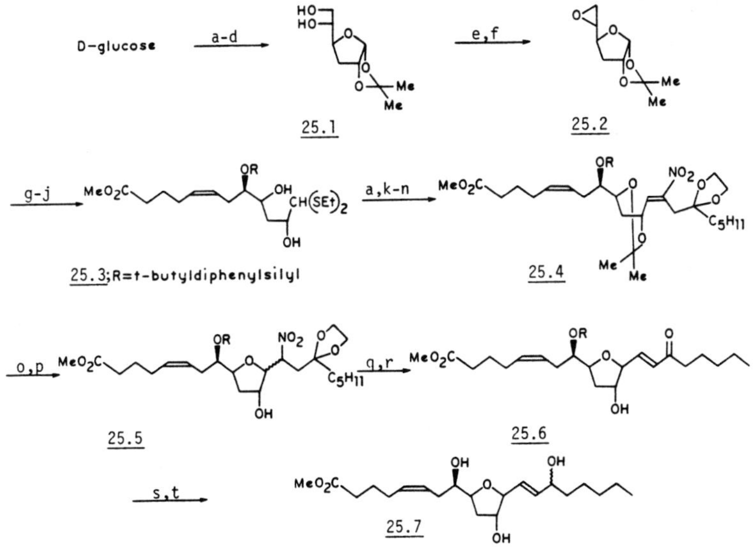

D-glucose $\xrightarrow{a-d}$ **25.1** $\xrightarrow{e,f}$ **25.2**

$\xrightarrow{g-j}$ **25.3**;R=t-butyldiphenylsilyl $\xrightarrow{a,k-n}$ MeO$_2$C ... **25.4**

$\xrightarrow{o,p}$ MeO$_2$C ... **25.5** $\xrightarrow{q,r}$ MeO$_2$C ... **25.6**

$\xrightarrow{s,t}$ MeO$_2$C ... **25.7**

a. acetone, H$^+$; b. CS$_2$, MeI, NaH; c. Bu$_3$SnH; d. AcOH; e. TsCl, pyr; f. NaOMe, MeOH;
g. HO$_2$C (CH$_2$)$_3$C≡CH, BuLi, HMPA; h. EtSH, ZnCl$_2$; i. P-2 Ni boride; j. Ph$_2$ t-BuSiCl,
imidazole, DMF; k. HgCl$_2$, HgO, aq. acetone; l. O$_2$N⟋ ... C$_5$H$_{11}$ R$_2$NH, DMF; m. Ac$_2$O
pyr. DMAP; n. K$_2$CO$_3$, benzene, 18-crown-6, reflux; o. MeOH, HCl;
p. Et$_3$N, THF; q. aq. TsOH; r. Et$_3$N, CHCl$_3$; s. NaBH$_4$; t. Bu$_4$NF.

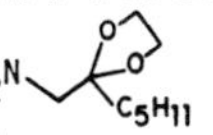

as found in the target. Note that the original stereochemistry at C-2,
C-4 and C-5 is carried over. A new center is created at the "anomeric"
carbon atom which is involved in ring formation via an intramolecular
Michael-type reaction of an α,β-unsaturated nitroolefin. In terms of
design, the key steps involve the generation of functionalized side-
chains which can be further elaborated to produce the prostaglandin-type
appendages.

Biogenetic considerations led the authors to a 9($\underline{S}$), 11($\underline{R}$)
stereochemistry as found in C-2 and C-4 of D-glucose. The easy access
to intermediate 25.1[23] with a predetermined additional chiral center at
C-5, and the possibility to elaborate at both termini, rendered the
synthetic approach practical. The "acid side-chain" was introduced
first in a critical chain-extension reaction involving the epoxide derivative 25.2. The product was then protected as the t-butyldiphenylsilyl
ether, and converted into the acyclic diethyl dithioacetal derivative
25.3, which was further elaborated into the nitroolefin 25.4. By a
judicious choice of O-protecting groups, it was possible to generate a
diol and induce a base-catalyzed intramolecular addition involving the
original C-4 hydroxyl and the anomeric carbon (the latter now being part
of the nitroolefin system). Further elaboration of the side-chain gave
a single enone (HPLC) with a stereochemistry at the C-12 of the ring
tentatively assigned as in expression 25.6 (*cis* to the other side-
chain). This synthesis illustrates an altogether different strategy in
the formation of tetrahydrofuran ring systems and demonstrates the
possibility of stereocontrol at the newly created center by intramolecular attack of a hydroxyl group on an α,β-unsaturated nitroolefin.

A conceptually different approach for the construction of the
chiral tetrahydrofuran ring present in 25.7 with complete control of
stereochemistry at three centers (C-9, C-11, C-12) has been communicated
previously[6,7] and took advantage of the symmetry present in D-mannitol
(Scheme 26). Thus, dehydration[5] of the readily available 1,6-di-O-
benzoyl-D-mannitol 26.1 proceeded via 26.2 to give the crystalline and
easily isolated (30%) tetrasubstituted tetrahydrofuran derivative 26.3
(2,5-anhydro-1,6-di-O-benzoyl-D-glucitol), in addition to isomeric
anhydro and dianhydro alditols. Note that the same product is formed
from cyclization of either C-2 or C-5 hydroxyl groups because of the *C2*
symmetry in the molecule. Conversion of 26.3 into the target containing
the desired substitution pattern and stereochemistry requires deoxygenation at C-4 and inversion of configuraton at C-3. It is also

Scheme 26

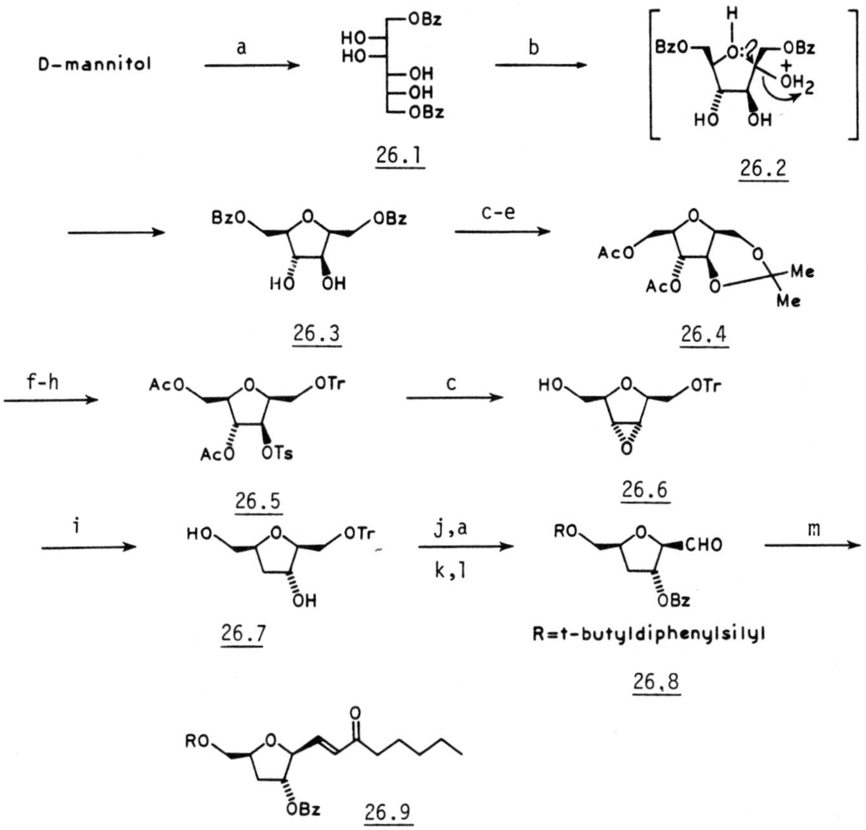

a. BzCl, pyr; b. TsOH, $Cl_2CHCHCl_2$, reflux; c. NaOMe, MeOH, d. acetone, H^+
e. Ac_2O, pyr; f. aq.AcOH; g. TrCl, pyr.; h. TsCl, pyr.; i. LAH, THF;
j. Ph_2t-BuSiCl, imidazole, DMF; k. HBr, AcOH, 0^0;

l. Collins; m. $Bu_3P=CHCOC_5H_{11}$

imperative that both hydroxymethyl groups be manipulated independently.
These requirements were met by 26.3. The *trans-* relationship of the
hydroxyl groups allowed preferential protection and manipulation of the
molecule to produce 26.5 which led to the epoxide 26.6. Reduction with
lithium aluminum hydride was highly regioselective (if not specific) to
give 26.7, presumably by virtue of coordination with the favorably
situated C-6 hydroxymethyl group. The combination of a trityl group at

C-1 and the t-butyldiphenylsilyl group at C-6 provided the required criteria for selective deprotection and further elaboration into 26.8 and 26.9. The ready accessibility of the chiral tetrasubstituted tetrahydropyran derivative 26.3 and the favorable "*cis*" orientation of the two side-chains makes it a useful precursor to other types of natural products containing such a ring system as in C-nucleosides[7] and, for example, the intriguing antibiotic aurodox[24] 27.4 (Scheme 27).

Scheme 27

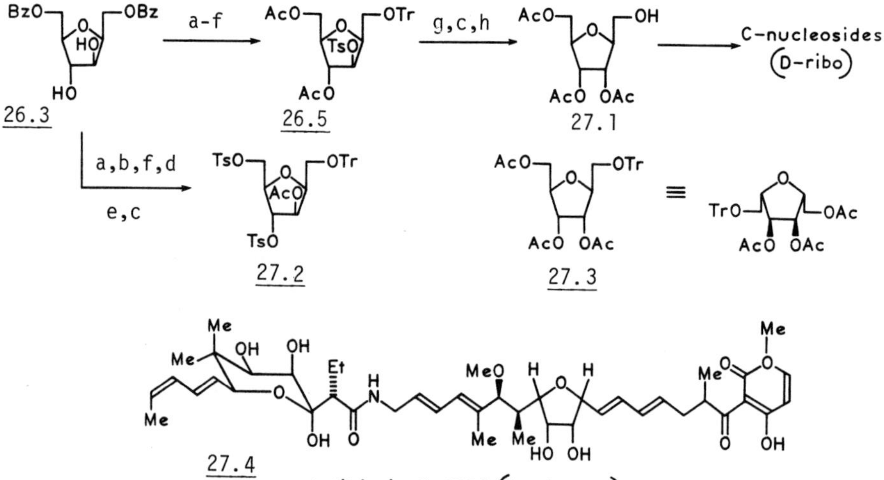

Antibiotic X-5108 (Goldinodox)

a. NaOMe, MeOH; b. acetone, H^+; c. Ac_2O, pyr.; d. H_3O^+; e. TrCl, pyr.; f. TsCl, pyr., 50^0; g. NaOAc, aq. DMF, reflux; h. HBr, AcOH, 0^0, 10 min.

Thus, 26.3 has been transformed into intermediates 27.1, 27.3 by standard reactions involving neighboring group participation of acetate or benzoate esters. Note that with preferential substitution at either hydroxymethyl group, one can elaborate each end individually and maintain the asymmetric character of otherwise symmetrical molecules. Chiral tetrahydrofurans with functionalized side-chains at C-2 and C-4 are available by other routes as well starting with carbohydrate precursors.[1]

g. 11-Oxaprostaglandin $F_2\alpha$ - Two independent syntheses of this prostaglandin analog start with different chiral precursors and employ different strategies for the construction of the trisubstituted tetrahydrofuran ring system (Scheme 28). Lourens and Koekemoer[25] started with D-glucose and utilized the furanose derivative 28.1 for the highly regioselective introduction of an acetate branch at C-3. Thus, the two side-chains in the target with the desired *trans*-orientation were present in 28.2 at an early stage. There remained to replace the anomeric hydroxyl group by a hydrogen atom thereby creating the tetrahydrofuran ring and to elaborate the side-chains by well known reactions in the prostaglandin field. Deoxygenation at C-1 of intermediate 28.3 was done via the corresponding phenylthio glycoside and treatment with Raney-nickel. The side-chain containing C-5 — C-6 served as a precursor to an aldehyde which, in turn, led to 28.6. The epimeric 11-oxaprostaglandins 28.8 and 28.9 were separated by chromatography.

Another approach[26] also utilizes D-glucose as a primary chiral precursor, but takes advantage of the ready availability of 1,4-anhydro-D-glucitol from the dehydration[27] of D-glucitol (Scheme 29). Thus, preferential protection of 29.2 gave intermediate 29.3 which was oxidized and homologated with high stereocontrol to give the desired acetate side-chain as in 29.5. The presence of the bulky t-butyldiphenylsilyl group at C-2 was probably responsible for the selectivity in the reduction of the olefinic intermediate 29.4. The sequence of reactions shown in Scheme 29 led to the C-15 epimeric 11-oxaprostaglandin $F_2\alpha$ methyl esters which were chromatographically separable. Synthesis of ($\pm$)-11-oxa and ($\pm$)-9-oxaprostaglandins have also been reported.[28]

h. 11-Oxaprostaglandin $F_2\beta$ - The first synthesis of an optically active 11-oxaprostaglandin also utilized 1,4-anhydro-D-glucitol as a chiral precursor, and led to the $F_2\beta$ analog 30.7[29] (Scheme 30). The critical acetate side-chain was introduced by opening of the epoxide intermediate 30.2, easily available via 30.1. Unfortunately the reaction was not regiospecific and gave a 1:1 mixture of isomers. The desired malonate derivative 30.3 was then converted via routine steps to the target 30.7 and its C-15 epimer 30.8 which could be separated by chromatography.

Scheme 28

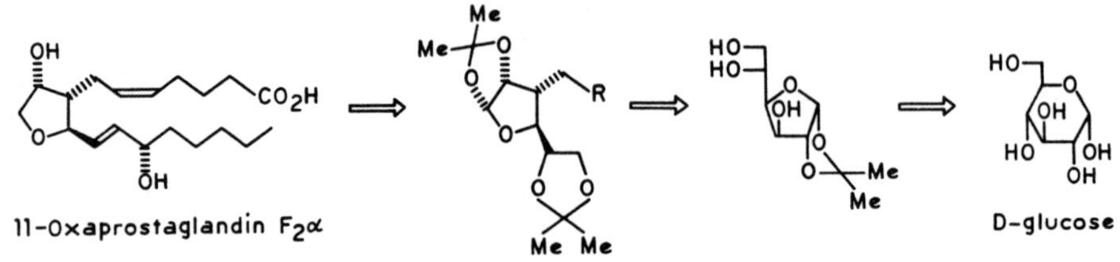

11-Oxaprostaglandin F$_{2\alpha}$

D-glucose

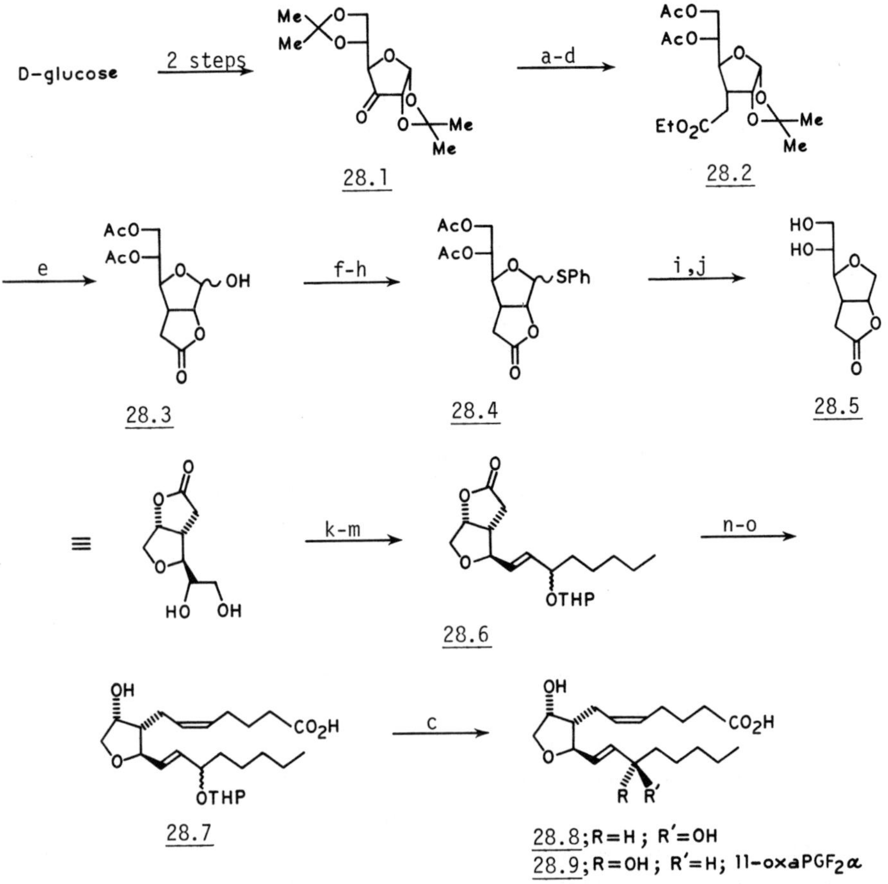

D-glucose $\xrightarrow{\text{2 steps}}$ 28.1 $\xrightarrow{\text{a-d}}$ 28.2

$\xrightarrow{e}$ 28.3 $\xrightarrow{\text{f-h}}$ 28.4 $\xrightarrow{\text{i,j}}$ 28.5

$\equiv$ $\xrightarrow{\text{k-m}}$ 28.6 $\xrightarrow{\text{n-o}}$

28.7 $\xrightarrow{c}$ 28.8; R=H; R'=OH

28.9; R=OH; R'=H; 11-oxaPGF$_{2\alpha}$

a. (MeO)$_2$P(O)CH$_2$CO$_2$Et, KH; b. Ra-Ni, H$_2$; c. aq. AcOH; d. Ac$_2$O, pyr.;
e. aq. AcOH reflux; f. O$_2$NC$_6$H$_4$COCl, pyr.; g. HBr, CH$_2$Cl$_2$; h. PhSK, EtOH;
i. Ra-Ni; j. aq. K$_2$CO$_3$, MeOH; k. NaIO$_4$; l. (MeO)$_2$P(O)CH$_2$COC$_5$H$_{11}$, then ZnBH$_4$,
NaH; m. dihydropyran, TsOH; n. DIBAL; o. Ph$_3$P=CH(CH$_2$)$_3$CO$_2$H, NaH, DMSO

Scheme 29

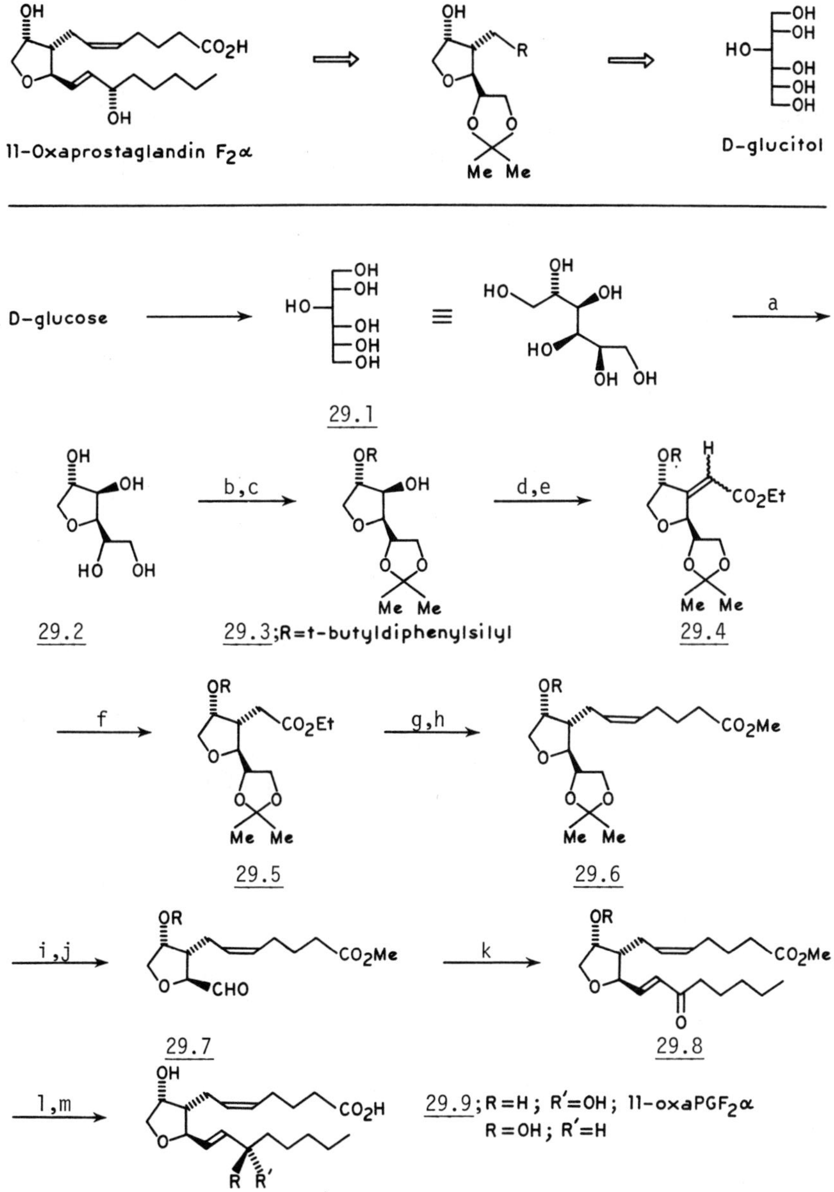

11-Oxaprostaglandin F$_2\alpha$

D-glucitol

D-glucose

29.1

29.2

b,c

29.3;R=t-butyldiphenylsilyl

d,e

29.4

f

29.5

g,h

29.6

i,j

29.7

k

29.8

l,m

29.9;R=H; R'=OH; 11-oxaPGF$_2\alpha$
R=OH; R'=H

a. aq. H$_2$SO$_4$; b. acetone, H$^+$; c. Ph$_2$t-BuSiCl, imidazole; d. EDAC, DMSO;
e. Ph$_3$P=CHCO$_2$Et; f. Pd/C, H$_2$; g. DIBAL; h. Ph$_3$P=CH$_2$(CH$_2$)$_3$CO$_2$HBr$^-$, then
CH$_2$N$_2$; i. aq. AcOH; j. NaIO$_4$; k. Bu$_3$P=CHCOC$_5$H$_{11}$; l. Zn(BH$_4$)$_2$, DMF; m. Bu$_4$NF,
THF.

Scheme 30

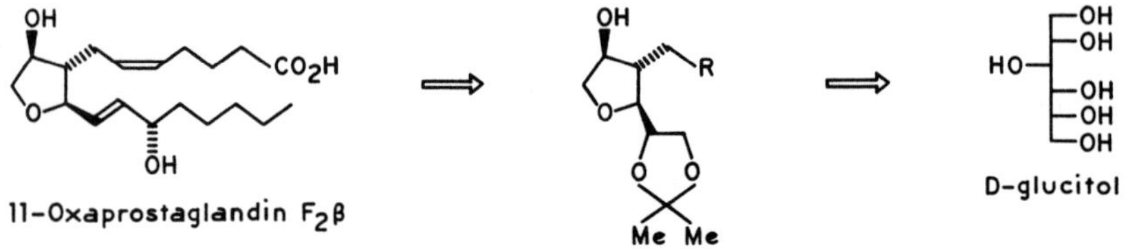

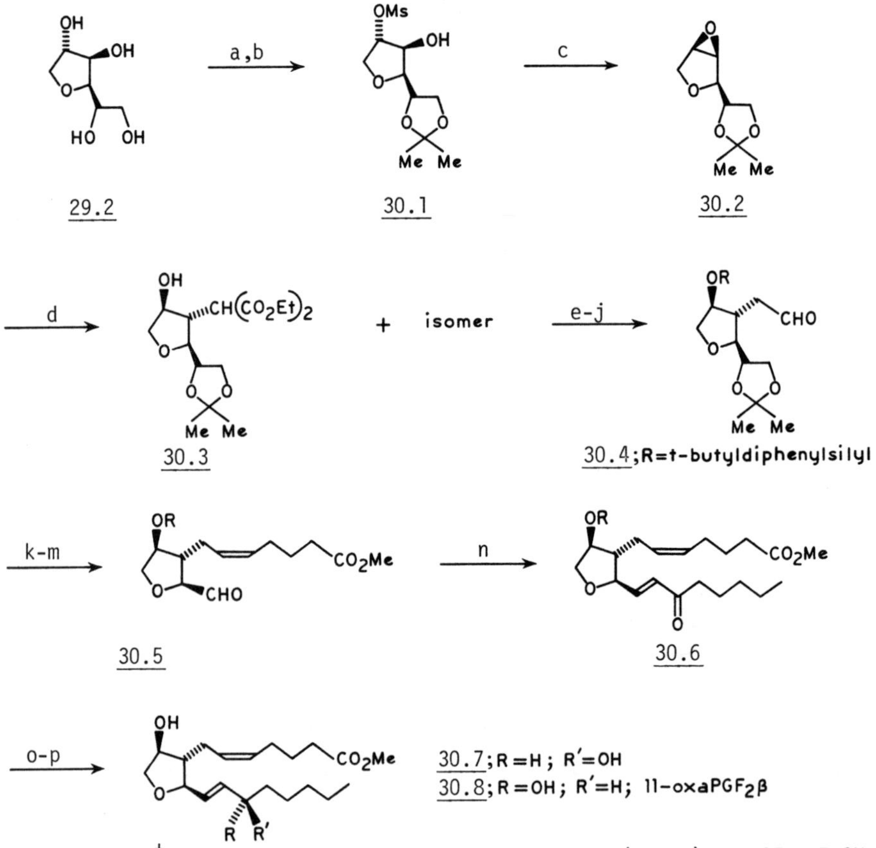

11-Oxaprostaglandin F$_2\beta$ $\Longrightarrow$ (intermediate with Me Me acetonide) $\Longrightarrow$ D-glucitol

29.2 $\xrightarrow{a,b}$ 30.1 $\xrightarrow{c}$ 30.2

$\xrightarrow{d}$ 30.3   + isomer $\xrightarrow{e-j}$ 30.4;R=t-butyldiphenylsilyl

$\xrightarrow{k-m}$ 30.5 $\xrightarrow{n}$ 30.6

$\xrightarrow{o-p}$ 30.7;R=H; R'=OH
30.8;R=OH; R'=H; 11-oxaPGF$_2\beta$

a. acetone, H$^+$; b. MsCl, pyr.; c. NaOMe, MeOH; d. CH$_2$(CO$_2$Et)$_2$, NaOEt, EtOH;
e. Ph$_2$t-BuSiCl, imidazole; f. aq. NaOH; g. Dowex-50(H$^+$); h. pyridine, reflux;
i. EtOH, H$_2$SO$_4$; j. DIBAL; k. Ph$_3$P$^+$CH$_2$(CH$_2$)$_3$CO$_2$H Br$^-$, then CH$_2$N$_2$; l. aq. AcOH;
m. NaIO$_4$; n. Bu$_3$P=CHCOC$_5$H$_{11}$; o. Zn(BH$_4$)$_2$, DMF; p. Bu$_4$NF, THF

i. (+)- and (-)-Nonactic acids - The syntheses of (+)- and
(-)-nonactic acids, components of the nonactins,[30] by Ireland and
Vevert[31] demonstrate the applicability of the enolate Claisen
rearrangement to the carbohydrate series and the possibility of
transferring chirality from one site to another. For the synthesis of
the (+)-acid 31.8 (Scheme 31), the commercially available D-gulono-γ-
lactone proved to be a useful chiral precursor. In essence it served as
a "template" for the enolate Claisen rearrangement which introduced a
carbon appendage having a *cis* orientation to an existing side-chain, and
circumvented the necessity of using an L-sugar such as L-mannose. Note
the favorable orientation of the side-chain in the starting lactone as
well as the   *cis*-disposition of the vicinal C-2/C-3 diol necessary for
glycal formation.[32] The strategy involved the generation of intermediate 31.4, a furanoid glycal containing a chain-extended and
protected side-chain with the desired sense of chirality at the ring
junction, and a C-3 hydroxyl group which served as an anchor for the
transfer of chirality. The major isomer 31.5 was then transformed to
the 8(S) and 8(R) (+)-nonactic acids, 31.7, and 31.8, respectively.
Unfortunately no stereochemical control was possible in the last organometallic addition step.

The synthesis of (-)-nonactic acid starts with D-mannose (Scheme
32) and proceeds by the same strategy as for the previous synthesis.
The choice of D-mannose as a chiral precursor is dictated by the availability of a furanoid derivative such as 32.3 which can be subjected to
elimination by the Ireland procedure to give the desired glycal 32.4.
Otherwise, the required *cis* orientation of the C-3 hydroxyl group and
the side-chain could also be found in the more readily available
D-glucose derivative, but the formation of a glycal in this series is
without precedent. The enolate Claisen rearrangement provided a mixture
of isomers 32.5 in which the major component was the desired one.
Further elaboration led to a mixture of (-)-nonactic acids 32.7 and
32.8, epimeric at C-8. Several other syntheses of the nonactic acids
are available[30,33] from carbohydrate as well as non-carbohydrate
precursors.   The stereocontrolled introduction of carbon chains at the
anomeric carbon atom in furanoid derivatives can also be accomplished by
other methods involving carbanion additions, Wittig-type reactions and
related methodologies.[7,34]

Scheme 31

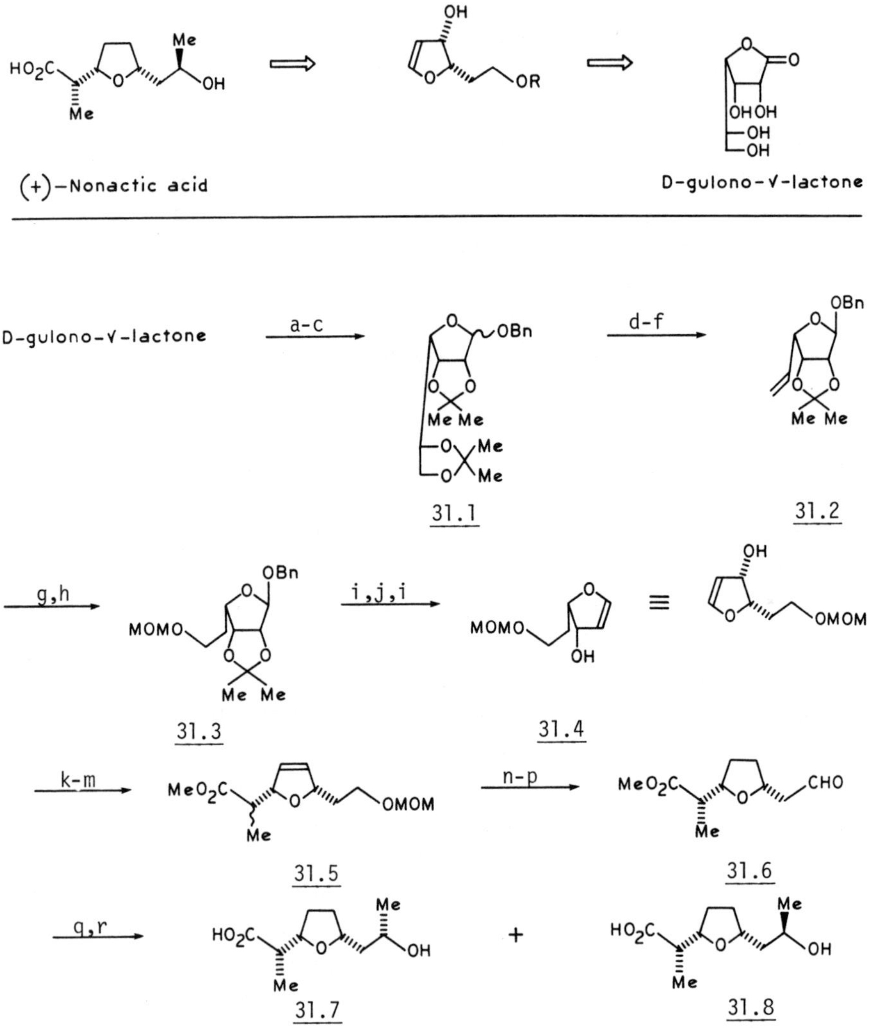

(+)-Nonactic acid   D-gulono-γ-lactone

D-gulono-γ-lactone $\xrightarrow{a-c}$ **31.1** $\xrightarrow{d-f}$ **31.2**

$\xrightarrow{g,h}$ **31.3** $\xrightarrow{i,j,i}$ **31.4**   ≡

$\xrightarrow{k-m}$ **31.5** $\xrightarrow{n-p}$ **31.6**

$\xrightarrow{q,r}$ **31.7**   +   **31.8**

a. acetone, H⁺; b. DIBAL, ether; c. BnCl, NaH, DMF; d. HCl, MeOH; e. Me₂
NCH(OMe)₂, CH₂Cl₂; f. Ac₂O, 130°; g. 9-BBN, NaOH, H₂O₂; h. CH₂(OMe)₂, P₂O₅;
i. Li, NH₃, -78°; j. Ph₃P, CCl₄; k. n-BuLi; l. EtCOCl, LDA, THF, -78°, TMSCl;
m. CH₂N₂; n. Pd/C, H₂; o. HCl, MeOH; p. (COCl)₂, DMSO, Et₃N; q. MeMgBr,
-78° → 10° ; r. aq. KOH

Scheme 32

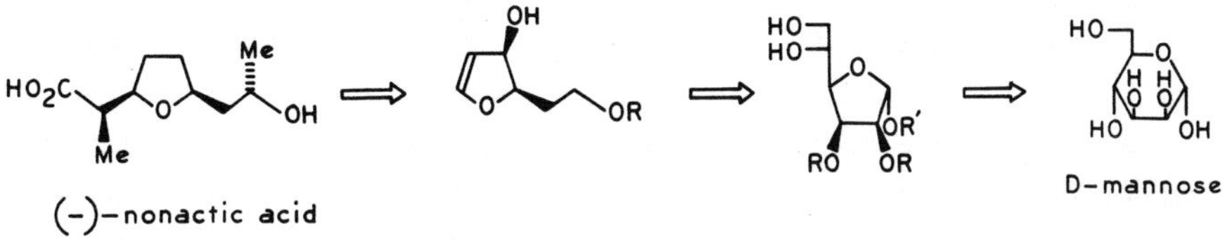

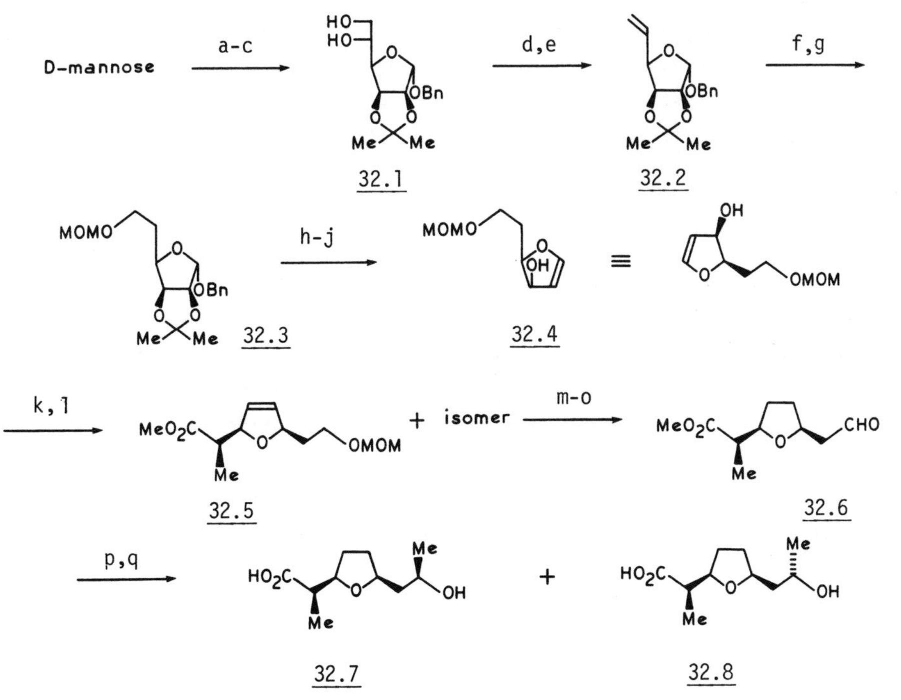

a. acetone, H$^+$; b. BnCl, NaH, DMF; c. HCl, MeOH; d. Me$_2$NCH(OMe)$_2$, CH$_2$Cl$_2$;
e. Ac$_2$O, 130^0; f. 9-BBN, NaOH, H$_2$O$_2$; g. ClCH$_2$OMe, KH; h. Li, NH$_3$, -78^0;
i. Ph$_3$P, CCl$_4$; j. n-BuLi; k. EtCOCl, LDA, THF -78^0, TMSCl; l. CH$_2$N$_2$; m. Rh/C,
H$_2$; n. MeOH, HCl; o. (COCl)$_2$, DMSO, Et$_3$N; p. Me$_2$CuLi; q. aq. KOH

j. Germacranolide Sesquiterpenes – Synthesis of the furenone
portion – The furenone portion of some germacranolide sesquiterpenes
such as eremantholide A and ciliarin has been constructed from a
furanoid sugar derivative[35] (Scheme 33). Examination of the target
structure reveals a highly branched dihydrofurenone system in which the
ring system and one of the side-chains can originate from a hexofuranose
derivative. The choice of D-mannose as starting material was not for
stereochemical reasons but for the operational convenience of having a
partially protected lactol in the furanose form, since D-mannose gives
the highly crystalline 2,3:5,6-di-O-isopropylidine derivative. The
ketol 33.2, easily obtained from the lactone 33.1 was subjected to a
Wittig reaction (note conditions) which led to a mixture of the
C-glycoside 33.4 and its isomer. Fortunately, it was possible to
equilibrate the minor isomer and obtain more of the desired 33.4, thus
rendering the synthesis quite efficient up to this point. Note the
feasibility of Wittig reactions with lactol precursors of the type
33.2[7,36] and the formation of cyclic tetrahydrofurans by intramolecular
Michael-type cyclizations to give 33.4. Introduction of the dienone
system was accomplished by first converting the C-5/C-6 side-chain into
the methyl ketone derivative 33.5 which was subjected to a β-elimination, followed by incorporation of the carbomethoxymethylene unit via
Wittig methodology and oxidation. Thus in the synthesis of the furenone
33.8, the choice of sugar precursor was an excellent one, based on sound
planning and taking advantage of inherent stereochemistry to introduce
carbon-branching with a high degree of regiocontrol. This illustrates
an example where destroying all the asymmetric centers of the starting
sugar derivative is advantageous and synthetically expedient.

Scheme 33

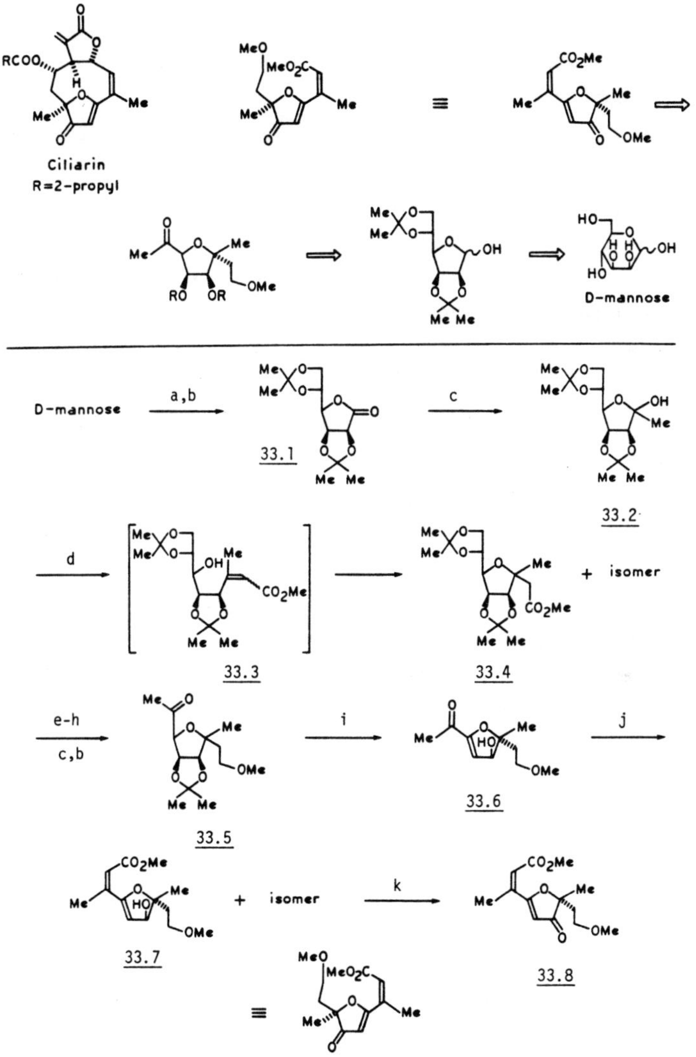

Ciliarin
R=2-propyl

D-mannose

33.1

33.2

33.3

33.4

33.5

33.6

33.7

33.8

a. acetone, H^+, b. Collins; c. MeLi; d. $Ph_3P=CHCO_2Me$, MeCN, 125 p.s.i., $160^°$; e. LAH, ether; f. MeI, NaH, DMF; g. aq. AcOH; h. $NaIO_4$, aq. MeOH; i. NaOMe, MeOH; j. $Ph_3P=CHCO$ Me, MeCN reflux; k. Ag_2CO_3, Celite

REFERENCES

1. E. Hardegger and F. Lohse, Helv. Chim. Acta, 40, 2383 (1957).

2. C.H. Eugster, Fortschr. Chem. Org. Naturst., 27, 288 (1969).

3. See for example, B.C. Bera, A.B. Foster and M. Stacey, J. Chem. Soc., 4531 (1956); J. Defaye, Advan. Carbohydr. Chem. Biochem., 25, 181 (1970).

4. A.M. Mubarak and D.M. Brown, J. Chem. Soc., Perkin I, 809 (1982); Tetrahedron Lett., 21, 2453 (1980).

5. T.A.W. Koerner, R.J. Voll and, E.S. Younathan, Carbohydr. Res., 59, 403 (1977); see also R.C. Hockett, H.G. Fletcher, Jr., E.L. Sheffield, R.M. Goepp, Jr., and S. Solzberg, J. Am. Chem. Soc., 68, 927 (1946) and references cited therein.

6. S. Hanessian and G. Rancourt, Abstracts 169th Natl. Am. Chem. Soc. Meeting, Philadelphia, Pa, 1975, CARB 26.

7. S. Hanessian and A.G. Pernet, Advan. Carbohydr. Chem. Biochem., 33, 111 (1976).

8. R. Amouroux, B. Gerin and M. Chastrette, Tetrahedron Lett., 23, 4341 (1982); S. Pochet and T. Huynh-Dinh, J. Org. Chem., 47, 193(1982); W.C. Still and J.A. Schneider, J. Org. Chem., 45, 3375 (1980); for other syntheses, see J. Whiting, Y.-K. Au-Young and B. Belleau, Can. J. Chem., 50, 3322 (1972); T. Matsumoto, A. Ichihara and N. Ito, Tetrahedron, 25, 5889 (1969).

9. P.-C. Wang, Z. Lysenko and M.M. Jouillé, Tetrahedron Lett., 1657 (1978); see also Heterocycles, 9, 753 (1978); P.C. Wang and M.M. Jouillé, J. Org. Chem., 45, 5359 (1980).

10. T. Ogawa, M. Matsui, H . Ohrui, H. Kuzuhara and S. Emoto, Agr. Biol. Chem., 36, 1449 (1972).

11. T. Ogawa, Y. Kikuchi, M. Matsui, H. Ohrui, H. Kuzuhara and S. Emoto, Agr. Biol. Chem., 35, 1825 (1971).

12. K. Katagiri, Y. Tori, Y. Kimura, T. Yoshida, T. Nagasaki and H. Minato, J. Med. Chem., 10, 1149 (1967).

13. T. Masamune and M. Ono, Chem. Lett., 625 (1975).

14. M.M. Jouillé, P.C. Wang and J.E. Semple, J. Am. Chem. Soc., 102, 887 (1980).

15. For a review see, N. R. Williams, Advan. Carbohydr. Chem. Biochem., 25, 155 (1970).

16. R. Urban and I. Ugi, Angew. Chem., Int. Ed. Engl., 14, 61 (1975).

17. H.R. Divanfard, Z. Lysenko, P.C. Wang and M.M. Jouillé, Synth. Commun., 269 (1978).

18. S. Mizuhara, H. Kodawa, S. Ohmori, K. Taketa and M. Ueda, Physiol. Chem. and Physics, $\underline{1}$, 87 (1974); $\underline{6}$, 91 (1974).

19. J. Yoshimura, S. Kondo, M. Ihara and H. Hashimoto, Carbohydr. Res., $\underline{99}$, 129(1982); Chem. Lett., 819(1979).

20. H. Ohrui, H. Kuzuhara and S. Emoto, Agr. Biol. Chem., $\underline{35}$, 754 (1971).

21. G. Just, C. Luthe and H. Oh, Tetrahedron Lett., $\underline{21}$, 1001, (1980).

22. C. Pace-Asciak, Biochemistry, $\underline{10}$, 3664 (1971); C. Pace-Asciak and L.S. Wolfe, Chem. Comm., 1234, 1235 (1970); C. Pace-Asciak, Biochemistry, $\underline{10}$, 3664 (1971); Chem. Commun., 1234, 1235 (1970); G. Bild, S.G. Bhat, C.S.Ramadoss and B. Axelrod, J. Biol.Chem., $\underline{253}$, 21 (1978).

23. D.H.R. Barton and S.W. McCombie, J. Chem. Soc., Perkin I, 1574 (1975).

24. H. Maehr, M. Leach, J.F. Blount and A. Stempel, J. Am. Chem. Soc., $\underline{96}$, 4034 (1974); C. Vos and P.E.J. Vermeil, Tetrahedron Lett., 5173 (1973).

25. G.J. Lourens and J.M. Koekemoer, Tetrahedron Lett., 3715, 3719 (1975).

26. S. Hanessian, P. Lavallée and Y. Guindon, unpublished results, Ph.D. Thesis, Y. Guindon, University of Montreal, 1981.

27. S. Soltzberg, R.M. Goepp, Jr., and W. Freudenberg, J. Am. Chem. Soc., $\underline{68}$, 919 (1946).

28. I.T. Harrison and V.R. Fletcher, Tetrahedron Lett., 2729 (1974); I. Vlattas and L. Dellavecchia, Tetrahedron Lett., 4267, 4455, 4459 (1974).

29. S. Hanessian, P. Dextraze, A. Fougerousse and Y. Guindon, Tetrahedron Lett., 3983 (1974).

30. J. Dominguez, J.D. Dunitz, H. Gerlach and V. Prelog, Helv. Chim. Acta, $\underline{45}$, 129 (1962); H. Gerlach and V. Prelog, Ann., $\underline{669}$, 121 (1963).

31. R.E. Ireland and J.-P. Vevert, J. Org. Chem., $\underline{45}$, 4259 (1980).

32. R.E. Ireland, S. Thaisrivongs, N. Vanier and C.S. Wilcox, J. Org. Chem., $\underline{45}$, 48 (1980).

33. See for example, K.M. Sun and B. Fraser-Reid, Can. J. Chem., $\underline{58}$, 2732 (1980); see also P.A. Barlett and K.K. Jernstedt, Tetrahedron Lett., 1607 (1980); M. J. Arco, M.H. Trammell and J.D. White,

J. Org. Chem., 41, 2075 (1976); H. Gerlach, K. Oertle, A. Thalmann
and S. Servi, Helv. Chim. Acta, 58, 2036 (1975); U. Schmidt, J.
Gombos, E. Haslinger and H. Zak, Chem. Ber., 109, 2628 (1976).

34. See for example, A.P. Kozikowski and K.L. Sorgi, Tetrahedron Lett.,
     24, 1563(1983); M.D. Lewis, J.K. Cha and Y. Kishi, J. Am. Chem.
     Soc., 104, 4976(1982); G.E. Keck and J.B. Yates, J. Am. Chem. Soc.,
     104, 5829(1982); T.L. Cupps, D.S. Wise and L.B. Townsend, J. Org.
     Chem., 47, 5115(1982); S. Danishefsky and J.F. Kerwin, Jr.,
     J. Org. Chem., 47, 3805(1982); M. Chmielewski, J.N. BeMiller and
     D.P. Cerretti, J. Org. Chem., 46, 3903(1981).

35. T.F. Tam and B. Fraser-Reid, J. Org. Chem., 45, 1344(1980).

36. See for example, Yu.A. Zhdanov, Y.E. Alexeev and V.G. Alexeeva,
     Advan. Carbohyd. Chem. Biochem., 27, 227(1972) and references cited
     therein, see also, J.G. Buchanan, A.R. Edgar, M.J. Power, and P.D.
     Theaker, Carbohydr. Res., 38, C22(1974); S. Hanessian, T. Ogawa and
     Y. Guindon, Carbohydr. Res., 38, C12(1974); H. Ohrui, G.H. Jones,
     J.G. Moffatt, M.L. Maddox, A.T. Christensen and S.K Byram, J. Am.
     Chem. Soc., 97, 4602(1975).

CHAPTER SEVEN

MOLECULES CONTAINING A TETRAHYDROPYRAN-TYPE RING

a. (-)-*cis*-Rose oxide - This substituted tetrahydropyran derivative has been isolated from certain flowers and shown to have structure 34.9 (Scheme 34).[1] The presence of a carbon chain next to the ring oxygen prompts consideration of a carbohydrate-precursor approach to its synthesis, particularly since the absolute configuration at that center is (R) and corresponds to a D-sugar. The main challenge resides in effecting practical deoxygenations, chain-branching and extension reactions from a suitable D-hexose precursor. A key step is the intro- duction of a C-methyl group at C-3 with the desired sense of chirality. A stereocontrolled synthesis by Ogawa and coworkers[2] was based on such an approach, starting with D-glucose and proceeding through the confor- mationally biased epoxide 34.4 by way of well known and readily avai- lable derivatives of 1,6-anhydro- β -D-glucopyranose.[3] Note how selective tosylation provided 34.1, which upon treatment with base gave the 3,4- epoxide 34.2, which in turn underwent regioselective reductive opening to give 34.3. Of interest also is the formation of the "down" epoxide 34.4 by refluxing 34.3 in methanolic sodium methoxide. This product arises by initial O-S cleavage of the 2-tosylate ester and subsequent intramolecular attack at C-3. With the partially deoxygenated deriva- tive 34.4 in hand, Ogawa and coworkers effected a regiospecific opening with lithium dimethylcuprate to give the desired C-methyl derivative 34.5 which, in a retrosynthetic sense can be considered as a suitable 'chiron'. Deoxygenation at C-1 was effected via the thioglycoside 34.6 which in turn was transformed into the aldehyde 34.8 and on to the target 34.9 by well known procedures. Thus, the overall process invol-

ved extensive deoxygenation of D-glucose but preservation of the original stereochemistry at C-5.   The introduction of the C-3 methyl group
took place regio- and stereospecifically by relying on the conformational bias afforded by the rigid dianhydro derivative 34.4.

Scheme 34

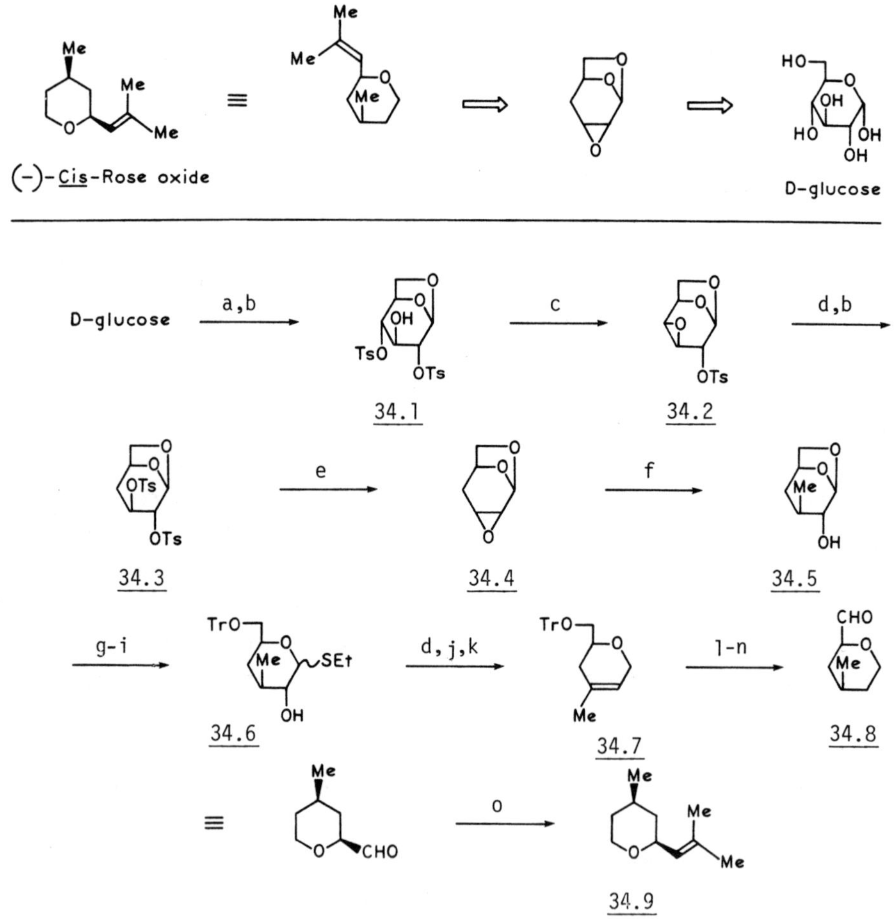

(-)-Cis-Rose oxide

D-glucose

a. pyrolysis of starch or base hydrolysis of phenyl-D-glucopyranoside;
b. TsCl, pyr.; c. NaOMe, MeOH; d. T-4 Ra-Ni, H_2; e. NaOMe, MeOH, reflux;
f. Me_2CuLi; g. Ac_2O, $BF_3.Et_2O$; h. $EtSSnBu_3$, $SnCl_4$, then NaOMe; i. TrCl, pyr;
j. MsCl pyr.; k. NaI, HMPA, $100°$; l. Pd/C, H_2; m. Amberlyst 15 (H^+), MeOH;
n. PCC; o. $Ph_3P=CMe_2$

b. (-)-Multistriatin - Examination of the structure of the
pheromone multistriatin[4] reveals its close resemblance to a highly
deoxygenated branched-chain 3,6-anhydro-octulose. The presence of the
tetrahydropyran-like ring makes the prospects of a stereocontrolled
synthesis from a hexose an interesting and logical goal. Indeed, a
stereoselective synthesis of (-)-α-multistriatin 35.8 (Scheme 35) has
been accomplished starting with D-glucose independently by Weiler and
Fraser-Reid and their coworkers.[5,6] From the retrosynthetic analysis it
is clear that C-methyl substitution must be achieved at C-2 and C-4 in a
D-hexopyranose structure with regio- and stereocontrol. The relative
orientation of the methyl groups is "up" (axial), and it is easy to
visualize how such an orientation can be achieved at C-2 from a conformationally biased epoxide intermediate (see Scheme 3). The C-4 axial
methyl group would have to be introduced by indirect means. The easily
available crystalline epoxide derivative 35.1,[7] does in fact undergo a
highly regioselective opening with lithium dimethylcuprate to give the
crystalline 2-C-methyl derivative[8] 35.2 having the desired sense of chirality. Deoxygenation at C-3 and subsequent manipulation of the remaining hydroxyl groups led to the 4-C-methylene derivative 35.4 which underwent stereoselective reduction with the Wilkinson catalyst to give
35.5. This 'chiron' embodies the basic six- membered ring structure of
the target, as well as the required three asymmetric centers with the
correct absolute stereochemistry. Elaboration of the ethyl side-chain
was done via the dithian derivative 35.6, and the product demercapta-
lated to give the target. Two comments are in order with regard to the
overall strategy shown in Scheme 35, namely, the sequence of introduction of C-methyl substitution, and the use of 35.6 as an acyl anion
equivalent. Considering the first point, it is of interest that rever-
sing the positions of methylene and C-methyl groups in 35.4 and subsequent hydrogenation would have led to mixtures of stereoisomers as has
been found in related work[9] and in other carbohydrate models.[10] Furthermore, the nature of the catalyst also appears to control the stereochemical outcome of the reduction.[5,9,10] With regards to 35.6, it is known
that generating synthetically useful carbanions from 1,3-dithian derivatives other than 1,3-dithian itself and some of its simpler 2-substi-
tuted analogs[11] may be unpredictable. Indeed, Sum and Weiler[5] were un-
successful in generating the anion of 35.6 using alkyl lithium reagents
in ether solvents, but alkylation with iodoethane could be achieved only
when the anion was generated with t-butyllithium in hexane.

Scheme 35

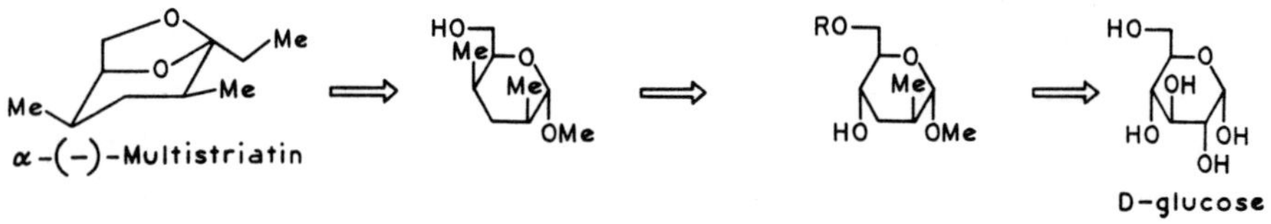

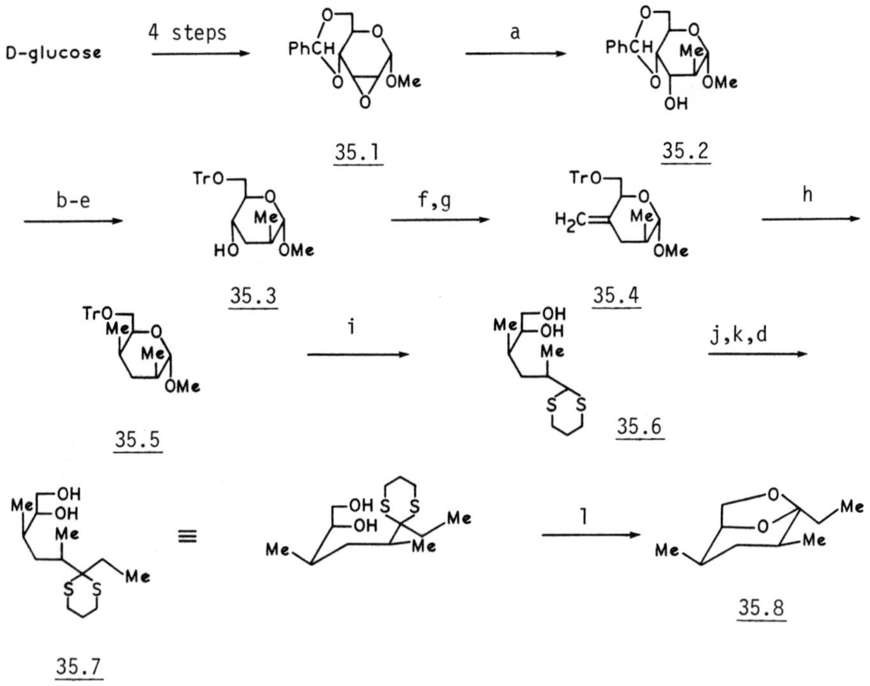

a. Me$_2$CuLi, ether; b. NaH, CS$_2$, MeI, ether; c. Bu$_3$SnH; d. TsOH, MeOH;
e. TrCl, pyr.; f. Collins; g. Ph$_3$P=CH$_2$, ether; h. (Ph$_3$P)$_3$RhCl, H$_2$, benzene;
i. HSCH$_2$CH$_2$CH$_2$SH, BF$_3$·Et$_2$O, CH$_2$Cl$_2$; j. Me$_2$C(OMe)$_2$, TsOH; k. t-BuLi, hexane,
then EtI, HMPA; l. HgCl$_2$, aq, MeCN

Other syntheses of (-)-multistriatin and three of its isomers have been accomplished starting from D-mannitol by way of (R)-glycer-aldehyde by K. Mori,[12] or from levo-D-glucosenone by M. Mori and coworkers.[12] The related natural serricornin was also made by the latter group.[12] Several other syntheses are also available.[13]

Other methods are also available for the synthesis of chiral pre-cursors to the multistriatins and related compounds that contain a simi-lar substitution pattern. Thus, Fraser-Reid and coworkers[14] have re-ported on the regiospecific conjugate addition of lithium dimethyl-cuprate to enones of type 36.1 (Scheme 36) available from D-glucose.

Scheme 36

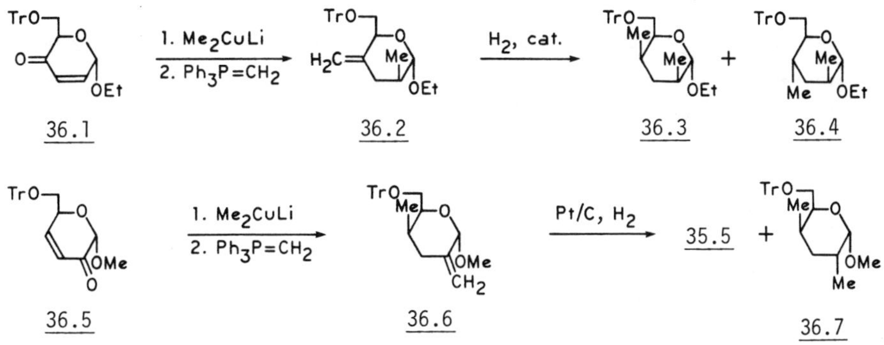

Reduction of the corresponding exocyclic methylene derivative 36.2 gave variable ratios of 36.3 and 36.4 depending on the catalyst used.[9] As expected, the Wilkinson catalyst favored 35.5 (7:3 ratio, but a better ratio (9:1) was obtained in the presence of Pd/C using the methylene derivative 36.6 which is derived from 36.5.

An expedient route to 'chirons' such as 36.3 and 36.7 is discussed on page 94 (Schemes 48,49).

   c. (+)-Thromboxane B_2,(TXB_2) - The cascade of biosynthetic products derived from arachidonic acid[15] consists of several oxygenated heterocycles among which are two intriguing structures, thromboxanes A_2 and B_2 (Scheme 37). Examination of the constitutional structure of the B_2 component (TXB_2), which is reported to arise from the hydrolytically unstable A_2 component reveals the presence of a deoxygenated hexose in which positions 4 and 6 are sites of chain-branching and extension respectively. Note that the *trans* olefinic side-chain extends from a position vicinal to the ring oxygen corresponding to C-6 of a D-hexose

Scheme 37/12

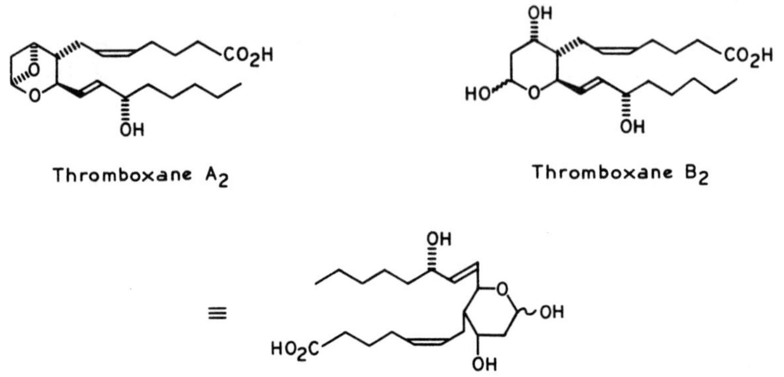

Thromboxane A$_2$

Thromboxane B$_2$

A carbohydrate-precursor approach therefore offers distinct operational
advantages. The retrosynthetic analysis shown in Scheme 38 illustrates
these points, and focuses the main challenge on the regio and stereocontrolled introduction of an acetate branch point at C-4 in a suitable
D-hexopyranose derivative.

Scheme 38

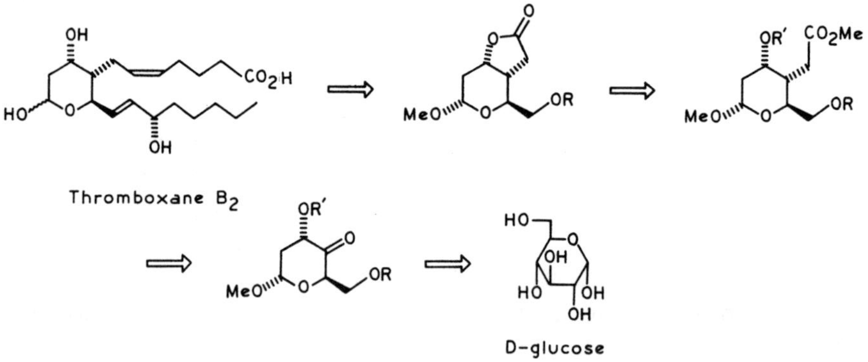

Thromboxane B$_2$

D-glucose

The first synthesis of TXB$_2$ by such a strategy was reported by Hanessian
and Lavallée.[16] Other syntheses of critical intermediates from carbohydrate precursors were reported soon thereafter (see below). Other
than the sites of chain-branching in TXB$_2$, one notes the presence of a
2-deoxy function and an axial C-3 hydroxyl group (carbohydrate numbering). This pattern is easily achieved by the well-known reductive
opening of conformationally biased epoxides (see Scheme 3). In the case
of 35.1 (Scheme 38A), such a reaction is practically regiospecific to

Scheme 38A

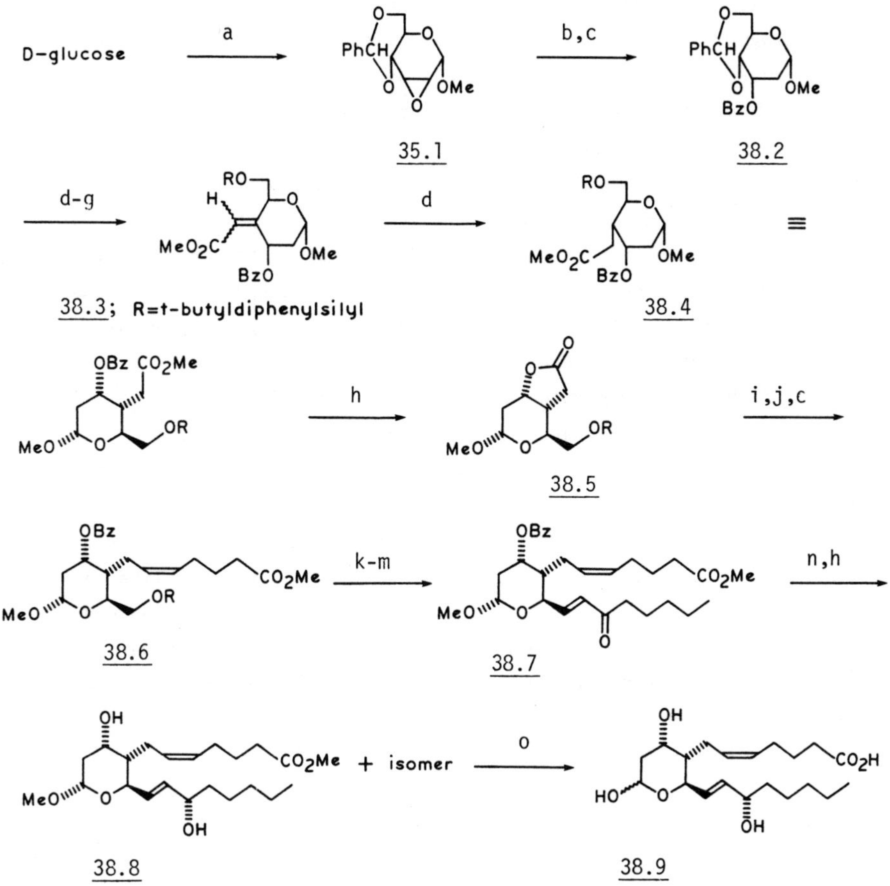

a. Ref. 76; b. LAH, ether; c. BzCl, pyr.; d. Pd(OH)$_2$/C, H$_2$; e. Ph$_2$t-BuSiCl, imidazole, DMF; f. 1-ethyl-3-(3'-dimethylaminopropyl)carbodiimide HCl, pyridinium trifluoroacetate, DMSO; g. (MeO)$_2$P(O)CH$_2$CO$_2$Me, t-BuOK, t-BuOH, toluene; h. K$_2$CO$_3$, MeOH, i. DIBAL, toluene; j. Ph$_3$P$^+$CH$_2$(CH$_2$)$_3$CO$_2$H Br$^-$, (Me$_3$Si)$_2$NLi, HMPA, then CH$_2$N$_2$; K. Bu$_4$NF, THF; l. Collins; m. Bu$_3$P=CHCOC$_5$H$_{11}$; n. Zn(BH$_4$)$_2$, DME, ether; o. aq. NaOH, then Dowex-50(H$^+$)

give the known axial, alcohol hence the benzoate 38.2. As in other
multistep syntheses of complex polyfunctional compounds, the choice of
protecting groups can be critical. The synthesis of TXB_2 proved to be
no exception and prompted the development of the 0-t-butyldiphenylsilyl
protecting group[17] which provided the elements of compatibility, stabi-
lity, crystallinity and possibly the steric bias needed in this synthesis. Thus, the acetal function in 38.2 was removed by hydrogenolysis
rather than hydrolysis because of the hydrolytic instability of 2-deoxy
glycosides in acidic media. Selective protection of the primary hydroxyl group as the t-butyldiphenylsilyl ether followed by oxidation with a
seldom used carbodiimide reagent,[18] and chain-extension gave 38.3. It
was ascertained by definitive chemical and spectroscopic methods that
the axially disposed 0-benzoyl group had   not undergone intramolecular
migration in the above operations.   Although each of the E and Z isomers of 38.3 was hydrogenated at different rates, the stereochemical
outcome was predictable, and gave the desired α-disposed side-chain as
in 38.4 preponderantly if not exclusively. The ensuing manipulations
are well known in the area of prostaglandin chemistry and will not be
discussed except for the fact that the introduction of the acid side-
chain with a Z-double bond was greatly facilitated by using lithium
hexamethyldisilazane in HMPA[19] rather than the more commonly used dimsyl
sodium-DMSO combination. Thus, with the recognition of an apparent
carbohydrate-type symmetry in TXB_2 , which is all the more evident by the
presence of a lactol function, a synthesis from a D-hexose precursor is
a logical first choice.

Corey and coworkers[20] have devised another synthesis of the bicyclic lactone precursor 39.5 to TXB_2 (Scheme 39). The key steps involved
the stereocontrolled introduction of the acetate side-chain at C-4 and
the vicinal axial C-3 hydroxyl group. This was achieved by transfer of
chirality via a [3,3]-sigmatropic rearrangement (Claisen) followed by
iodolactonization. The retrosynthetic analysis shows how this can be
accomplished from the appropriate allylic alcohol derived from a carbohydrate. By a known sequence of reactions, D-glucose was transformed
into the 2,6-dibenzoate derivative 39.1, hence to the desired diol
39.3. Application of the dimethylamide variant of the orthoester
Claisen rearrangement[21] led to the desired dimethylacetamide derivative
39.4, which after hydrolysis, iodolactonization and reduction gave the
key intermediate 39.5. In Corey's approach, the hydroxyl groups in D-
glucose were efficiently manipulated to achieve a high level of regio

Scheme 39

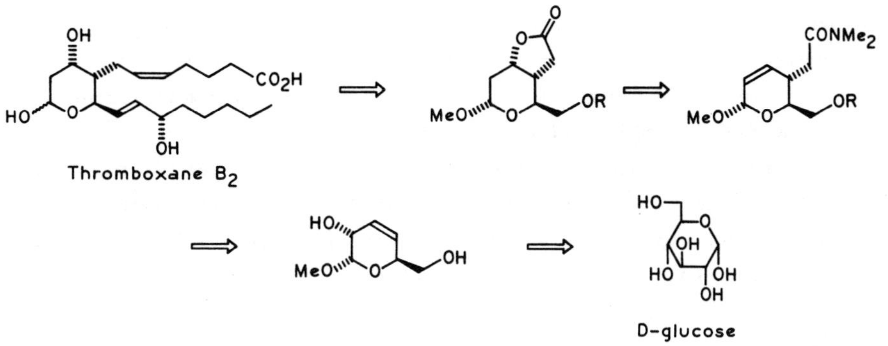

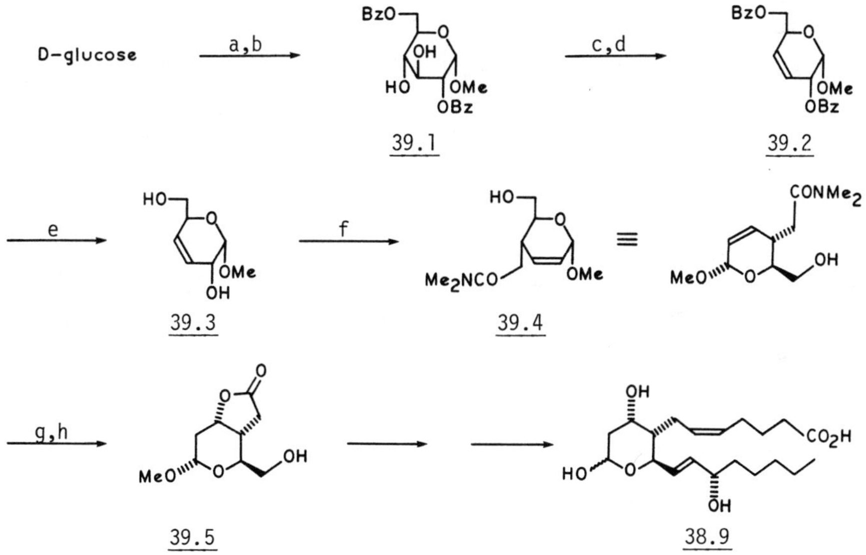

a. MeOH, H^+; b. BzCl, pyr.; c. MsCl, pyr.; d. NaI, DMF, Zn; e. NaOMe, MeOH; f. $Me_2NCMe(OMe)_2$, diglyme, $25° \rightarrow 160°$, over 2h; g. I_2, THF, H_2O; h. Bu_3SnH.

and stereocontrol in generating functionality at two contiguous centers. Such allylic rearrangements could also be useful in generating C–C linkages at C-2 in hexopyranosides via the corresponding 4-allylic alcohols with complete control of regio and stereochemistry. Furthermore, the availability of a set of epimeric alcohols provides a great deal of

flexibility and latitude to the application of sigmatropic
rearrangements in the carbohydrate series. Precedents have already been
recorded by Ferrier and coworkers[22]. A second synthesis of the trityl
ether derivative of the lactone intermediate 39.5 was subsequently
reported[23] and follows essentially the Corey approach, except for the
application of the classical orthoester Claisen rearrangement rather
than the dimethylamide variation.

Finally, a practical entry into the TXB$_2$ ring system was reported
by Kelly and Roberts[24] who used the rigid bicyclic framework of 1,6-
anhydro-β-D-glucopyranose as a means of achieving regio and stereocontrol (Scheme 40). The 2-deoxy function (carbohydrate numbering) was

Scheme 40

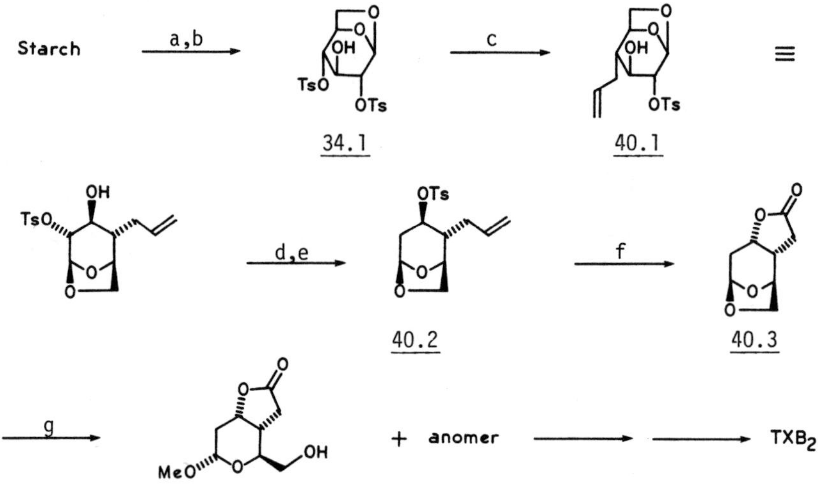

a. 3 steps; b. see Scheme 34; c. allylmagnesium chloride, CuI, inverse
addn., THF; d. LiEt$_3$BH, THF; e. TsCl, pyr.; f. RuO$_2$, NaIO$_4$, aq. acetone;
g. MeOH, Amberlyst H$^+$.

introduced by a regioselective deoxygenation via an epoxide intermediate
to give 40.2. Regiospecific introduction of the two-carbon side-chain
was achieved via the reaction of the epoxide intermediate 34.2 (Scheme

34) with an organoallyl derivative. Oxidative cleavage of the double
bond was followed by intramolecular displacement of the suitably
disposed *trans* - tosylate to give 40.3 in high yield. With the three
contiguous centers on the ring having the required substituents and
oxidation/reduction states, there remained to unfold the "template"
40.3, which could lead to the lactone 39.5 (Scheme 39) or its derivative
38.5 (Scheme 38). This approach is practical and synthetically
expedient since it uses conformational bias as a means of achieving
regio and stereocontrol.It is however necessary to separate 39.5 from
its β-anomer (1.5:1 ratio).

   d. Ambruticin — Synthesis of ring A — Ambruticin, a novel
antifungal agent has been assigned its stucture (Scheme 41) based on
X-ray diffraction studies.[25] A synthesis of the ring A portion from a
carbohydrate precursor by Just and Potvin[26] established its absolute
stereochemistry. The target 41.6 is a tetrasubstituted tetrahydropyran
which can be formally related a 3,7-anhydro-octonic acid. Since eight
carbon sugars are rare, it would be practical to select an appropriate
five- or six-carbon sugar precursor and to effect chain-extensions from
either end depending on the nature of the configuration of the
asymmetric centers and the site of deoxygenation. With the arbitrary
choice of L-arabinose (rather than the D-isomer) as starting material
the configurations of the three contiguous carbon atoms bearing the two
hydroxyl groups and the hydroxymethyl group were secured at the outset.
Note that chain extension was done at the aldehyde end via a nitro-
methane-type condensation which further allowed construction of the ring
system and introduction of the deoxy group. The selection of methyl 3-
nitropropionate as the three-carbon appendage was therefore an excellent
choice. The required selectively O-protected aldehyde 41.2 was prepared
by standard procedures from the readily available L-arabinose. Condensation with methyl 3-nitropropionate afforded a mixture of nitroalcohols
41.3 which were transformed to the β,γ-unsaturated nitro ester and then
to the α,β-unsaturated ester derivative 41.4 by DBU-catalyzed elimination and hydrolysis of the acetal function. Intermediate 41.4 can be
considered as an appropriate 'chiron' in this synthetic scheme.
Treatment with refluxing sodium methoxide effected intramolecular
cyclization to give the bicyclic lactone 41.5 in 51% yield which ensured
the *cis*-disposition of the two side-chains. Removal of the benzyl
ether groups by catalytic hydrogenation was not uneventful, but could be
achieved by treatment with trimethylsilyl iodide, whereupon the

Scheme 41

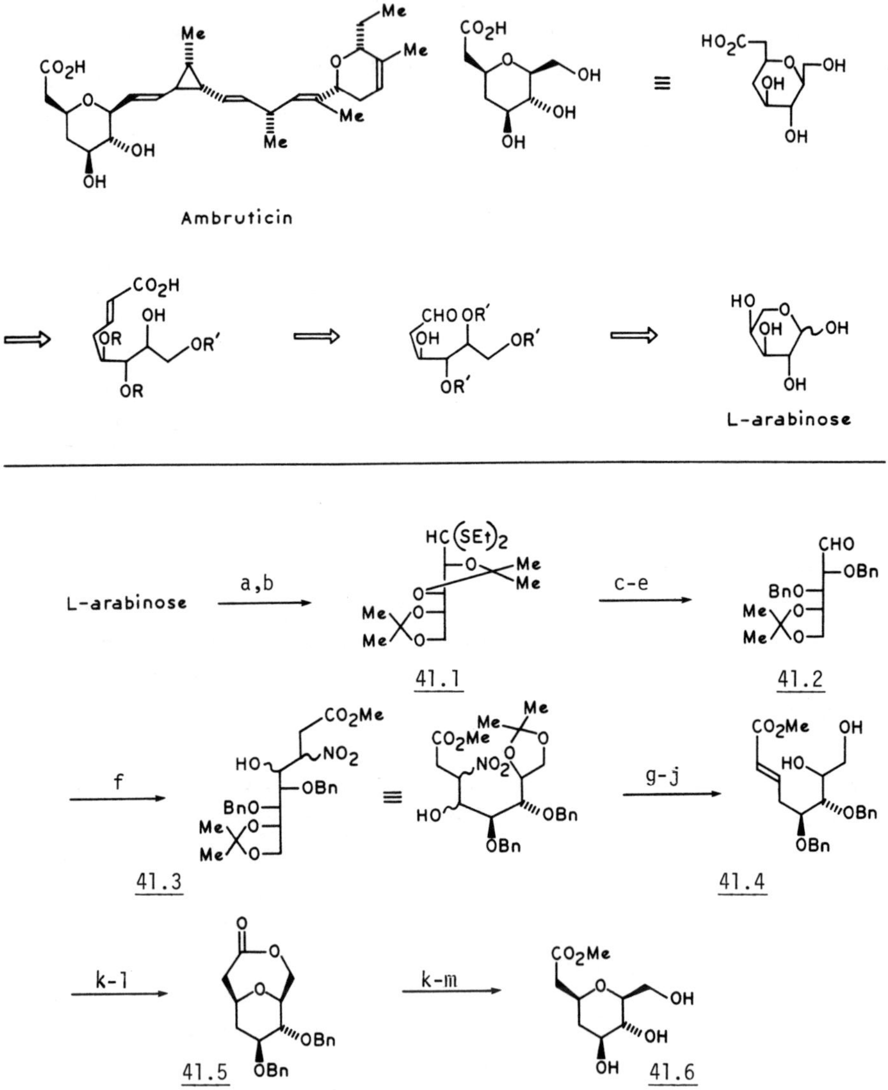

Ambruticin

L-arabinose

L-arabinose $\xrightarrow{a,b}$ 41.1 $\xrightarrow{c-e}$ 41.2

$\xrightarrow{f}$ 41.3 $\equiv$ $\xrightarrow{g-j}$ 41.4

$\xrightarrow{k-1}$ 41.5 $\xrightarrow{k-m}$ 41.6

a. EtSH, H^+; b. acetone, H^+; c. aq. AcOH; d. BnBr, NaH, THF; e. aq. $NaHCO_3$, I_2; f. $O_2NCH_2CO_2Me$, t-BuOK, THF, $0°$; g. Ac_2O, DMAP, ether; h. Et_3N, $NaBH_4$; i. DBU; j. aq. AcOH; k. NaOMe, MeOH; l. Dowex-50(H^+); m. TMSI, CH_2Cl_2

trihydroxy ester <u>41.6</u> was formed in modest yield and converted into the corresponding crystalline tribenzoate ester. The latter was found to be identical with a sample obtained by degradation of ambruticin. The absolute stereochemistry of the cyclopropane unit was also secured in this study by synthesis of an appropriate derivative from (<u>R</u>)-Feist's acid and correlation with a similar product isolated by degradation of ambruticin.

     e. <u>Pseudomonic acid C</u> - Pseudomonic acid C[27] (Scheme 42) is an antibiotic of novel structure isolated from fermentation sources. It occurs as a minor component, along with pseudomonic acid A in which the disubstituted olefinic function is replaced by an epoxide. Pseudomonic acid A has been reconstructed from a degradation product arising from ozonolysis of the trisubstituted double bond. A synthesis of such an intermediate corresponding to pseudomonic acid C (Scheme 43) would

Scheme 42

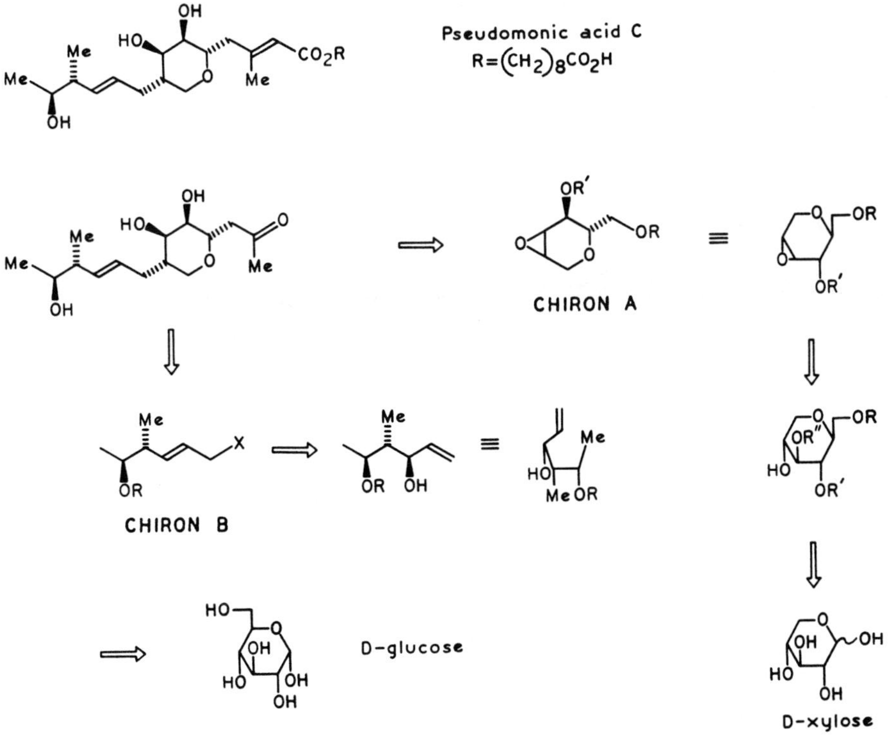

constitute a useful immediate synthetic precursor to the antibiotics in
question.

Inspection of the structure of the target ketone reveals the presence of a tetrasubstituted tetrahydropyran which can be related to a
1,5- or a 2,6-anhydro-alditol. The orientation of the hydroxyl groups
is *cis* (D-erythro), as is that of the two C-branch points. The introduction of the "left hand appendage" would thus require a regioselective
C-branching, and the *trans*-disposition with regard to the vicinal hydroxyl group should lead to consideration of epoxide opening as a viable
strategy. The "right hand appendage" can be introduced by methodology
known in the formation of C-glycosides.[28] Considering a 2,6-anhydro-
hexitol as a viable precursor leads to the selection of a starting
carbohydrate with inherent "all-trans" stereochemistry which would
ensure the formation of an epoxide intermediate as well as a 1,2-*trans*-
C-glycoside. Sinäy[29] and coworkers have completed the synthesis of the
target ketone from the readily available D-xylose and D-glucose as
starting materials for the ring portion and the "left hand appendage"
respectively. Bond disconnection in their strategy can be related to
the creation of 'chirons' A and B. Scheme 43A illustrates the synthesis

Scheme 43 A

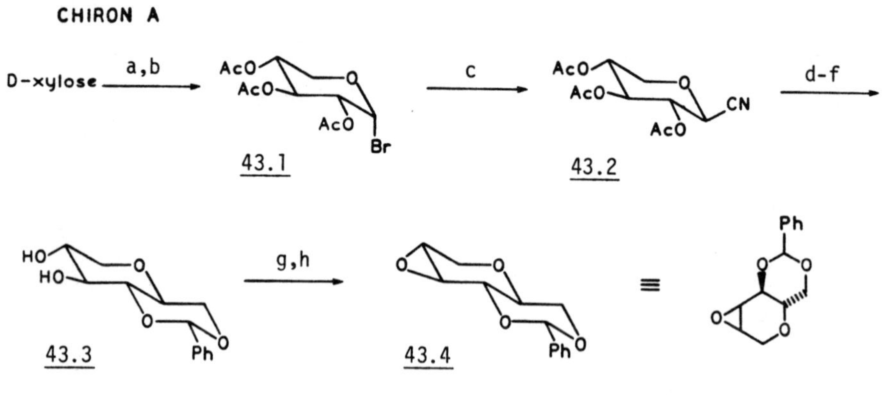

CHIRON A

a. Ac_2O, pyr.; b. HBr, AcOH; c. NaCN; d. NaOMe, MeOH, then 25% aq. NaOH,
reflux; MeOH, HCl, reflux 10 min.; e. LAH, THF; f. $PhCH(OMe)_2$, TsOH, DMF;
g. TsCl, pyr.; h. NaOMe, CH_2Cl_2

of 'chiron' A 43.4, a chiral epoxide obtained from D-xylose. The key
reaction is the formation of the known[30] β-D-xylopyranosyl cyanide derivative 43.2 (considered as a C-glycoside). By standard procedures it
was transformed into the acetal derivative 43.3 which could be regio-
selectively tosylated at the C-4 hydroxyl group, thus facilitating the
formation of the desired epoxide 43.4. This selectivity is not entirely
evident and must rely on the kinetic acidity of the C-4 hydroxyl group
compared to the apparently more accessible C-5 hydroxyl group.

The synthesis of the chiral "left hand appendage" ('chiron' B,
Scheme 43B) shows a carbohydrate-based strategy for the preparation of

Scheme 43 B

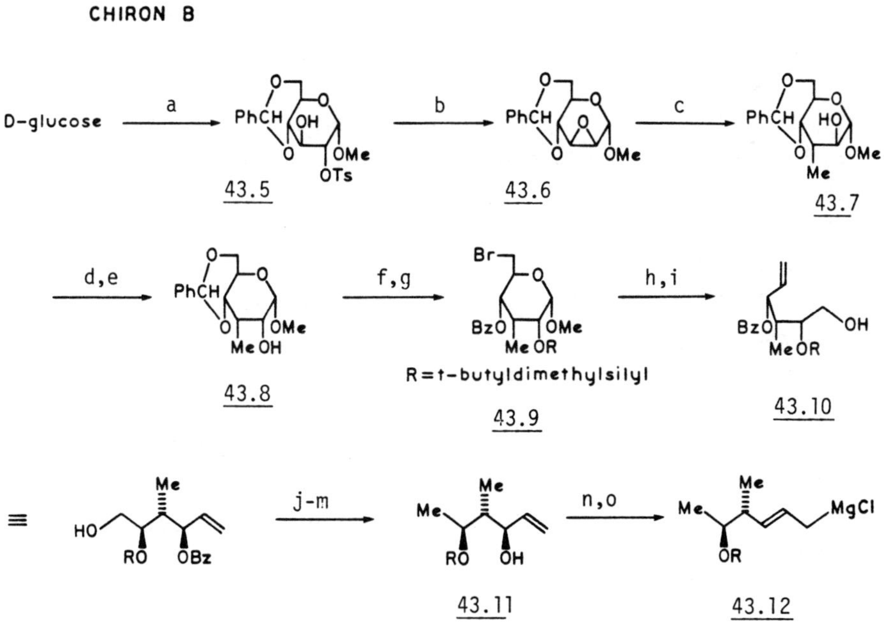

CHIRON B

a. Ref. 7; b. NaOMe, MeOH; c. MeMgI, ether; d. DCC, DMSO, TFA, pyr.;
e. LAH, ether, $0°$; f. t-BuMe$_2$SiCl, imidazole, DMF; g. NBS, BaCO$_3$,
CCl$_4$, $80°$; h. Zn, PrOH, H$_2$O, $80°$, 30 min.; i. NaBH$_4$, EtOH, $-35°$;
j. TsCl, pyr.; k. NaI, methylethyl ketone, $80°$; l. NaBH$_4$, DMSO; m.
NaOMe, MeOH; n. SOCl$_2$, pyr., $0°$; o. Mg, ether

the desired allylic Grignard derivative 43.12. Note that the hydroxyl
group is five carbons removed from a terminal center in a six-carbon
E-olefinic derivative. The synthesis of such an intermediate hinges on
the choice of a carbohydrate precursor in which the requisite
functionality can be introduced. The hydroxyl group could correspond to
C-5 (or C-2) of a hexose depending on which end of the chain is taken as
C-1 ("rule of five"). Sinaÿ and coworkers utilized an approach in which
the two ends of a hexose were interchanged, so that the hydroxyl and
C-methyl groups in 43.12 would correspond to C-2 and C-3 respectively.
This strategy takes advantage of the facile and regiospecific
introduction of a C-methyl group by opening of an epoxide in a conformationally biased derivative. The *cis*-orientation of the hydroxyl and
methyl groups can be achieved by a sequence involving oxidation, and
reduction. The alternative approach would have required the hydroxyl-
bearing carbon atom of 43.12 to be C-5 of an L-hexose and ensuring
regiospecific C-methylation at C-4. Scheme 43 shows the steps leading
to the required derivative 43.8 with the *cis*-arrangement of hydroxyl and
C-methyl groups from D-glucose. Protection and cleavage of the
benzylidene acetal[31] gave a critical intermediate 43.9 which, when
subjected to a zinc-mediated fragmentation[32] led to the acyclic
structure 43.10. The remainder of the sequence was straightforward,
although the interesting transposition of the secondary allylic alcohol
to the primary analog 43.12 is noteworthy as it introduces the
E-olefinic linkage.

With the availability of the two crucial precursors, union was
attempted and successfully achieved based on a copper-mediated[33] opening
of the epoxide 43.4 with the chiral allylic Grignard reagent 43.12
(Scheme 44). This reaction proceeded in 97% yield and with very high
regioselectivity to give the desired substitution pattern. Manipulation
of functional groups then led to the chiral ketone 44.3 in very good
overall yield.

The construction of this important chiral precursor to the
pseudomonic acids from readily available carbohydrates such as D-xylose
and D-glucose was possible in major part because of anticipated high
regioselectivity in crucial C-C bond forming reactions. Conformational
bias was a built-in asset which was nicely capitalized upon.
Pseudomonic acid C has been recently synthesized by a different route in
which a 3,4-dihydro-2 H-pyran derivative was the precursor of the
central ring.[34]

Scheme 44

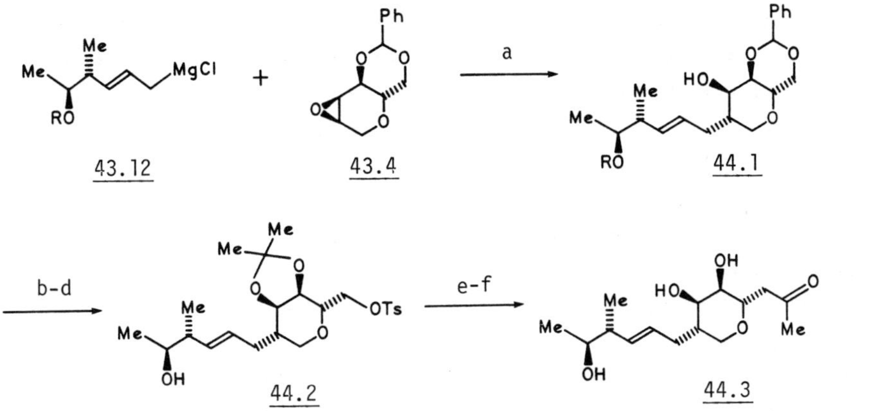

43.12   43.4   44.1

b-d   e-f

44.2   44.3

a. CuI, THF, -30°; b. aq. HCl,diox.; c. acetone, H$^+$; d. TsCl, Et$_3$N, DMAP, CH$_2$Cl$_2$; e. KCN, HMPA, 18-crown-6; f. MeLi, Et$_2$O, 0°, then HCl, diox.

f. A-23187 – Synthesis of chiral intermediates – The divalent cation ionophore A-23187 (calcimycin) has a unique structure[35] in comparison with other ionophores. The 1,7-dioxaspiro-[5.5]-undecane entity encompasses all of the chiral centers in the antibiotic (Scheme 45), and in acyclic perspective, it shows a pattern of substitution that

Scheme 45

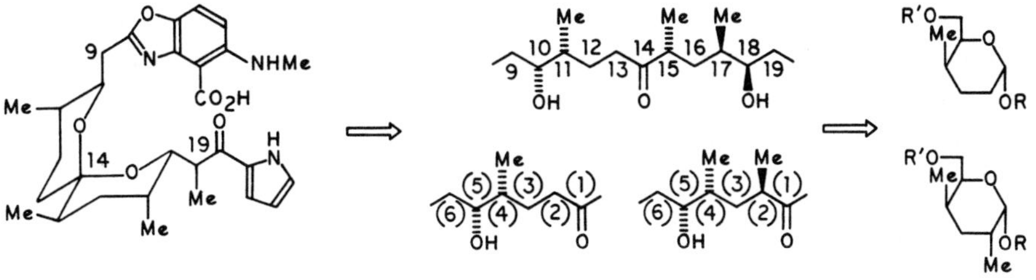

corresponds to a vicinal and alternating arrangement of hydroxyl and methyl groups (C-9 — C-19). Unlike segments of polyketide-derived molecules such as the macrolides and ansa antibiotics, which can be chemically assembled by a series of aldol condensations, stereocontrolled access to the carbon chain containing alternating C-methyl substituents as in A-23187 is not as general. Examination of the target structure shows that the two tetrahydropyran units are joined by a carbon atom common to both rings and that the absolute configuration and nature of substituents at C-10, C-11 and C-7, C-18 respectively are the same. The "rule of five" applies if one considers C-14 as the equivalent of the anomeric carbon in a suitable hexose. There emerge two possible 'chirons' corresponding to C-14 — C-19 and C-9 — C-14, and in each case a D-hexose is required. The challenge in obtaining such intermediates resides in the regio- and stereocontrolled introduction of C-methyl substituents. Each intermediate could then be used independently to construct the remainder of the chain, as the presence of the C-14 carbon atom precludes the utilization of the two six-carbon units as such since it would be common to both. The solution to this problem could reside in using the more substituted intermediate as a six-carbon chain and to degrade the other to a four-carbon chain encompassing the vicinal C-methyl and hydroxyl substituents. Union of two such segments in a C-C bond forming reaction would produce the C-9 — C-19 chiral portion of A-23187. This strategy also offers the advantage of building up the required heterocyclic units at the terminal primary carbon atoms of the two intermediates independently. Only the stereochemistry at C-19 would have to be elaborated independently in such an approach.

Access to 'chirons' corresponding to the C-14 — C-19 portion can be considered via the reduction of appropriate C-2 methylene carbohydrate derivatives (Scheme 36). However the stereoselectivity is low and favors a *cis*-dimethyl derivative (9:1 ratio). Ogawa and coworkers[36] have prepared the chiral intermediate 46.4 (Scheme 46) which corresponds to the C-9 — C-13 segment. They have also reported on an elegant method for the stereocontrolled introduction of the C-19 methyl group in a model intermediate in which the C-methyl groups are *cis* rather than *trans* as required in A-23187 (Scheme 47). The key reactions in this scheme are the oxidative transformation of a nitrile into an amide (47.2 →47.4), and the stereocontrolled alkylation of the conformationally biased iminoether derivative 47.4 . Standard transformations led to the dithian derivative 47.9 which corresponds to the C-14 — C-20 segment of A-23187, except for the stereochemistry at the C-15 position. This center is subject to epimerization, and can presumably be inverted during the intramolecular acetalization step.

Scheme 46

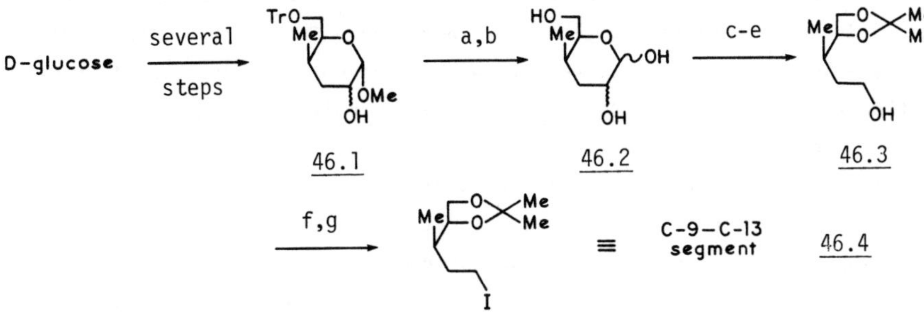

46.1 46.2 46.3

46.4

a. Ac_2O, $BF_3 \cdot EtO_2$; b. NaOMe, MeOH; c. $NaIO_4$; d. $NaBH_4$; e. $Me_2C(OMe)_2$, H^+;
f. TsCl, pyr.; g. NaI, acetone, reflux

Scheme 47

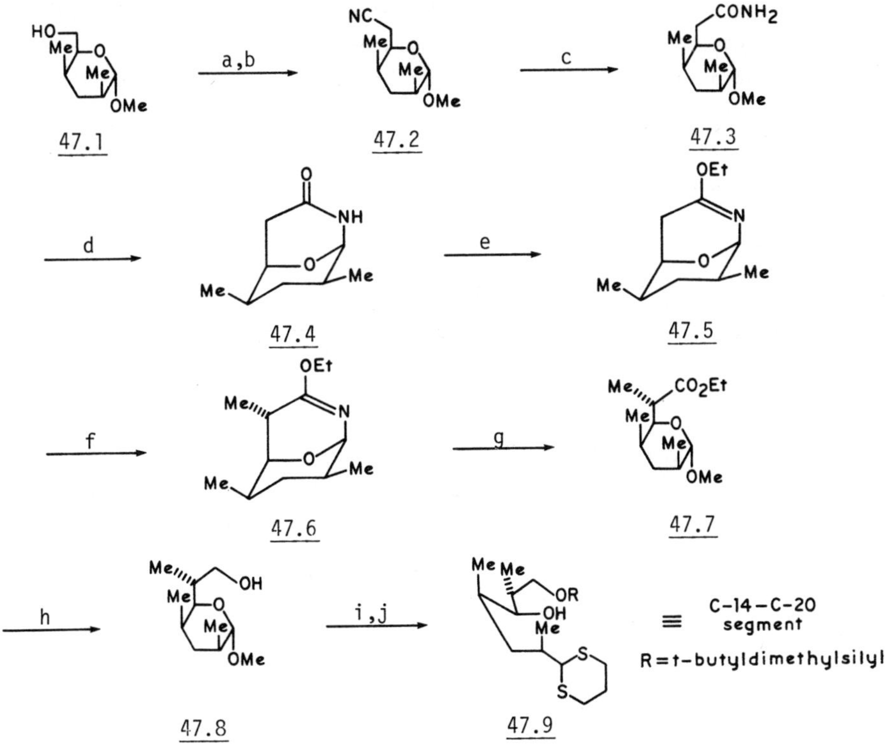

47.1 47.2 47.3

47.4 47.5

47.6 47.7

47.8 47.9

a. TsCl, pyr.; b. Bu_4NCN, DMF; c. H_2O_2, Na_2CO_3, acetone; d. H^+;
e. $Et_3O^+BF_4^-$; f. LDA, MeI; g. aq. MeOH, H^+; h. LAH; i. 1,3-propane
dithiol, BF_3; j. $t\text{-}BuMe_2SiCl$

An expedient route[37] to chiral precursors representing the two tetrahydropyran rings in A-23187 relies on the preparation of intermediate
48.3 in four steps from D-glucose (Scheme 48). Treatment of the readily
available D-glucal derivative 48.1 with t-butanol and boron trifluoride
etherate according to Ferrier[38] gave the crystalline glycoside 48.2,
which, when treated with lithium dimethylcuprate afforded the C-methyl
derivative 48.3 directly. This transformation involves the conversion
of a β-acyloxy enol ester into a β-alkyl ketone and presumably proceeds
by conjugate 1,4-methylation of an intermediate enone derivative. The
behavior of such alkyl 2-acyloxy- α-D-hex-2-enopyranosides toward other
nucleophiles appears to follow a similar course (see page 109). Reduction of the exocyclic methylene function produced a 1:1 mixture of the
desired precursor 48.5 and its C-2 epimer. The choice of the t-butyl
group as the aglycon was to alter the stereochemistry of reduction in
favor of 48.5 (equatorial C-2 methyl), compared to reductions of related
methyl glycosides.[9,10] The corresponding neopentyl glycoside gave a
similar ratio indicating that the limits of stereoselective reductions
had been attained. The analogous methyl glycoside derivative led to a
4:1 mixture of reduced products of which the major isomer corresponds to
the ring structure in multistriatin (see Scheme 35). Variation in the

Scheme 48

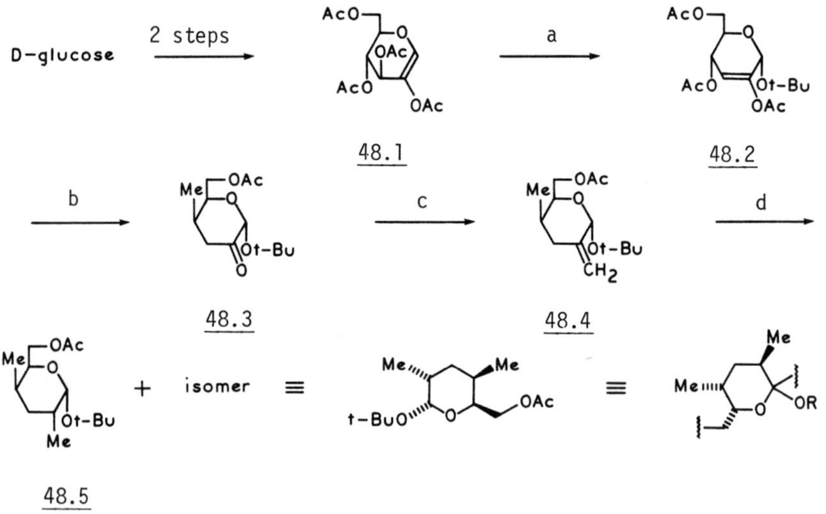

a. t-BuOH, BF_3Et_2O; b. Me_2CuLi, THF ; c. $Ph_3P=CH_2$; d. $(Ph_3P)_3RhCl$, H_2

catalysts and possibly the nature of the C-6 substituent (compare Scheme 36) may account for the differences in the ratio of C-2 isomers.

Intermediate 48.3 can also serve as a precursor to the C-9—C-14 portion of A-23187 or to a segment containing the two chiral centers, by suitable chain-shortening processes (Scheme 49) (see also Scheme 46). Scheme 49/8 lgns

Scheme 49

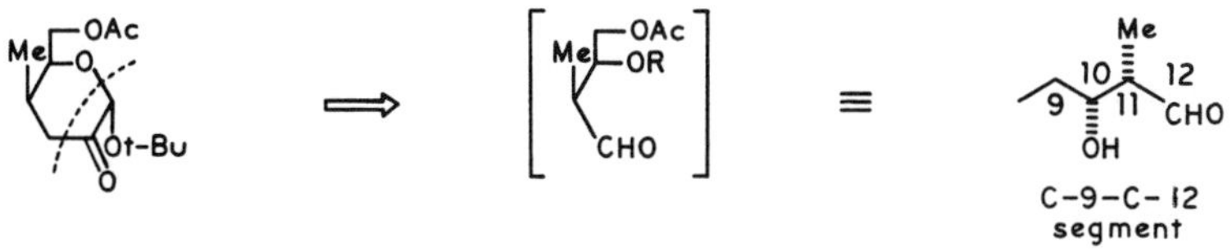

C-9-C-12 segment

A total synthesis of A-23187 as well as the synthesis of an immediate precursor (a formal synthesis) have been reported based on different strategies.[39]

g. Tirandamycic acid - Tirandamycin is a member of structurally unique antibiotics in the 3-acyl tetramic acid group.[40] The synthesis of tirandamycic acid (Scheme 50) from D-glucose has been accomplished by the Ireland group.[41] In spite of the complexity of the structure, the carbohydrate-type of symmetry is apparent in the bicyclic segment. By virtue of the substitution pattern α- to the ring oxygen it is evident that stereocontrolled C-C bond formation with a chiral center containing a C-methyl group could arise via an enolate Claisen rearrangement which has been pioneered by Ireland and coworkers.[42] The retrosynthetic analysis provided in Scheme 50, illustrates the elegant strategy that starts with D-glucose and proceeds by the systematic introduction of functional groups and chiral centers. Manipulation of the readily available tri-0-acetyl-D-glucal 50.1 afforded 50.2 in which the propionate ester was strategically anchored at C-3. Enolate Claisen rearrangement followed by iodolactonization gave 50.3 and its epimer. Deoxygenation and dehydration led to the enol ether 50.4 which was functionalized and chain-extended to give 50.5. Note that the correct oxygenation pattern is present at sites destined to become the hemiacetal and ketone functions. Elaboration of the double bond and the side-chain led to the bicyclic acetal 50.7 with the correct anomeric stereoselection. The enone system in 50.9 was incorporated by a series

Scheme 50

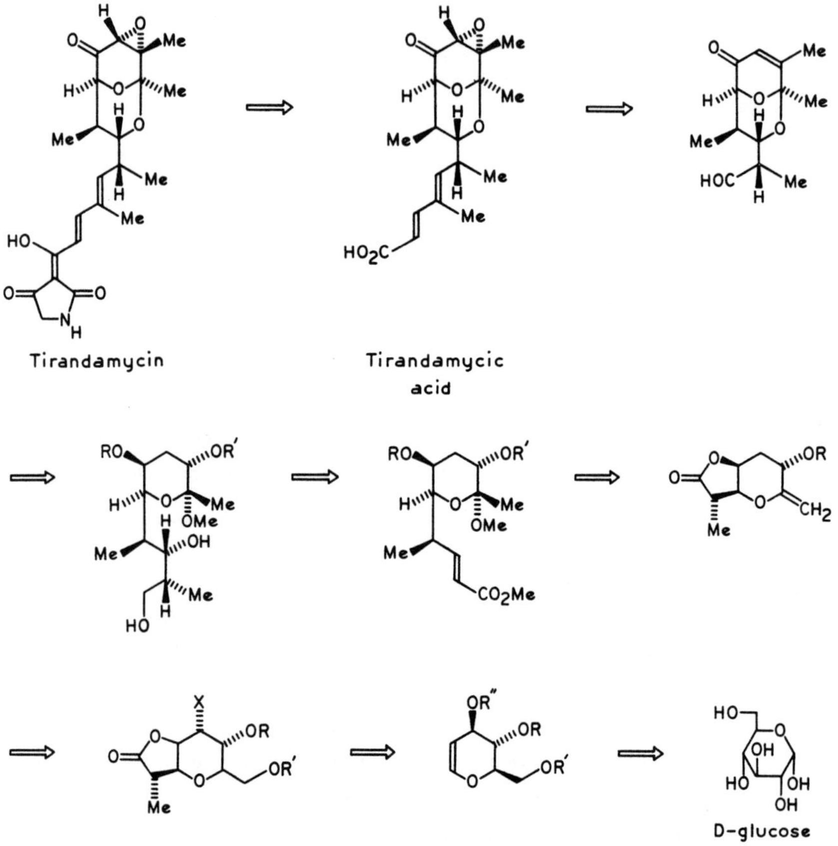

Tirandamycin

Tirandamycic
acid

D-glucose

of reactions which were initially plagued by the inability to cleave the
MEM group. Having overcome these difficulties, the enone was further
elaborated and chain-extended to give optically active tirandamycic
acid. It should be noted that the Ireland strategy uses the
six-membered dihydropyran structure not only as a 'chiron' that overlaps
with a portion of the target, but also to direct the stereochemistry of
functional groups and dictate the pattern of substituents. Except for
the ring oxygen atom, none of the original hydroxyl groups present in
D-glucose can be found in the target. The carbon framework of the
starting carbohydrate was therefore used to construct the entire
molecule around it, together with complete regio- and stereocontrol of
substituents.

Scheme 50A

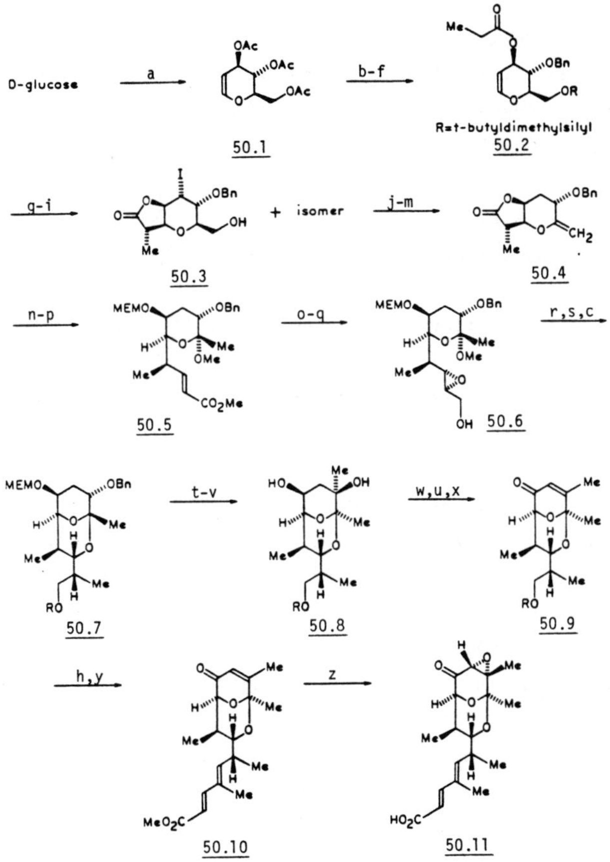

a. See for example scheme 99; b. NaOMe, MeOH; c. t-butyldimethylsilyl
chloride, pyr., then BzCl, pyr.; d. BnBr, KH, THF; e. NaOH, EtOH;
f. MeCH₂COCl, pyr.; g. hexamethyldisilazane, BuLi, THF, -78°; R₃SiCl,
HMPA, benzene, reflux; h. aq. HCl, THF; i. KI, I₂, NaHCO₃; j. n-Bu₃SnH,
EtOH; k. TsCl, pyr.; l. NaI, MEK, reflux; m. AgF, pyr., MeCN; n. TsOH,
MeOH, reflux; o. Dibal, -78°; then Ph₃P=CHCO₂Me, THF; p. MEMCl, R₂NH;
q. m-CPBA; r. CuBr, Me₂S, MeLi, ether; s. TsCl, CHCl₃; t. Pd/C, H₂;
u. PDC, CH₂Cl₂; v. MeMgBr, Et₂O; w. n-BuLi, benzene then Hg(OAc)₂,
aq. THF; x. TsOH, pyr., DMSO, reflux; y. DMSO, (COCl)₂; then Wittig;
z. t-BuO₂H, Trition, B, benzene, reflux, then aq. KOH, MeOH.

REFERENCES

1. C.F. Seidel and M. Stoll, Helv. Chim. Acta, 42, 1830 (1959); Y.R.
    Naves, D. Lamparsky and P. Ochsner, Bull. Soc. Chim. France, 645
    (1961).
2. T. Ogawa, N. Takasaka and M. Matsui, Carbohydr. Res., 60, C4 (1978).
3. L. Vargha and E. Kasztreiner, Chem. Ber., 93, 1608 (1960); see also
    S. Solzberg, Advan. Carbohydr. Chem. Biochem., 25, 229 (1970); M.
    Černý and J. Staněk, Jr., Advan. Carbohydr. Chem. Biochem., 34, 24
    (1977).
4. G. T. Pearce, W.E. Gore, R.M. Silverstein, J.W. Peacock, R.A.
    Cuthbert, G.N. Lanier and J.B. Simeone, J. Chem. Ecol., 1, 115
    (1975).
5. P.E. Sum and L. Weiler, Can. J.Chem., 60, 327 (1982); 56, 2700
    (1978).
6. D.E. Plaumann, B.J. Fitzsimmons, B.M. Ritchie and B. Fraser-Reid,
    J. Org. Chem., 47, 941 (1982).
7. N.K. Richtmyer, Methods Carbohydr. Chem., 1, 107 (1962); L.F.
    Wiggins, Methods Carbohydr. Chem., 2, 188(1963).
8. N.L. Holder and B. Fraser-Reid, Can. J. Chem., 51, 3357 (1973); see
    also S. Hanessian and G. Rancourt, Can. J. Chem., 55, 1111 (1977).
9. B.J. Fitzsimmons, D.E. Plaumann and B. Fraser-Reid, Tetrahedron
    Lett., 3925 (1979).
10. M. Miljković and D. Glišin, J. Org. Chem., 40, 3357 (1975).
11. D. Seebach Angew. Chem. Int. Ed. Engl., 18, 239(1979) and
    references cited therein.
12. K. Mori, Tetrahedron, 32, 1979(1976); M. Mori, T. Chuman, K. Kato
    and K. Mori, Tetrahedron Lett., 23, 4593 (1982).
13. P.A. Bartlett and J. Myerson, J. Org. Chem., 44, 1625 (1979);
    W.E. Gore, G. T. Pearce and R.M. Silverstein, J. Org. Chem., 40,
    1705 (1975); G.T. Pearce, W.E. Gore and R.M. Silverstein, J. Org.
    Chem., 41, 2797 (1976); G. J. Cernigliaro and P.J. Kocienski,
    J. Org. Chem., 42, 3622 (1977).
14. M. B. Yunker, D.-E. Plaumann and B. Fraser-Reid, Can. J. Chem., 55,
    4002 (1977).
15. For some reviews, see, K.C. Nicolaou, G. P. Gasic and W.E. Barnette,
    Angew. Chem. Int. Ed. Engl. 17, 293 (1978); K. H. Gibson, Chem.
    Soc. Rev., 6, 489 (1977); Science, 196, 1072 (1977).

16. S. Hanessian and P. Lavallée, Can. J. Chem., 55, 562 (1977).

17. S. Hanessian and P. Lavallée, Can. J. Chem., 53, 2975 (1975).

18. J.C. Sheehan, P.A. Cruickshank and G.L. Boshart, J. Org. Chem., 26, 2525(1961).

19. P. Rosen, G.W. Holland, J.L. Jernow, F. Kienzle and S. Kwoh, Abstracts 9th IUPAC Conference on Natural Products, Ottawa, Ont., June 24-28,1974, p. 57.

20. E.J. Corey, M. Shibasaki and J. Knolle, Tetrahedron Lett., 1625 (1977).

21. A.E. Wick, D. Felix, K. Steen and A. Eschenmoser, Helv. Chim. Acta, 47, 2425 (1964); 52, 1030 (1969).

22. R.J. Ferrier and N. Vethaviyasar, J. Chem. Soc., C, 1907 (1971); see also R.J. Ferrier in, MTP International Review of Science, Vol. 7, Carbohydrates, G.O. Aspinall, Ed., University Park Press, Baltimore, Md., 1973.

23. O. Hernandez, Tetrahedron Lett., 219 (1978).

24. A.G. Kelly and J.S. Roberts, J. Chem. Soc., Chem. Comm., 228 (1980).

25. D.T. Connor, R.C. Greenough and M. von Strandtmann, J. Org. Chem., 42, 3664 (1977).

26. G. Just and P. Potvin, Can. J. Chem., 58, 2173 (1980).

27. J.P. Clayton, P.J. O'Hanlon and N. H. Rogers, Tetrahedron Lett., 881 (1980).

28. S. Hanessian and A.G. Pernet, Advan. Carbohydr. Chem. Biochem., 33, 111(1976).

29. J.-M. Beau, S. Aburaki, J.-R. Pougny and P. Sinaÿ, J. Am. Chem. Soc., 105, 621(1983), see also S. Aburaki, J.-M. Beau, J. R. Pougny and P. Sinaÿ; Abstracts 178th Natl. Am. Chem. Soc., Washington, D.C., Sept. 1979, CARB. 29.

30. B. Helferich and W. Ost, Chem. Ber., 95, 2612(1962).

31. S. Hanessian and N.R. Plessas, J. Org. Chem., 34, 1035 (1969); S. Hanessian, Methods   Carbohydr. Chem., 6, 183 (1972), and· references cited therein.

32. B. Bernet and A. Vasella, Helv. Chim. Acta, 62, 1990 (1979); M. Nakane, C.R. Hutchinson and H. Gollman, Tetrahedron Lett., 1213 (1980).

33. C. Huynh, F. Derguini-Boumechal and G. Linstrumelle, Tetrahedron Lett., 1503 (1979); see also M. Schlosser and M. Stähle, Angew. Chem., Int. Ed. Engl., 19, 487 (1980).

34. A.P. Kozikowski, R.J. Schmiesing and K.L. Sorgi, J. Am. Chem. Soc.,
    102, 6577 (1980); R.K. Boeckman, Jr., and E.W. Thomas, J.Am. Chem.
    Soc., 101, 987 (1979).

35. M.O. Chaney, P.V. De Marco, N.D. Jones and J.L. Occolowitz, J. Am.
    Chem. Soc., 96, 1932 (1974).

36. Y. Nakahara, K. Beppu and T. Ogawa, Tetrahedron Lett., 22,
    3197 (1981).

37. S. Hanessian, P.C. Tyler and Y. Chapleur, Tetrahedron Lett., 22,
    4583 (1981).

38. R.J. Ferrier, N. Prasad and G.H. Sankey, J. Chem. Soc., C,
    587(1969); see also, R.J. Ferrier, Methods Carbohydr. Chem., 6,
    307(1972); L.M. Lerner and D.R. Rao, Carbohydr. Res., 22,
    345(1972).

39. D. A. Evans, C.E. Sacks, W.A. Kleschick and R.R. Taber, J. Am.
    Chem. Soc., 101, 6789 (1979), see also G.R. Martinez, P.A. Grieco,
    E. Williams, K. Kanai and C.V. Srinivasan, J. Am. Chem. Soc., 104,
    1436 (1982).

40. D.J. Duchamp, A.R. Branfman, A.C. Button and K.L. Rinehart, Jr.,
    J. Am. Chem. Soc., 95, 4077 (1973); F.A. Mackellar, M.F. Grostic,
    E.C. Olson, R.J. Wnuk, A.R. Branfman and K.L. Rinehart, Jr., J.
    Am. Chem. Soc., 93, 4943 (1971).

41. R.E. Ireland, P.G.M. Wuts and B. Ernst, J. Am. Chem. Soc., 103,
    3205 (1981).

42. R.E. Ireland, R.H. Mueller and A.K. Willard, J. Am. Chem. Soc., 98,
    2868 (1976).

CHAPTER EIGHT

MOLECULES CONTAINING TETRAHYDROFURAN AND TETRAHYDROPYRAN-TYPE RINGS

a. Lasalocid A - The synthesis of lasalocid A by Ireland and coworkers[1] demonstrates the utility of furanoid and pyranoid glycals derived from carbohydrates in the total synthesis of ionophores. The ingenious application of the enolate Claisen rearrangement is once again a pivotal step in the control of stereochemistry in C-C bond-forming reactions at a position vicinal to a ring oxygen (equivalent to an anomeric carbon atom). Scheme 51 shows the overall strategy which calls

Scheme 51

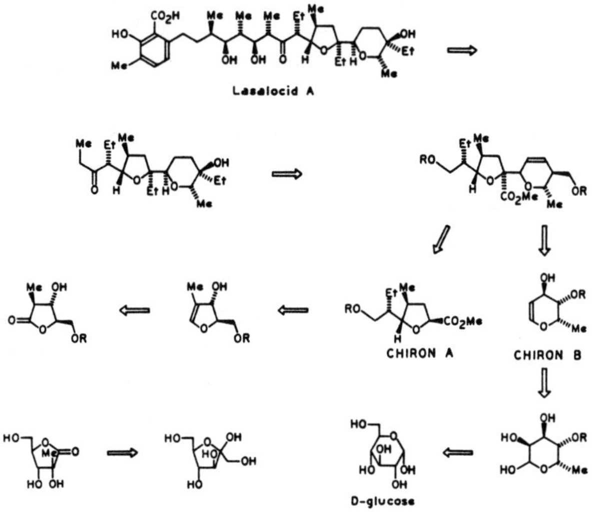

for the formation of two precursors, A and B. The first is a 2,4,5-
trisubstituted tetrahydrofuran in which the C-2 substituent (a CO_2Me
group) orginally belongs to the sugar precursor (as a CH_2OH group). The
proper choice of precursor provides the desired "<u>template</u>" for the
transfer of chirality from C-3 to C-1 (sugar numbering), hence, to
generate the functionalized two carbon side-chain in stereospecific
manner. The C-methyl branch in the ring is nicely incorporated via the
well-known but seldom exploited saccharinic acid rearrangement of
ketoses which leads to the so called "α-saccharinolactones".[2] The
starting sugar for 'chiron' A is D-fructose, while 'chiron' B is a
pyranoid glycal which originates from 6-deoxy-L-gulose, a seemingly rare
sugar.[3] In essence, the preparation of this sugar from D-glucose would
simply involve deoxygenation at C-1, and oxidation of C-6 to an aldehyde
(see Scheme 4). These formal transformations were conveniently achieved
in five steps and 52% overall yield starting not from D-glucose itself,
but from the configurationally related and readily available D-glucurono-
γ-lactone. The preparation of 'chiron' A, <u>52.5</u>, is illustrated in
Scheme 52. The perspective drawings show that it can be related to a
furanoid-type precursor in which the terminal hydroxymethyl group serves
as a latent carboxyl group. D-Fructose provided the branched-chain
intermediate <u>52.1</u> (α-D-glucosaccharino 1,4-lactone) which in turn was
a precursor to the glycal <u>52.2</u>. Enolate Claisen rearrangement and·
reduction led to a mixture in which the major component corresponded to
the desired isomer <u>52.3</u>. Manipulating the extremities of both side-
chains then provided 'chiron' A as represented by structure <u>52.5</u>. The
preparation of 'chiron' B <u>53.6</u> is shown in Scheme 53. The key transformation involved the overall conversion of D-glucose into a derivative
of L-gulose by a head-to-tail-type inversion. The readily available
D-glucurono- γ-lactone <u>53.1</u> was converted into the diethyl dithioacetal
derivative <u>53.2</u>, which, after reduction with Raney-nickel, gave the
corresponding 1-deoxy derivative <u>53.3</u>. As such, the 6-deoxy-L-gulose
structure is in hand except for the oxidation state at C-1. The preparation of the pyranoid glycal intermediate ('chiron' B, <u>53.6</u>) necessitated conversion into the appropriate acetal derivative <u>53.5</u> which
contained the required *cis*-diol unit at C-2/C-3. The completion of the
synthesis involved the union of 'chirons' A and B via a critical and
unique enolate Claisen rearrangement in which the "acid unit" consisted
of 'chiron' A itself (Scheme 54). The required ester <u>54.1</u> was thus
subjected to the Claisen rearrangement to give the desired product <u>54.2</u>

Scheme 52

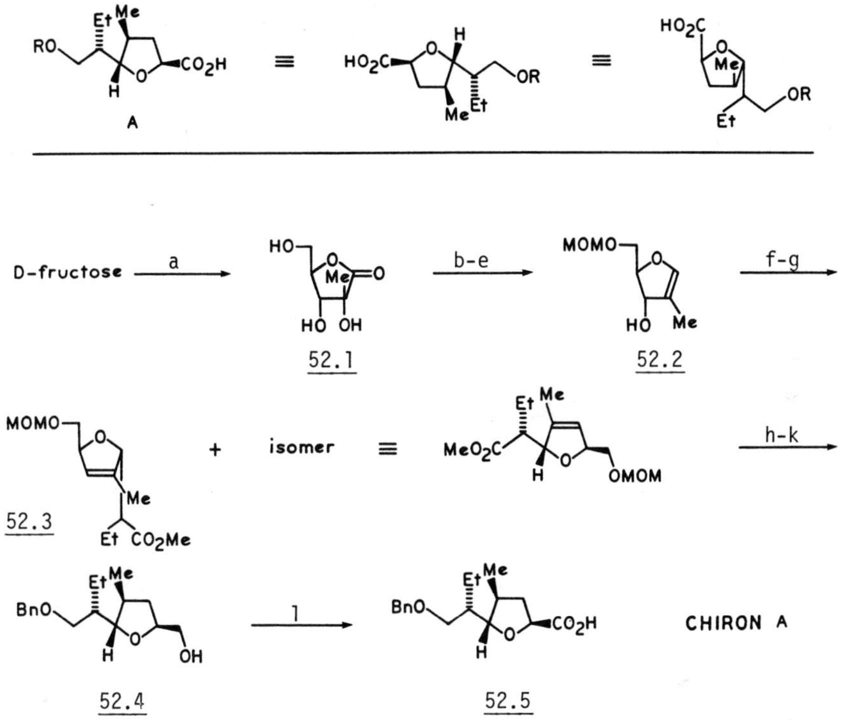

a. Ca(OH)$_2$; b. acetone, H$^+$; c. ClCH$_2$OMe , KH; d. DIBAL, Et$_2$O; e. P(NMe$_2$)$_3$, CCl$_4$, Li/NH$_3$; f. n-BuLi, n-C$_3$H$_7$COCl, LDA, HMPA/THF, TMSiCl; g. CH$_2$N$_2$; h. Pd/C, H$_2$, EtOAc; i. LAH, Et$_2$O; j. BnBr, KH; k. H$_3$O$^+$; 1. Pt/C, O$_2$

as the major component. This optically pure intermediate, derived almost entirely from carbohydrate precursors, was then further elaborated via 54.4 → 54.5 to ultimately give 54.6 which had previously been converted into lasalocid A in another landmark synthesis of this compound by Kishi and coworkers.[4]

Scheme 53

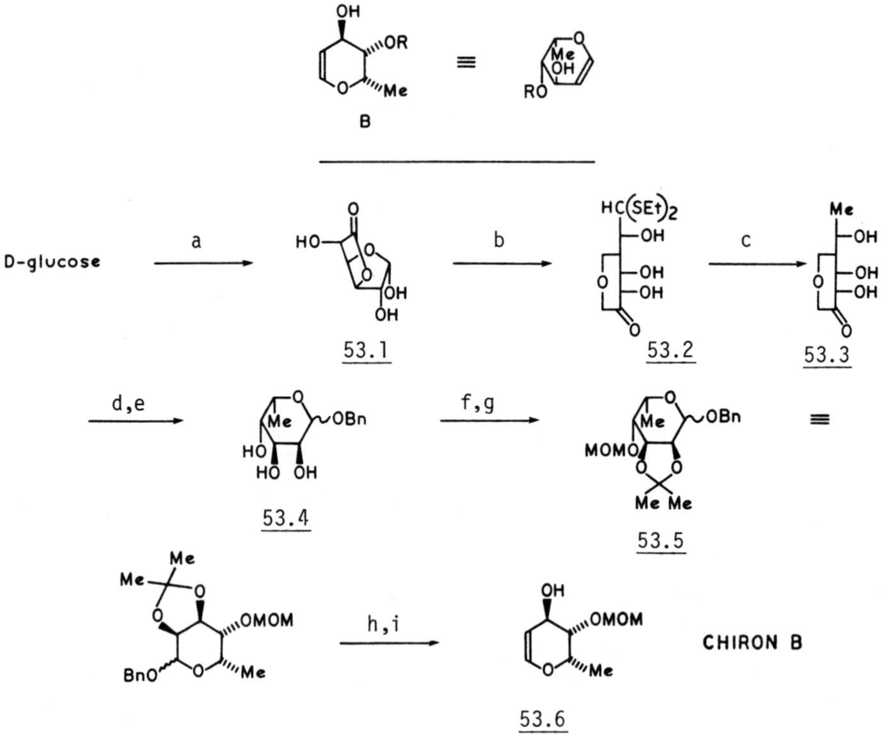

B

53.1   53.2   53.3

53.4

53.5

53.6

CHIRON B

a. oxid; b. EtSH, H$^+$; c. Ra-Ni; d. Na/Hg, H$_3$O$^+$; e. BnOH, H$^+$; f. acetone, H$^+$;
g. ClCH$_2$OMe, KH; h. Pd/C, H$_2$; i. P(NMe$_2$)$_3$, CCl$_4$, Li/NH$_3$

   b. Boromycin - Synthesis of the 'upper' and 'lower' halves -
Boromycin (Scheme 55) is a unique macrodiolide antibiotic containing
boron, whose constitutional structure was determined by single crystal
X-ray analysis.[5] The only other known structurally related natural
product also containing boron is aplasmomycin[6] which consists of two
identical halves, hence its *C2* symmetry. The constitutional structures
of boromycin and aplasmomycin are the same except for the acyclic portion,
the opposite senses of chirality at the neopentylic alcohol centers, and
the type of unsaturation. In fact the C-10 — C-17 aliphatic portion
in boromycin appears to be the open-chain counterpart of the C-10'—
C-17' tetrahydrofuran ring portion. Limited degradative studies[5] have

Scheme 54

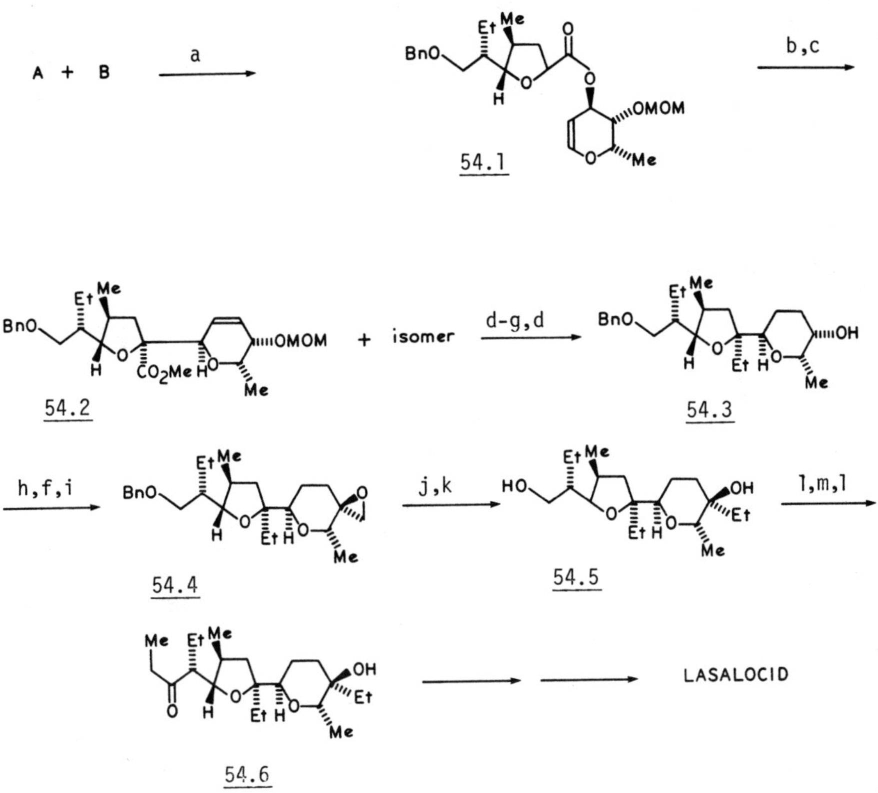

54.1

54.2

54.3

54.4

54.5

54.6

LASALOCID

a. $(COCl)_2$; b. n-BuLi, THF, LDA, TMSCl; c. CH_2N_2; d. Ra-Ni, H_2; e. DIBAL;
f. $Ph_3P=CH_2$, THF; g. ROH, H^+; h. DMSO, $(COCl)_2$; i. m-CPBA, CH_2Cl_2;
j. Me_2CuLi, pentane; k. Li/NH_3; l. PCC; m. EtMgBr

produced two fragments for which the structures of a $C_{18}H_{32}O_5$ lactone
(Scheme 55, structure A) and an acidic $C_{13}H_{24}O_4$ fragment have been
suggested.   Schemes 55 and 56 illustrate retrosynthetic analyses which
create logical precursors such as A-E representing the tetrahydrofuran,
tetrahydropyran and acyclic C-10 — C-17 segments, all of which can be
related to sugar precursors with a D-configuration.   In such a strategy,

Scheme 55

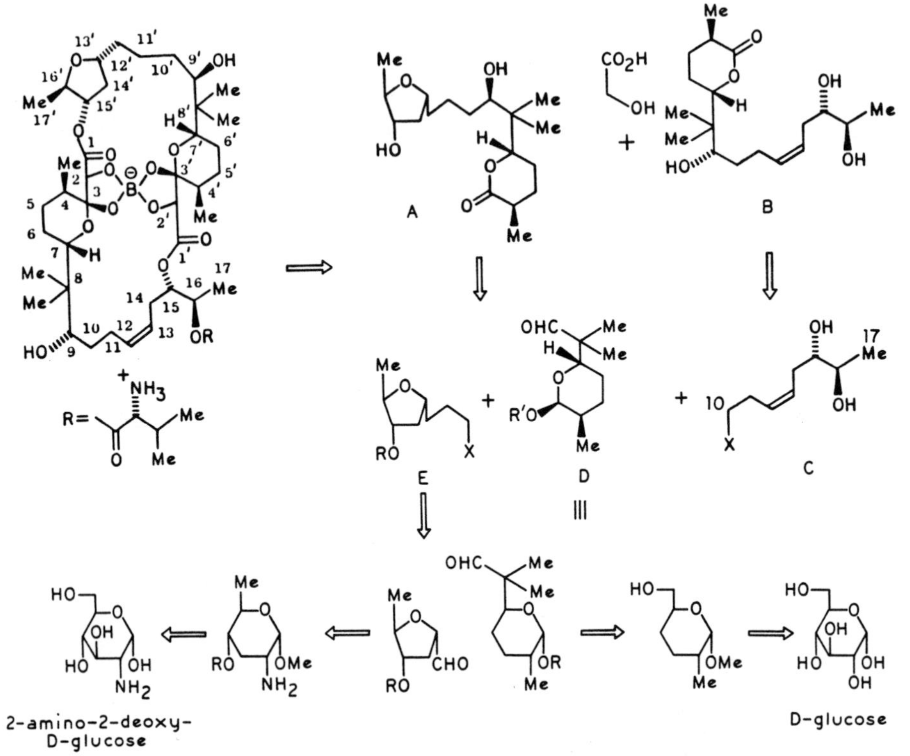

2-amino-2-deoxy-
D-glucose

D-glucose

precursors C and E can be considered as nucleophilic counterparts (X=Mg
halide, -SPh, etc) while D would be a common electrophilic partner.
Carbon-carbon bond formation would thus create the neopentylic center.
The prospects of achieving stereoselective condensations favoring one
isomer can be appreciated particularly if a metal-assisted reaction were
successful and if some aspect of "chelation control"[7] could be expected
by virtue of the presence of the ring oxygen. Access to the
trisubstituted tetrahydrofuran portion with a three-carbon appendage
(precursor E) was visualized via a 2,5-anhydro-D-aldose derivative,
which in turn could be obtained by stereocontrolled ring contraction of
a 2-amino-2,3,6-tri-deoxy-D-hexopyranose derivative. Precursor D on the
other hand was envisaged to arise from D-glucose by systematic deoxygenations. Precursor C (Scheme 56) can be related to a 2,5-dideoxy-D-
erythro-pentose and the latter can be prepared by an adaptation of a
recent efficient synthesis of 2-deoxy-D-erythro-pentose (2-deoxy-D-

Scheme 56

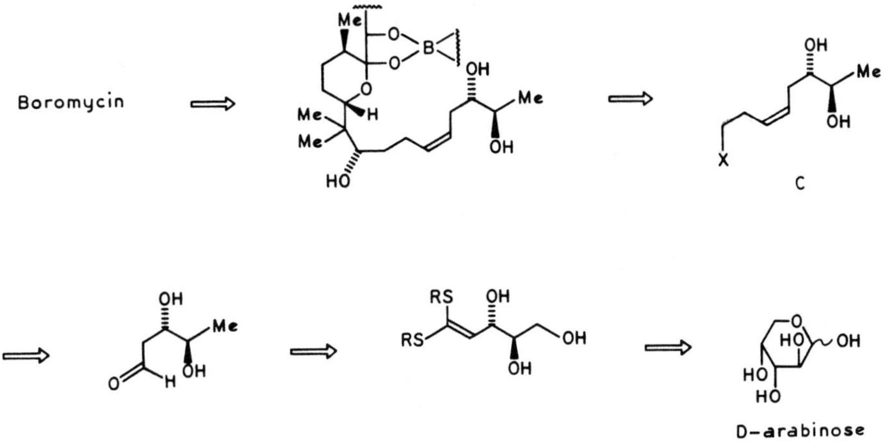

D-arabinose

ribose) from the readily available D-arabinose.[8] With the retrosynthetic analysis shown in scheme 55, it is clear that the main
challenges in the synthesis of lactones A and B, representing the
'upper' and 'lower' segments of boromycin, are the creation of
appropriately functionalized chiral ring systems and joining them. The
synthesis of lactone A ($C_{18}H_{32}O_5$) has been accomplished using D-glucose
and 2-amino-2-deoxy-D-glucose as chiral building blocks.[9] Scheme 57
illustrates the steps leading to the synthesis of the sulfoxide
derivative 57.8 (precursor E) from the readily available 2-amino-2-
deoxy-D-glucose. The key step in this strategy was the anticipated[10,11]
stereocontrolled ring contraction of the 2-amino sugar derivative 57.4
to give the trisubstituted tetrahydrofuran 57.5. Access to 57.5 was
based on systematic deoxygenations at C-3 and C-6 by standard
methodology. Note however the efficient displacement at C-3 by iodide
(see Scheme 3) using the versatile O-imidazolylsulfonyl leaving group,[12]
and the tributyltin hydride-mediated reduction[13] of a 3,6-dihalo
derivative in one step (conversion 57.2 → 57.3). Treatment of 57.4 with
dinitrogen trioxide[14] led to an aldehyde isolated as the dimethyl acetal
in very good yield. Elaboration of the side-chain followed standard
procedures to give the sulfoxide 57.8. Note the direct conversion[15] of
the alcohol 57.7 into the corresponding thioether. Alcohol 57.7 was
also transformed into the corresponding bromide in high yield, but all

Scheme 57

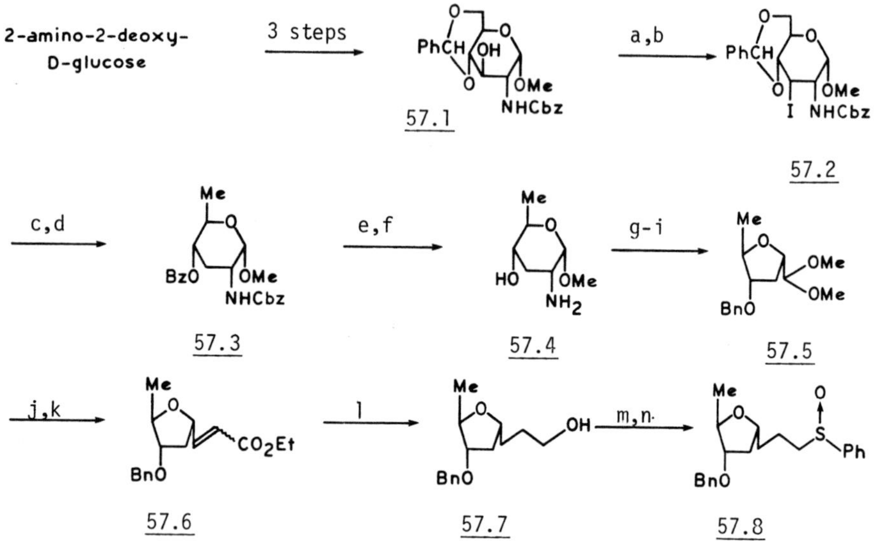

a. SO_2Cl_2, imidazole, DMF; b. Bu_4NI, benzene, reflux; c. NBS, CCl_4, reflux;
d. Bu_3SnH, AiBN, toluene, 80^0 ; e. NaOMe, MeOH; f. Pd/C, H_2; g. N_2O_3, aq.
MeOH; h. MeOH, HCl; i. BnBr, NaH, DMF; j. aq. AcOH, THF; k. $Ph_3P=CHCO_2Et$;
1. LAH ; m. PhSSPh, Bu_3P; n. m-CPBA

attempts to form the corresponding Grignard derivative were not
successful. Scheme 58 shows an extremely efficient route to the tetrahydropyran ring form as in 58.2, and its subsequent modification to
incorporate the neopentyl side-chain to give precursor D. A key
reaction was the finding that treatment of the "enoside" derivative 48.2
containing a β-acyloxy enol ester function, with methylenetriphenyl-
phosphorane afforded the diene 58.1 in one step and in high yield.
Presumably reaction takes place by initial formation of an enone,
followed by a Wittig reaction, reminiscent of the reaction of 48.2 with
lithium dimethylcuprate[16] (Scheme 48). The presence of the t-butoxy
group in 58.1 ensured a high degree of stereoselection in the catalytic
hydrogenation to give a high preponderance of the desired isomer 58.2
(9:1 ratio), which was converted to the corresponding methyl ketone
58.3. Elaboration of the quaternary center with the aldehydic appendage
was effected by first preparing an enone derivative followed by
conjugate addition of lithium dimethylcuprate, trapping the enolate as

the trimethylsilyl enol ether 58.4, and ozonolysis. This sequence
provided the desired aldehyde 58.5 (precursor D) in good overall
yield.   Methods for preparing functionalized carbon compounds
containing quaternary centers are not abundant in the literature[17] and
the sequence used for the preparation of 58.5 seemed admirably suited,
particularly because of the uncommon type of appendage required.
Initially, the quaternary center was to have been introduced by the
elegant method developed by Martin,[18] and Wender[19] and their coworkers
(Scheme 59).

Scheme 58

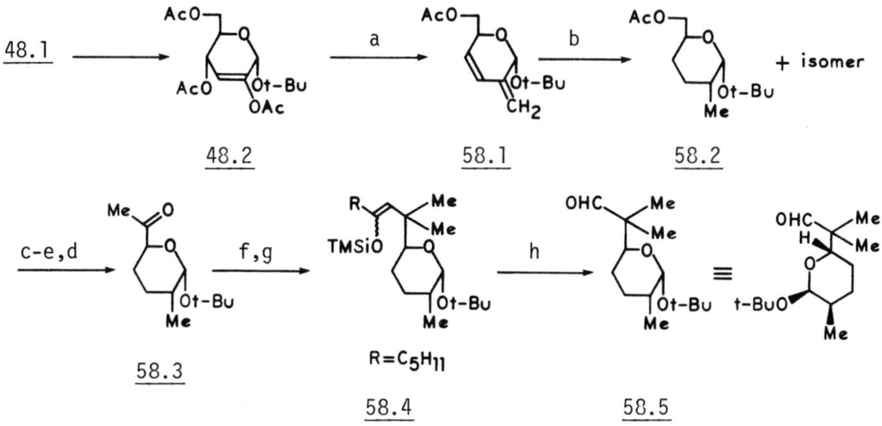

48.1

48.2

58.1

58.2

c-e,d

f,g

58.3

$R=C_5H_{11}$

58.4

h

58.5

a. $Ph_3P=CH_2$, THF; b. Pd/C, H_2, EtOAc; c. NaOMe, MeOH; d. Collins; e. MeMgBr,
ether; f. $(MeO)_2P(O)CH_2COC_5H_{11}$, NaH, DME, reflux; g. Me_2CuLi, ether, then
TMSiCl; h. O_3, CH_2Cl_2, 1% pyr.

The ketone 58.3 was transformed into the corresponding azadiene
derivative 59.1 by treatment with diethyl benzylideneaminomethylphospho-
nate[20] . Unfortunately the formation of the metalloenamine 59.2 was
complicated by concomitant ring opening. The route outlined in Scheme
59 also offered the attractive possibility of constructing the intended
target lactone A from the ketone 58.3 and a benzylideneaminophosphonate
derivative such as 59.4 (prepared from the bromide corresponding to
57.7). However, the behavior of the metalloenamine 59.2, precluded the
pursuit of such a route. Having generated the required precursors 57.8
and 58.5, the stage was set for attempted coupling (Scheme 60). Indeed
treatment of 58.5 with the carbanion generated from 57.8 led to 60.1 in
good yield as a mixture of four stereoisomers. Desulfurization

CHAPTER EIGHT

Scheme 59

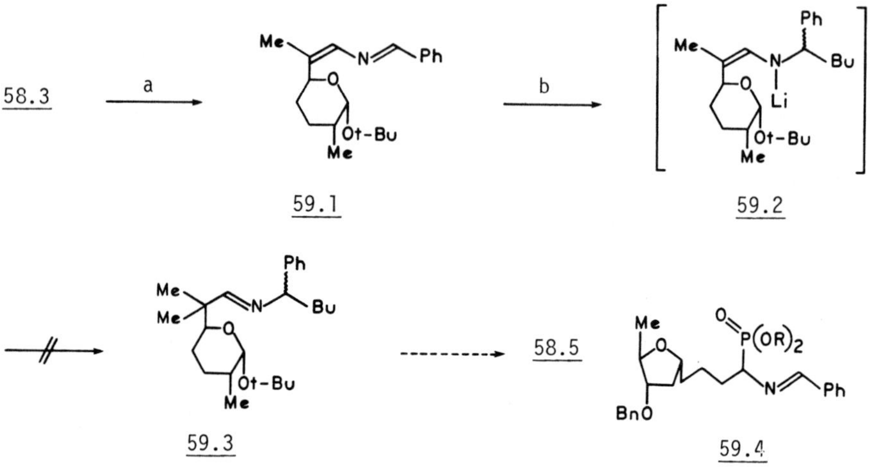

a. $(MeO)_2P(O)CH_2N=CHPh$, THF, BuLi; b. BuLi, $-78°$, then MeI

with Raney-nickel gave the expected mixture of C-9' (boromycin numbering) epimers which after chromatographic separation, followed by routine manipulations afforded the desired lactone A, 60.4. This product was found to be identical with the lactone[5] prepared by degradation of boromycin. The other isomer (epimeric at C-9') could be recycled by oxidation and hydride reduction which gave a mixture of alcohols rich in the desired isomer. Attachment of the glycolic acid chain led to 60.5 then 60.6 which represents the entire upper-half of boromycin.

The synthesis of lactone B[21] utilized the aldehyde 58.5 and acyclic derivatives (precursors C and D) prepared from D-arabinose[22] (Scheme 61). A key step is the formation of the ketene dithioacetal derivative 61.1[8] and its regiospecific reduction with lithium aluminum hydride. The introduction of the 5-deoxy function was achieved via reductive methods, and the three-carbon extension containing a Z-olefinic linkage was introduced by Wittig procedures to give 61.5. The strategy for uniting 61.5 with aldehyde 58.5 was based on a sulfone carbanion technology which led to efficient coupling.[21]

Scheme 60

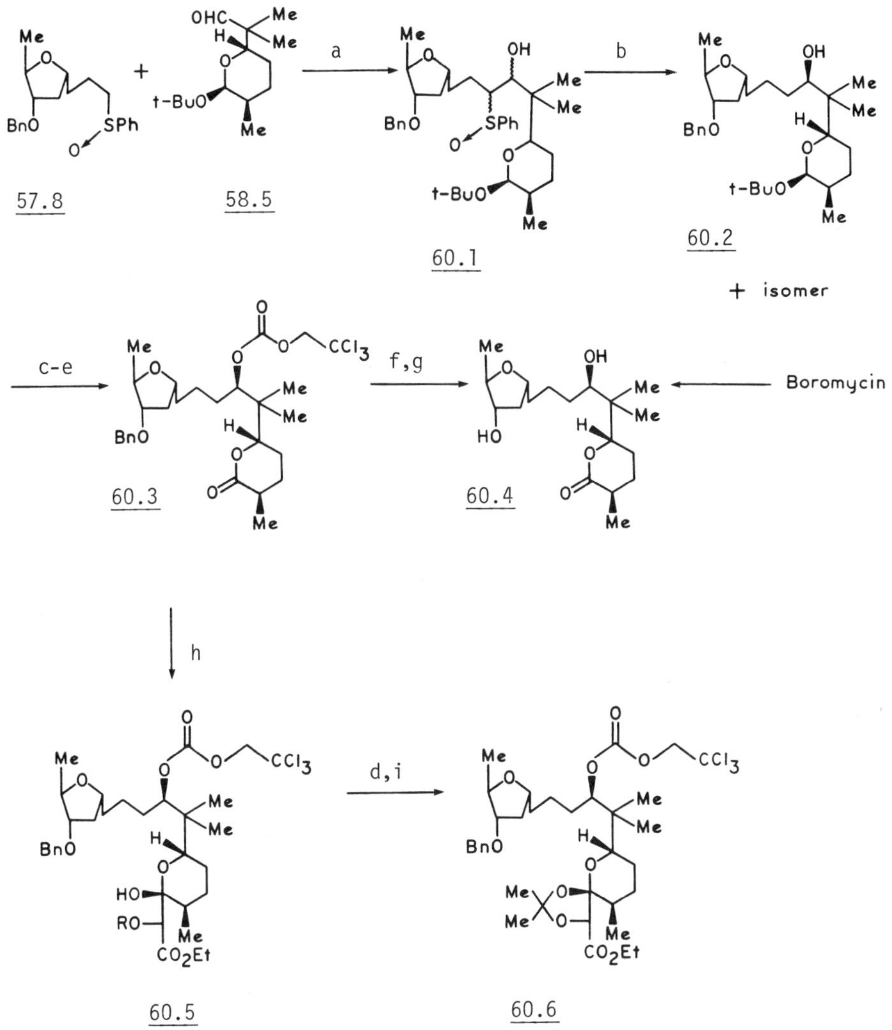

a. LDA, THF, 10% HMPA, $-78°$; b. Ra-Ni, hexanes then Pd/C, H_2, EtOAc;
c. $ClCO_2CH_2CCl_3$, pyr.; d. aq. HCl, THF; e. PCC; f. Zn, aq. KH_2PO_4;
g. Pd/C, H_2; h. ethyl 2-(methoxy-isopropyl) glycolate, LDA; i. acetone,
H^+.

Scheme 61

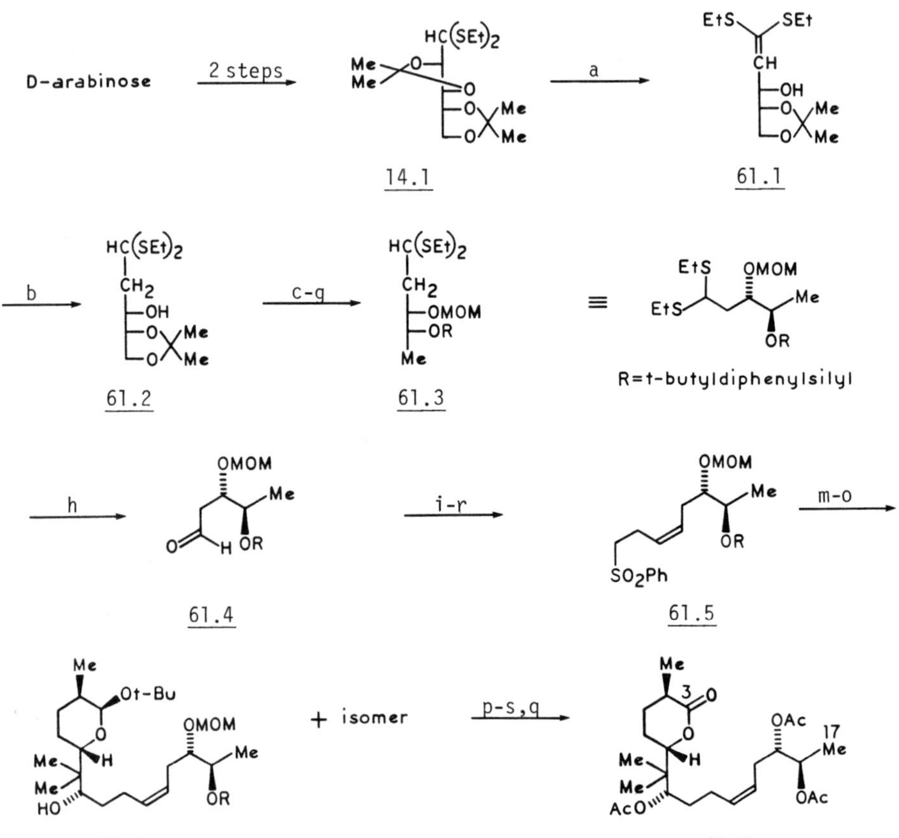

D-arabinose — 2 steps → 14.1 — a → 61.1

EtS   SEt

b → 61.2 — c-g → 61.3 ≡

EtS   OMOM

EtS   Me

OR

R=t-butyldiphenylsilyl

h → 61.4 — i-r → 61.5 — m-o →

61.6 + isomer — p-s,q → 61.7

a. t-BuOK, t-BuOH, THF, DMSO; b. LAH, ether; c. ClCH$_2$OMe, Et$_3$N; d. aq. AcOH;
e. TsCl, pyr.; f. LiEt$_3$BH; g. Ph$_2$t-BuSiCl, imidazole; h. Br$_2$, aq. NaHCO$_3$;
i. Ph$_3$P=CHCH$_2$CH$_2$OTHP, THF; j. HCl, MeOH; k. PhSSPh, Bu$_3$P; l. KHSO$_5$, aq. MeOH;
m. aldehyde 58.5, LDA, THF, HMPA, -78°; n. Collins; o. Na/Hg, then NaBH$_4$,
aq. EtOH; p. Bu$_4$NF; q. Ac$_2$O, pyr.; r. aq. AcOH; s. PCC; t. aq. HCl, THF.

Conversion to a β-ketosulfone, desulfonylation and reduction led to the desired 61.6 and its epimer. The latter could be recycled via an oxidation-reduction sequence. The stereochemical and constitutional identity of 61.6 was unambiguously established by its conversion to the lactone derivative 61.7 which was found to be identical to a sample obtained by the oxidative cleavage of desboron desvaline boromycin. Having thus gained access to lactone B which constitutes the lower half of boromycin, the stage is set for chain-extension with a glycolic acid unit to provide the C-1—C-2 chain and to effect two critical ester forming bonds to give the macrolide. Glycolate extension has been successfully accomplished as for the upper half (Scheme 60) to give the C-1 — C-17 segment of boromycin.[23] The boron atom can be inserted in desboron desvaline boromycin as shown in Scheme 62.[24,25] This has also been described independently by White and coworkers [25] (Scheme 62), who have developed stereoselective syntheses of segments of boromycin using a different approach.[26]

Scheme 62

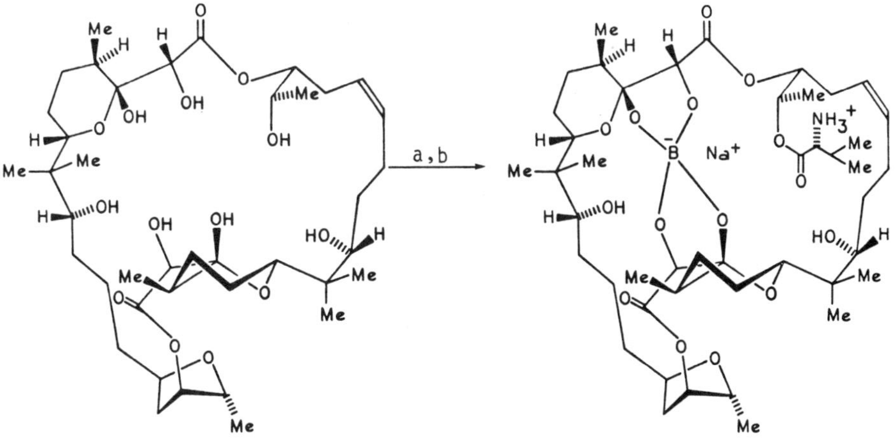

a. (MeO)$_3$B, MeOH, reflux, then silica column chromatography (ref.24); or H$_3$BO$_3$, THF, reflux, then NaH, benzene (ref.23); b. t-BOC-D-val, DCC, DMAP, CH$_2$Cl$_2$ (ref.25).

REFERENCES

1. R.E. Ireland, R.C. Anderson, R. Badoud, B.J. Fitzsimmons, G.J.
    McGarvey, S. Thaisrivongs and C.S. Wilcox, J. Am. Chem. Soc.,
    105, 1988 (1983); R.E. Ireland, S. Thaisrivongs and C.S. Wilcox, J.
    Am. Chem. Soc, 102, 1155 (1980); see also R.E. Ireland, G.J.
    McGarvey, R.C. Anderson, R. Badoud, B.J. Fitzsimmons and S.
    Thaisrivongs, J. Am. Chem. Soc., 102, 6178 (1980).

2. R.L. Whistler and J.N. BeMiller, Methods Carbohydr., Chem., 2, 484
    (1963).

3. R.E. Ireland and C.S. Wilcox, J. Org. Chem., 45, 197 (1980).

4. T. Nakata, G. Schmid, B. Vranesic, M. Okigawa, T. Smith-Palmer and
    Y . Kishi, J. Am. Chem. Soc., 100, 2933 (1978).

5. R. Hütter, W. Keller-Schierlein, F. Knusel, V. Prelog, G.C.
    Rodgers, Jr., P. Suter, G. Vogel, W. Voser and H Zähner, Helv.
    Chim. Acta, 50, 1533 (1967); J.D. Dunitz, D.M. Hawley, D. Miklos,
    D.N.J. White, Yu. Berlin, R. Marusic and V. Prelog, Helv. Chim.
    Acta, 54, 1709 (1971); W. Marsh, J. D. Dunitz and D.N.J. White,
    Helv. Chim. Acta, 57, 10 (1974).

6. H. Nakamura, Y. Iitaka, T. Kitahara, T. Okazaki and Y. Okami, J.
    Antibiotics, 30, 714 (1977); K. Sato, T. Okazaki, K. Maeda and Y.
    Okami, J. Antibiotics, 31, 632 (1978); for biosynthetic studies,
    see T.S.S. Chen, C.J. Chang, and H.G. Floss, J. Am. Chem. Soc.,
    101, 5826 (1979).

7. See for example, W.C. Still and J.H. McDonald, III, Tetrahedron
    Lett., 21, 1031 (1980); M.L. Wolfrom and S. Hanessian, J. Org.
    Chem., 27, 1800 (1962); D.J. Cram and K.R. Kopecky, J. Am. Chem.
    Soc., 81, 2748 (1959).

8. M.Y.H. Wong and G. R. Gray, J. Am. Chem. Soc., 100, 3548 (1978).

9. S. Hanessian, P.C. Tyler, G. Demailly and Y. Chapleur, J. Am. Chem.
    Soc., 103, 6243 (1981); S. Hanessian, D. Delorme, P.C. Tyler, G.
    Demailly and Y. Chapleur, Current Trends in Organic Synthesis, H.
    Nozaki, ed., Pergamon Press, Oxford, 1983, p.205.

10. See for example, B.C. Bera, A.B. Foster and M. Stacey, J. Chem.
    Soc., 4531 (1956); J. Defaye, Advan. Carbohydr.Chem. Biochem., 25,
    181 (1970).

11. See for example G.E. McCasland, J. Am. Chem. Soc., 73, 2293
    (1951); D.Y. Curtin and S. Schmukler, J. Am. Chem. Soc., 77, 1105
    (1955); A. Streitwieser, Jr., J Org. Chem., 22, 861 (1957);

Taguchi, T. Matsuo and M. Kojima, J. Org. Chem., 29, 1104 (1964);
M. Chérest, H. Felkin, J. Sicher, F. Sipos and M. Tichy, J. Chem.
Soc., 2513 (1965).

12. S. Hanessian and J.-M. Vatèle, Tetrahedron Lett., 22, 3579 (1981).

13. H. Arita, N. Ueda and Y. Matsushima, Bull. Chem. Soc. Japan. 22,
    3579 (1972).

14. P. Angibeaud, J. Defaye and H. Franconie, Carbohydr. Res., 78, 195
    (1980).

15. I. Nakagawa and T. Hata, Tetrahedron Lett., 1409 (1975).

16. S. Hanessian, P.C. Tyler and Y. Chapleur, Tetrahedron Lett., 22,
    4583 (1981).

17. S.F. Martin, Tetrahedron, 36, 419 (1980).

18. S.F. Martin and G. W. Phillips, J. Org. Chem., 43, 3792 (1978).

19. P.A. Wender and M.A. Eissenstat, J. Am. Chem. Soc., 100, 292 (1978).

20. R.W. Ratcliffe and B.G. Christensen, Tetrahedron Lett., 4645
    (1973); K. Yamauchi, Y. Mitsuda and M. Kinoshita, Bull. Chem.
    Soc. Japan, 48, 3285 (1975); A. Dehnel, J.P. Firet and G.
    Lavielle, Synthesis, 474 (1977).

21. S. Hanessian, D. Delorme, P. Tyler, G. Demailly and Y. Chapleur,
    Can. J. Chem., 61, 634(1983).

22. See for example N.A. Hughes and R. Robson, J. Chem. Soc., C, 2366
    (1966).

23. S. Hanessian and D. Delorme, unpublished results.

24. S. Hanessian and P.C. Tyler, unpublished results.

25. M.A. Avery, J.D.White and B.H. Arison, Tetrahedron Lett., 22, 3123
    (1981).

26. J. D. White, Abstracts 179th National Meeting Am. Chem. Soc.,
    Houston, March 23-28, 1980, ORGN. 48.

CHAPTER NINE

MOLECULES CONTAINING A
BUTYROLACTONE-TYPE RING

The apparent carbohydrate-type symmetry is quite evident in the structure of a number of monocyclic and bicyclic natural products containing a butyrolactone-type ring. Thus, the nature of the five membered ring, the oxidation level at C-1 and the presence of oxygen substituents, particularly the one in the ring clearly indicate a close structural as well as stereochemical relationship to a furanoid carbohydrate derivative. In fact with few exceptions, the recorded syntheses of compounds in this class from carbohydrate precursors are based on this premise. Since the sense of chirality of the oxygen-bearing carbon atoms in the individual target molecules will dictate the choice of starting carbohydrate, the main challenge in the synthesis of such molecules by this approach resides in the regio- and stereocontrolled introduction of carbon branching. This can be accomplished rather easily by preferential protection of hydroxyl groups and controlling the stereochemistry of C-C bond forming reactions based on considerations of steric hindrance and other parameters.

a. <u>(-)-Avenaciolide</u> - Avenaciolide is a naturally-occurring antifungal compound whose constitutional structure was determined by degradative studies and by n.m.r.[1] Anderson and Fraser-Reid,[2] and Ohrui and Emoto[3] have independently described the synthesis of (-)-avenaciolide from D-glucose by essentially similar routes. This is not as a result of coincidence; rather it is an illustration of how once a carbohydrate approach is considered, there remains essentially one logical route to follow. Scheme 63 shows that the sense of chirality of the two lactone

116

ring oxygens can be directly related to C-2 and C-4 of D-glucose. Utilizing a readily available furanoid derivative such as 1,2:5,6-di-O-isopropylidene-α-D-glucofuranose[4] offers unique operational advantages as it allows regiospecific chain-branching at C-3 and provides the opportunity to control stereochemistry at that carbon in introducing

Scheme 63

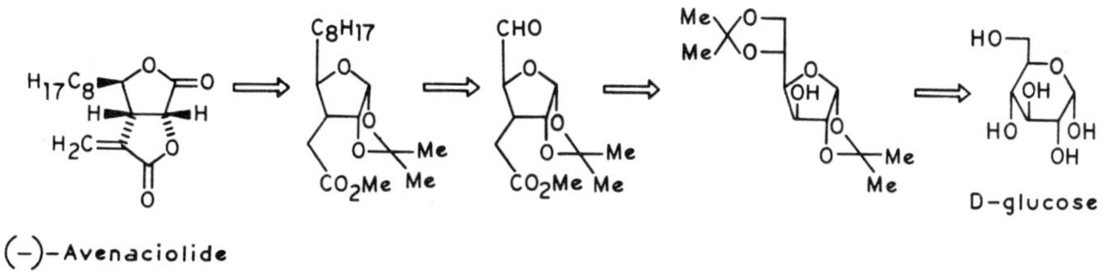
D-glucose
(–)-Avenaciolide

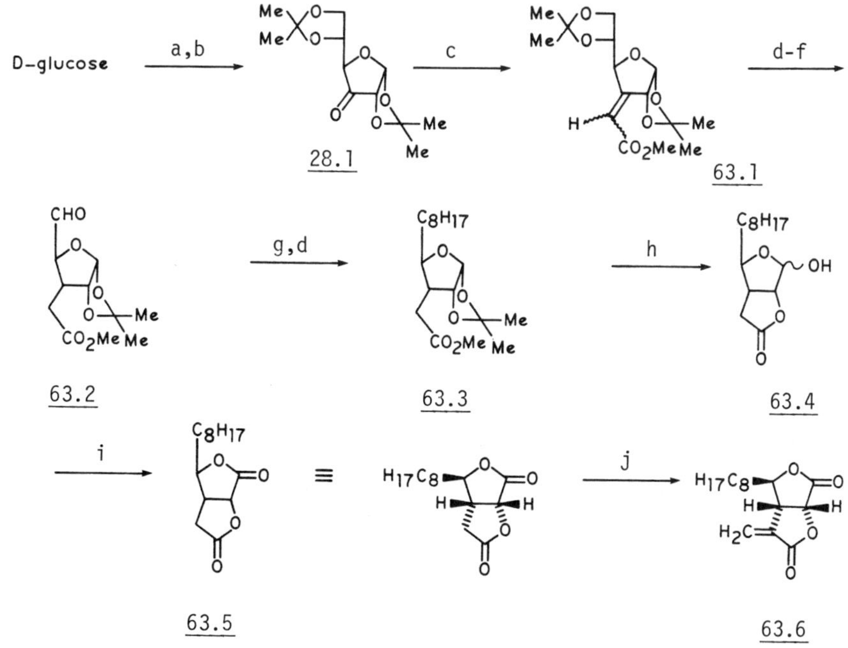
D-glucose
a,b
c
d-f
28.1
63.1
g,d
h
63.2
63.3
63.4
i
j
63.5
63.6

a. acetone, H^+; b. RuO_4, CCl_4; c. $(MeO)_2P(O)CH_2CO_2Me$, t-BuOK; d. Pd/C, H_2; e. aq. AcOH; f. aq. $NaIO_4$; g. $Ph_3P^+C_7H_{15}Br^-$, THF, BuLi; h. aq. H_2SO_4, diox. reflux; i. CrO_3; j. aq. CH_2O, NaOAc, AcOH; Et_3N, 100°, 5 min., Ref. 6.

an α-acetate chain. The C-5 — C-6 side chain has the correct orientation and serves as a latent aldehyde function which can be utilized in chain-extension. Once the acetate branch and the C_8H_{17} appendage are secured, the stage is set for lactone formation, oxidation and introduction of the methylene group. Analysis of the structures in Scheme 63 shows how these steps were systematically and predictably executed to arrive at the target 63.6. Note that the phosphonate modification of the Wittig reaction as applied to 28.1[5] and subsequent stereocontrolled reduction of the carbomethoxymethylene derivative 63.1 gave the desired α-acetate branch. The subsequent steps were straightforward as illustrated in the scheme. The synthesis of (-)-avenaciolide from D-glucose also allowed the correct stereochemical assignment of the natural product as being 3a(R), 4(R), 6a(R) and not the enantiomeric form as previously suggested.[1] The Ohrui and Emoto paper[3] also described a synthesis of the enantiomer. There are several reports of the synthesis of racemic avenaciolide.[6]

b. (-)-Isoavenaciolide - Another antifungal metabolite in the family of butyrolactone-type bicyclic structures is isoavenaciolide[7], which differs from avenaciolide simply in the inverted orientation of the C_8H_{17} side-chain. An approach from a D-glucofuranose derivative is also feasible provided that the C-5 — C-6 side-chain can be inverted to an α-orientation. This was accomplished by Anderson and Fraser-Reid[8] in their synthesis of (-)-isoavenaciolide from D-glucose (Scheme 64). The key C-C bond forming reactions were the same as in the case of avenaciolide (Scheme 63) except that the orientation of the side-chain was inverted early in the sequence via a well-known elimination reaction[9] to provide the cyclic enol ether derivative 64.2. Hydroboration-oxidation[10] afforded the desired intermediate 64.3 which was processed as indicated in the scheme. Note that in this case the reduction of the carbomethoxymethylene derivative 64.5 was virtually stereospecific. A synthesis of racemic isoavenaciolide is also available.[11]

c. (-)-Canadensolide - The antifungal metabolite, canadensolide[12] differs from avenaciolide in being a positional as well as configurational isomer. Thus, scheme 65 shows it to be a bicyclic α-methylene lactone structure in which chain-branching is at C-2 (carbohydrate numbering). The convenience of utilizing a D-glucofuranose

Scheme 64

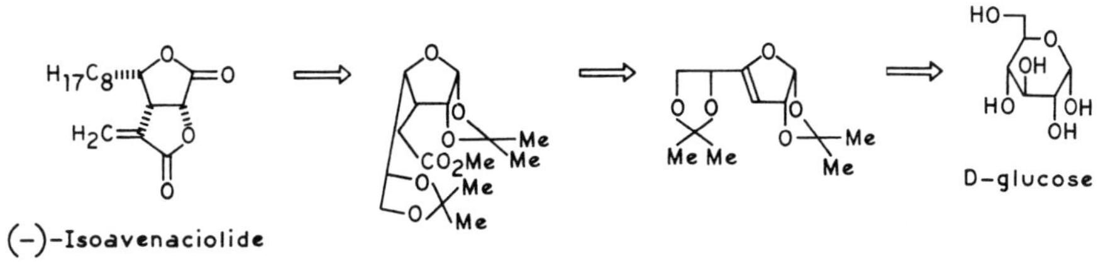

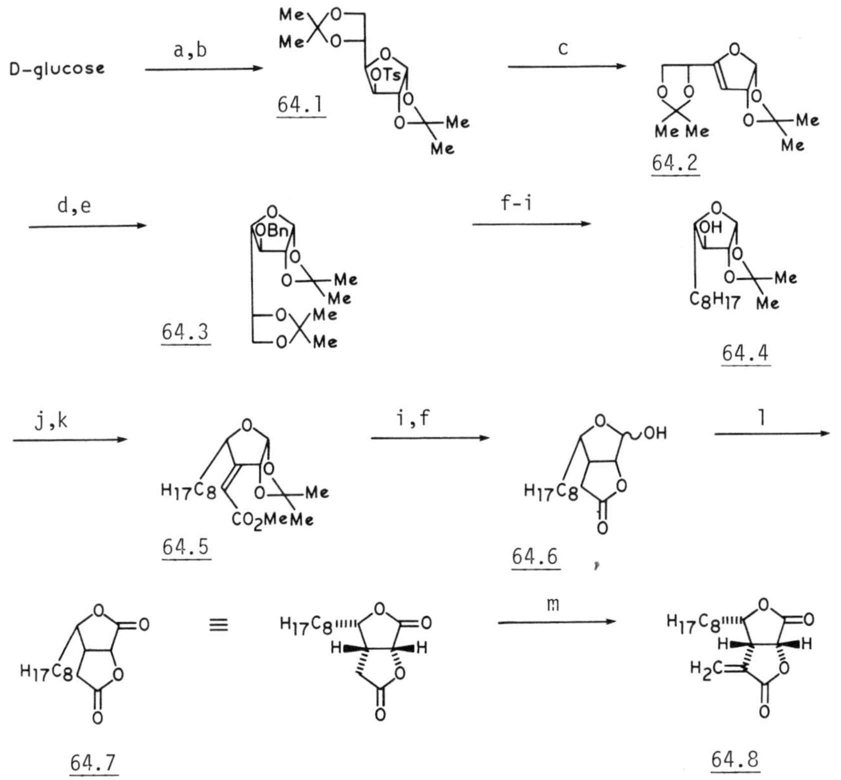

(−)-Isoavenaciolide

a. acetone, H^+; b. TsCl, pyr.; c. soda lime, 220^0; d. B_2H_6, THF, then H_2O_2, NaOH; e. BnCl, NaH, Bu_4NI, THF; f. H_3O^+; g. aq. $NaIO_4$; h. $Ph_3P^+C_7H_{14}Br^-$; i. Ra-Ni 50 p.s.i., then Pd/C H_2; j. Collins; k. $Ph_3P=CHCO_2Et$; l. CrO_3; m. Ref.6.

type   precursor is inescapable since in addition to ring size and
favorable oxidation states, the sense of chirality at C-3 and C-4 is
identical with corresponding centers in the target. The main challenge
here is to introduce the acetate chain at C-2 with a β-configuration.
Scheme 65 illustrates the route adopted by Anderson and Fraser-Reid[13]
in their synthesis of canadensolide from D-glucose. Note that the
order of C-C bond forming reactions were changed here compared to avenaciolide, and the C_8H_{17} chain was introduced first via the readily available aldehyde 65.1,[14] which in turn, was prepared from 1,2:5,6-di-0-
isopropylidene-α-D-glucofuranose. In order to introduce the acetate
side-chain at C-2, the furanoid glycoside, which now contains an
unprotected hydroxyl group at C-2, was generated by methanolysis.
Oxidation, Wittig reaction and reduction of the resulting methylene
derivative 65.3 gave a mixture of the desired 65.4 and its epimer which
could be separated after hydrolysis and lactone formation,as in 65.5.
Note the influence of the orientation in the anomeric methoxyl group in
controlling the stereochemistry of reduction at C-2. A β-D-glycoside
would have most assuredly provided a much larger preponderance of the
desired product 65.4. The remainder of the sequence leading to the
target 65.7 followed precedents known in the literature. A synthesis of
racemic canadensolide has been reported also.[15]

      d.   Terpenoids containing α-methylene-γ-butyrolactone rings
- Carbohydrate models - The α-methylene-γ-butyrolactone ring system is a
common structural unit in a large number of natural products belonging
to the sesquiterpene family.[16] Many of these have marked cytotoxic,
antitumor or other biological activities,[17] and their structure elucida-
tion and subsequent total syntheses are cornerstones of elegant studies
in modern natural product chemistry. The preceding syntheses of bicyclic antifungal metabolites such as avenaciolide, etc.demonstrated the
utility of furanoid derivatives of carbohydrates as chiral synthetic
precursors to this class of compounds. Model 2-C-methylene lactones in
the carbohydrate series have been prepared in several laboratories[18,19]
by standard procedures involving a Wittig reaction (Scheme 66). In view
of the fact that the α-methylene-γ-butyrolactone unit is a likely target
for thiols and thiol-containing enzymes,[17] the 2-deoxy-2-C-methylene-D-
threo-pentono-1,4-lactone model 66.1 was found to form adducts with cysteine and glutathione.[19] Structure 66.2 is a model[18] for an α-methylene-

Scheme 65

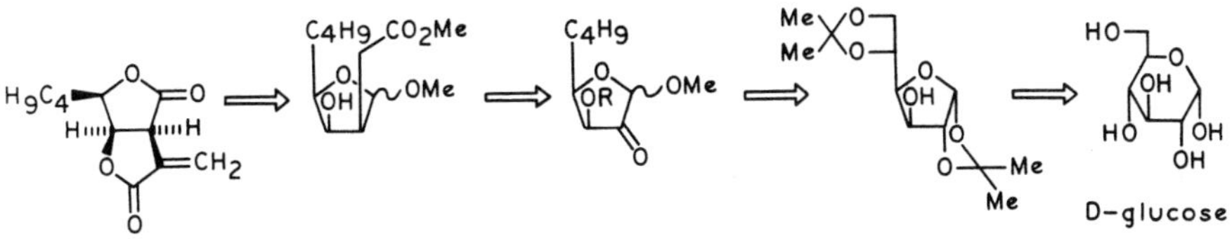
D-glucose

(−)-Canadensolide

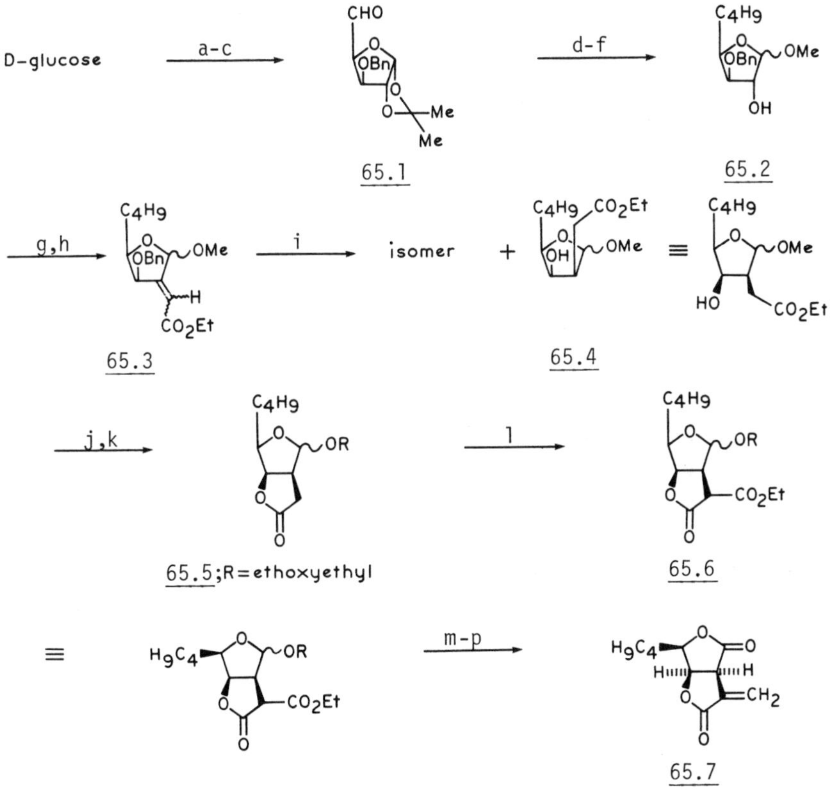
D-glucose
a-c
d-f
g,h
i
isomer
j,k
l
m-p
65.1
65.2
65.3
65.4
65.5;R=ethoxyethyl
65.6
65.7

a. acetone, H^+; b. BnBr, KOH, toluene; c. H_3O^+, then $Pb(OAc)_4$, benz.; d. $Ph_3P^+C_3H_7Br^-$, BuLi, THF; e. Ra-Ni, H_2; f. MeOH, HCl; g. PCC, benz., reflux; h. $Ph_3P{=}CHCO_2Et$; i. Pd/C, H_2; j. PPTS; k. ethylvinyl ether, PPTS, CH_2Cl_2; l. $ClCO_2Et$, LDA; m. H_3O^+; n. CrO_3; o. aq. HCl, diox., reflux; p. Et_2NH, CH_2O, AcOH

γ-butyrolactone portion of bicyclic sesquiterpenes in which the β- and
γ-carbon atoms of the lactone bear carbon substituents. It can be seen
how furanoid sugars can prove to be useful synthetic precursors to
targets such as helenalin[20] and confertin.[21] Expression 66.3 depicts the
structure of a synthetic bicyclic α-methylene lactone which formally
belongs to a perhydrofuropyran class of compounds.[22,23] Compounds 66.4[24]
and 66.5[25] are readily available by regiospecific opening of appropriate
epoxide derivatives with malonate anion. Two features are worthy of
comment with regard to these structures. Firstly, when viewed in a
different perspective (66.4A and 66.5A), a remarkable congruence of

Scheme 66

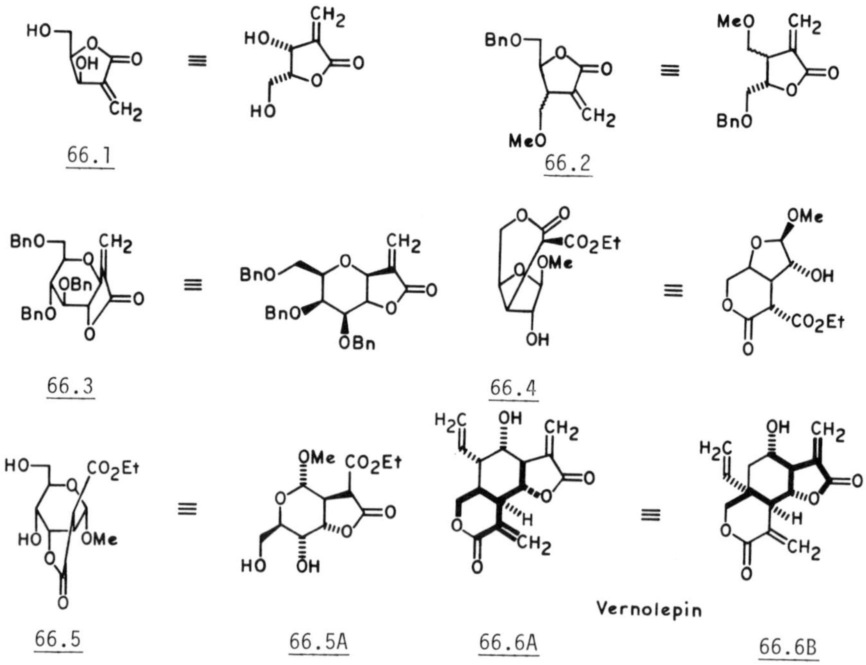

some structural and stereochemical features emerges with vernolepin,
66.6.[26] Thus, expression 66.6A and 66.6B show the extent of overlap
with 66.4A and 66.5A respectively. Interestingly, the crucial C-C bonds
required to construct the vernolepin skeleton can be generated from

sites occupying oxygen functionality. Secondly, an X-ray crystallographic study[27] of 66.4 showed it to be a single diastereoisomer with the
carbethoxy group in an exo-orientation. Such derivatives can therefore
be considered as containing a "chiral malonyl carbon atom" by virtue of
the preferred orientation of the carboethoxy group either in the transition state leading to 66.4 or after base-catalyzed equilibration(thermodynamic product). It is not unlikely that 66.5 also consists of a pure
diastereoisomer.

e. (+)-Blastmycinone – Saponification of the antibiotic
antimycin A$_3$ (blastmycin)[28] affords (+)-blastmycinone (Schemes 67,68).

Scheme 67

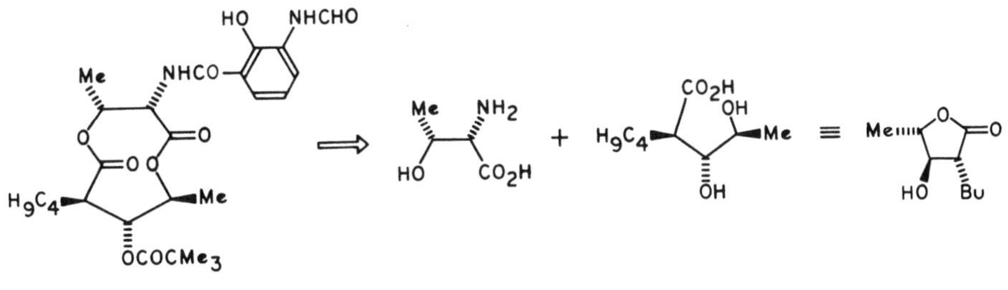

Antimycin

Correlation of its structure with that of a carbohydrate derivative and
consideration of a possible synthetic approach reveals the necessity to
generate an L-configuration and to introduce a C-butyl group with an
"α"-orientation at C-2 in a pentofuranose derivative. At first glance,
this seems feasible starting with a 5-deoxy-L-lyxofuranose or L-ribo-
furanose derivative since they each secure the functionality and desired
sense of chirality at C-3 and C-4 of the target. Only the stereocontrolled introduction of a 2-C-butyl group remains as a key step. A more
cursory examination shows that this option is operationally not very
practical since the L-sugars in question are not readily available and
the preferential protection of the C-3 hydroxyl group in the cis -diol
unit is not easily achieved. In this regard the 1,2-0-isopropylidene
derivatives of other sugars such as arabinose xylose, glucose, galac-
tose,etc. are excellent precursors for five membered oxygen heterocycles
with substituents at C-2 and/or C-3 (see for example Schemes 63, etc).
The solution to the problem turns out to be simple but necessitates a

Scheme 68

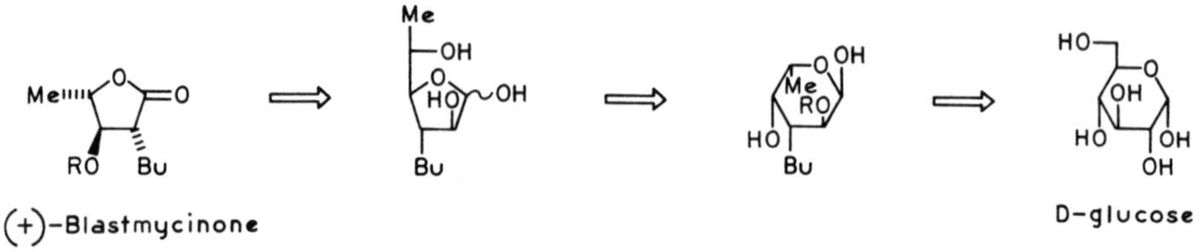

(+)-Blastmycinone

D-glucose

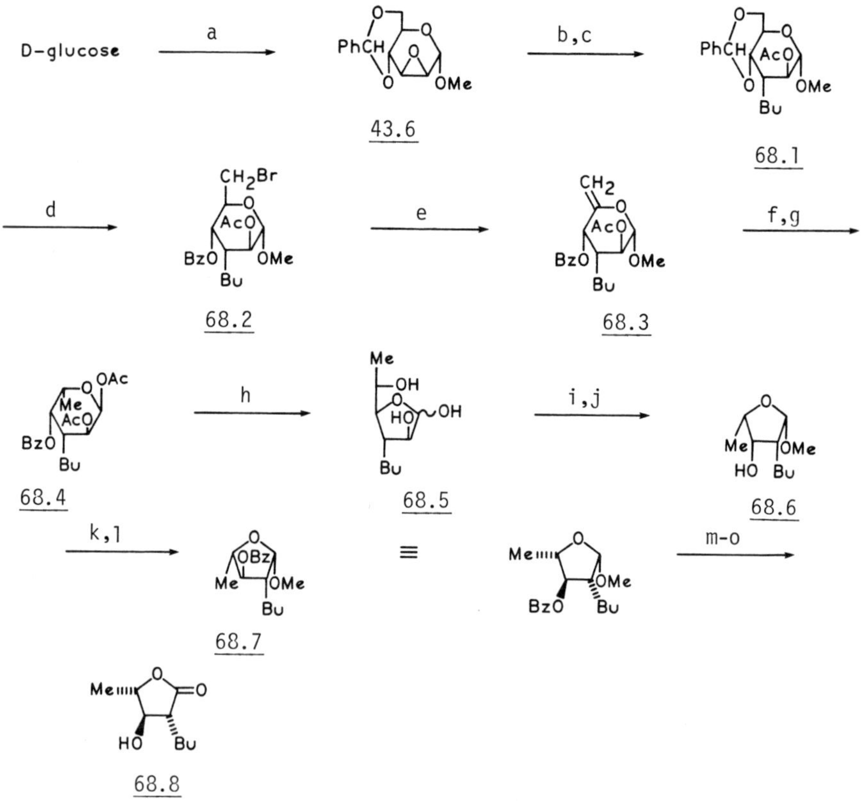

D-glucose →(a)→ PhCH ... 43.6 →(b,c)→ PhCH ... OMe 68.1

→(d)→ 68.2 →(e)→ 68.3 →(f,g)→

68.4 →(h)→ 68.5 →(i,j)→ 68.6

→(k,l)→ 68.7 ≡ →(m-o)→

68.8

a. see Scheme 43; b. BuMgCl, ether; c. Ac$_2$O, pyr.; d. NBS, CCl$_4$, reflux;
e. AgF, pyr.; f. Pd/black, H$_2$; g. Ac$_2$O, AcOH, H$_2$SO$_4$; h. MeOH, NaOH;
i. NaIO$_4$; j. MeOH, HCl; k. MsCl, pyr.; l. NaOBz, DMSO, reflux; m. NaOMe,
MeOH; n. aq. HCl; o. Br$_2$ water

few extra steps. If a pentose can accommodate the structural but not
functional requirements of the target, then a hexose derivative could be
considered with the option of excising one carbon atom from the reducing
or non-reducing end. For example 1,2:5,6-di-0-isopropylidene-D-galacto-
furanose would be an ideal precursor to blastmycinone (note stereochemistry at C-3 and C-4, the latter corresponding to the desired L-<u>glycero</u>
configuration). Preferential protection of the C-3 hydroxyl group and
voluntary manipulation of the C-5 — C-6 diol would lead to a useful intermediate for the elaboration of the C-2 chain in the form of a methyl
glycoside. Unfortunately, the galactofuranose derivative in question is
not the thermodynamic product when D-galactose is treated with acetone
under acid catalysis. Rather, 1,2:3,4-di-0-isopropylidene-α-D-galacto-
pyranose[29] is the major product. There are other procedures to obtain
the furanose derivative in question, but they may not be suitable for
what is required of a starting material. Kinoshita and coworkers[30] have
devised a route to the target <u>68.8</u> which starts with D-glucose and takes
advantage of conformational bias in the highly regioselective ring
opening of the readily available epoxide derivative <u>43.6</u>.[31] Having made
this choice they were committed to effect inversions of configuration at
C-4 and C-5 and to excise C-1 in order to achieve congruence with the
target. Key steps involved indirect functionalization at C-6 via the
NBS reaction,[32] elimination and stereocontrolled reduction of the exocyclic methylene derivative <u>68.3</u>. Note that this procedure allows inversion of configuration at C-5 and provides access to an L-sugar derivative. There remained to invert the configuration at C-4 in <u>68.5</u> and
to excise C-1. These operations were done by standard methods, although
it should be noted that the displacement of a sulfonate ester in a furanose derivative, particularly by benzoate ion (<u>68.6</u> → <u>68.7</u>) is a
difficult process, even in a virtually unhindered approach. By manipulating the C-4 and C-5 centers and following essentially the same route
it was possible to prepare the three possible diastereoisomers of (+)-
blastmycinone. A total synthesis of antimycin A₃ has been reported.[33]

  f. <u>(+)-Anhydromyriocin - Synthesis of the enantiomer</u> - Anhydromyriocin[34] is the γ-lactone form of myriocin[35] (thermozymocidin), whose
absolute configuration was unknown until the synthesis of an enantiomeric anhydromyriocin by Just and Payette.[36] Examination of the structure of the target (Scheme 69) and taking the lactone carbonyl as C-1 of
a pentose or a hexose, reveals that anhydromyriocin could be derived

Scheme 69

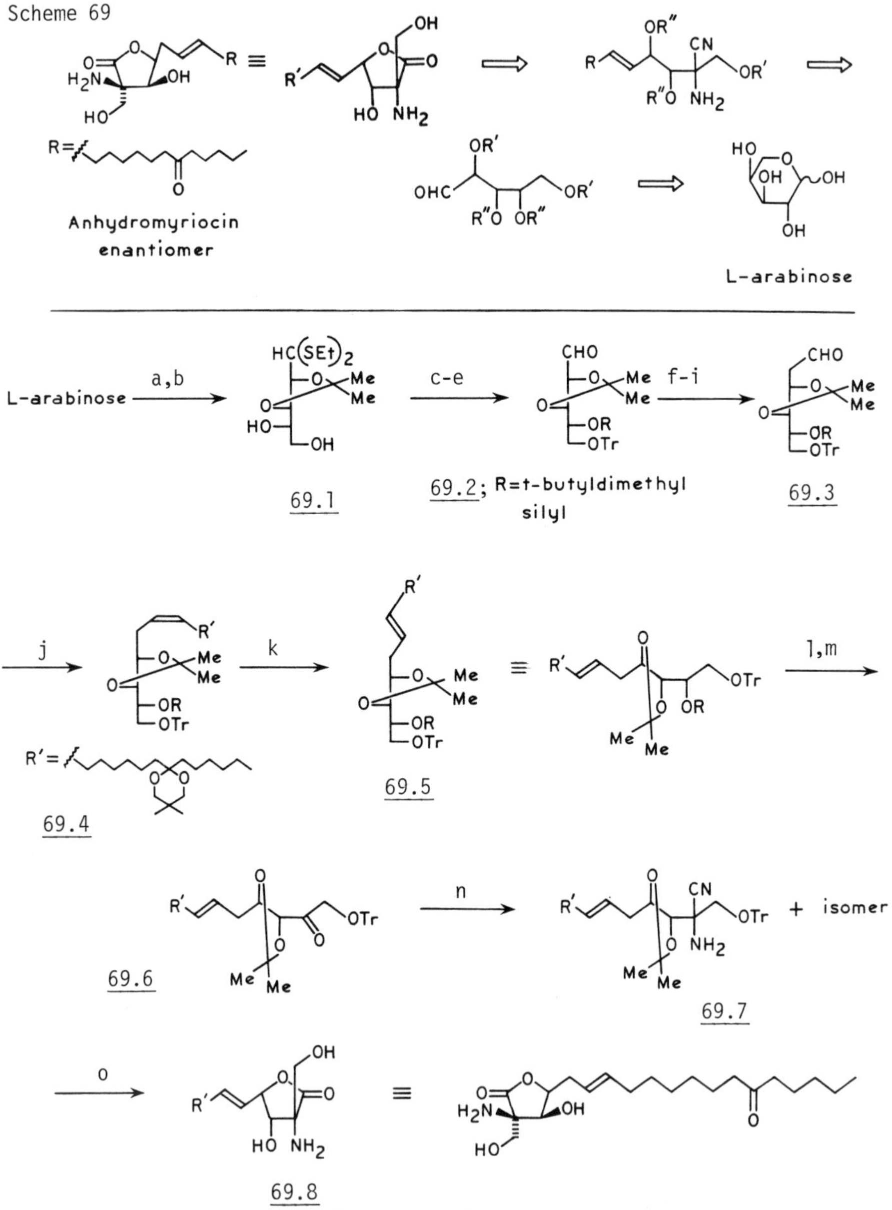

Anhydromyriocin
enantiomer

L-arabinose

L-arabinose $\xrightarrow{a,b}$ **69.1** $\xrightarrow{c-e}$ **69.2**; R=t-butyldimethyl silyl $\xrightarrow{f-i}$ **69.3**

$\xrightarrow{j}$ **69.4** $\xrightarrow{k}$ **69.5**

69.6 $\xrightarrow{n}$ **69.7** + isomer

$\xrightarrow{o}$ **69.8**

a. EtSH, HCl; b. acetone, H^+, then H_3O^+; c. TrCl, pyr.; d. t-BuMe$_2$SiCl, imidazole, DMF; e. HgCl$_2$, H$_2$O, acetone; f. CH$_3$NO$_2$, Et$_3$N, then NaOEt; g. O$_3$; h. NaBH$_4$, DMSO; i. PCC; j. Ph$_3$P$^+$CH$_2$R Br$^-$, BuLi, hexanes; k. m-CPBA then Ph$_2$PLi, THF and MeI; l. Bu$_4$NF; m. DMSO, Ac$_2$O; n. NaCN, NH$_4$Cl, NH$_3$, MeOH; o. MeOH, HCl

from an appropriate C-2 branched-chain derivative in which the C-3 and
C-4 hydroxyl groups have a *trans*(threo)relationship. Note that the
branch point involves a *trans*-hydroxymethyl group and a tertiary amine.
The arabinose configuration provides the desired *trans*-orientation of
hydroxyl groups at C-2 and C-3 and lends itself to easy selective
protection and manipulation compared to other pentoses which may offer
the same stereochemical criterion (at C-2/C-3 or C-3/C-4). Note that
once this *trans* diol relationship is found, the adjoining center should
eventually serve as a point of chain-extension (site of the allylic
double bond). Therefore, if arabinose were to be used in the present
case, the original C-1 becomes part of the allylic double bond system.
In a series of maneuvers, Just and Payette used L-arabinose as an
arbitrary form and interchanged both ends such that the hydroxymethyl
group became the carboxyl group in the final product. The readily
available partially protected dithioacetal derivative 69.1[37] was
homologated to 69.3, then extended with the desired C_{15} appendage. The
cyanoamine isomers could be separated and the desired isomer 69.7
carried through to the target 69.8 which was found to be identical to,
but enantiomeric with, (+)-anhydromyriocin.

g. Chiral butyrolactones – intermediates for sesquiterpene
lactones – D-Mannitol has been shown to be a source of a three carbon
'chiron' which can be transformed into chiral γ-butyrolactones.[38] These
in turn could be potential intermediates for the synthesis of
secologanin and sesquiterpene lactones.

REFERENCES

1. D. Brookes, B.K. Tidd and W.B. Turner, J. Chem. Soc., 5385 (1963);
    D. Brookes, S. Sternhell; B.K. Tidd and W.B. Turner, Aust. J.
    Chem., 18, 373 (1965).

2. R.C. Anderson and B. Fraser-Reid, J. Am. Chem. Soc., 97, 3870
    (1975).

3. H. Ohrui and S. Emoto, Tetrahedron Lett., 3657 (1975).

4. O. Th. Schmidt, Methods Carbohydr. Chem., 2, 318 (1963).

5. A. Rosenthal and L. Nguyen, J. Org. Chem., 34, 1029 (1969).

6. W.L. Parker and F. Johnson, J. Org. Chem., 38, 2489 (1973);
    J.L. Hermann, M.H. Berger and R.H. Schlessinger, J. Am. Chem. Soc.,
    105, 1544 (1979); 95, 7923 (1973).

7. D.C. Aldridge and W. B. Turner, J. Chem. Soc., C. 2431 (1971).

8. R.C. Anderson and B. Fraser-Reid, Tetrahedron Lett., 2865 (1977).

9. H. Zinner, G. Wulf and R. Heinatz, Chem. Ber., 97, 3536 (1964).

10. Compare, H . Paulsen and H. Behre, Carbohydr. Res., 2, 80 (1966).

11. K. Yamada, M . Kato, M.I. Iyoda and Y. Hirata, J. Chem. Soc.,
    Chem. Commun., 499 (1973); R.E. Damon and R.H. Schlessinger,
    Tetrahedron Lett., 4551 (1975).

12. N.J. McCorkindale, J.L. C. Wright, P.W. Brian, S.M. Clarke and
    S.A. Hutchinson, Tetrahedron Lett., 727 (1968).

13. R.C. Anderson and B. Fraser-Reid, Tetrahedron Lett., 3233 (1978).

14. M.L. Wolfrom and S. Hanessian, J. Org. Chem., 27, 1800 (1962).

15. M. Kato, M. Kageyama, R. Tanaka, K. Kuwahara and A. Yoshikoshi,
    J. Org. Chem., 40, 1932 (1975).

16. See for example, C.H. Heathcock,in The Total Synthesis of Natural
    Products, J. ApSimon, ed., Wiley-Interscience, New York, N.Y., 1973,
    p.197.

17. S.M. Kupchan, Pure and Appl. Chem., 21, 227 (1970).

18. T.F. Tam and B. Fraser-Reid, J. Chem. Soc., Chem. Commun. 556
    (1980).

19. V. Nair and A.K. Sinhababu, J.Org. Chem., 45, 1893 (1980).

20. Y. Ohfune, P.A. Grieco, C.L.J. Wang and C. Majetich, J. Am. Chem.
    Soc., 100, 5946 (1978).

21. J.A. Marshall and R.H. Ellison, J.Am. Chem. Soc., 98, 4312 (1976).

22. S. Hanessian, T.J. Liak and D.M. Dixit, Carbohydr. Res., 88, C14
    (1981).

23. S. Hanessian, D.M. Dixit and T.J. Liak, Pure Appl. Chem., 53, 129
    (1981).

24. S. Hanessian and P. Dextraze, Can. J. Chem., 50, 226 (1972);
    Chem. Ind., 958 (1971).

25. S. Hanessian, P. Dextraze and R. Massé, Carbohydr. Res., 26, 264
    (1973).

26. For further discussion and references to total syntheses of
    vernolepin, see S. Hanessian, Acc. Chem. Res., 12, 159 (1979).

27. F. Brisse, P. Dextraze and S. Hanessian, J. Cryst. Mol. Struct., 11,
    173 (1981).

28. K. Watanabe, T.Tanaka, K. Fukuhara, N. Miyairi, H. Yonehara and
    H. Umezawa, J. Antibiotics, 10, 39 (1957); see also E.E. van
    Tamelen, J. P. Dickie, M.E. Loomans, R.S. Dewey and F.M. Strong,
    J.Am. Chem. Soc., 83, 1639 (1961); A.J. Birch, D.W. Cameron, Y.

Harada and R.W. Rickard, J. Chem. Soc., 889 (1961); H. Yonehara and S. Takeuchi, J. Antibiotics, 11, 254 (1958).

29. O.Th. Schmidt, Methods Carbohydr. Chem., 2, 318 (1963).

30. S. Aburaki, N. Konishi and M. Kinoshita, Bull. Chem. Soc. Japan, 48, 1254 (1975).

31. L.F. Wiggins, Methods Carbohydr. Chem., 2, 188 (1963).

32. S. Hanessian and N.R. Plessas, J. Org. Chem., 34, 1035 (1969); S. Hanessian, Methods Carbohydr. Chem., 6, 183 (1972) and references cited therein.

33. M. Kinoshita, S. Aburaki, M. Wada S. Umezawa, Bull. Chem. Soc. Japan, 46, 1279 (1973).

34. F. Aragozzini, P.L. Manachini, R. Graveri, B. Rindone and C. Scholastico, Tetrahedron, 28, 5493 (1972).

35. D. Kluepfel, J.F. Bagli, H. Baker, M.-P. Charest, A. Kudelski, S.N. Sehgal and C. Vézina, J. Antibiotics, 25, 109 (1972); J.F. Bagli, D. Kluepfel and M. St-Jacques, J. Org. Chem., 38, 1253 (1973); R. Craveri, P.L. Manachini and F. Aragozzini, Experentia, 28, 867 (1972); C.H. Kuo and N.L. Wendler, Tetrahedron Lett., 211 (1978).

36. C. Just and D.R. Payette, Tetrahedron Lett., 21, 3219 (1980); Can. J. Chem., 59, 269 (1981).

37. H. Zinner, G. Rembarz and H.-P. Klöcking, Chem. Ber., 90, 2688 (1957).

38. T. Kametani, Heterocycles, 19, 205 (1982).

CHAPTER TEN

MOLECULES CONTAINING A VALEROLACTONE-TYPE RING

a. "Asperlin" – Synthesis of the enantiomer – The stereochemistry of "asperlin", a fungal metabolite isolated from Aspergillus nidulans[1] has been deduced by comparison with a synthetic sample obtained from D-galactose (Scheme 70).[2] The α-pyrone structure can be derived from an appropriate hexose in which the configuration at C-4 and C-5 is known (threo). Scheme 70 illustrates the route which involves chain extension of the readily available 70.1 via the corresponding aldehyde and the derivative with a cis-propenyl side chain.[3] Photochemical isomerization of the latter provided a 1:4 mixture of cis and trans isomers which was converted to the lactone derivative 70.6 by well known procedures. Epoxidation of the mixture of olefins afforded a cis-epoxide (not shown) and a 3:2 mixture of the trans-epoxides 70.8 and 70.9. The major isomer was found to be identical but enantiomeric wich "asperlin". A large number of natural products contain α,β-unsaturated lactone structures and their synthesis can be envisaged from carbohydrate precursors as shown in the case of "asperlin".

b. Prelog-Djerassi lactone – This lactonic acid (Scheme 71) is a key degradation product[4] of some macrolide antibiotics. Although there are several syntheses of this lactone (see later), it is evident that access to the chiral ring system can be gained from manipulation of pyranoid carbohydrate derivatives. Scheme 71 shows such a route based on the Ireland strategy[5] of forming critical carbon-carbon bonds by application of the Claisen enolate rearrangments. Examination of the

130

Scheme 70

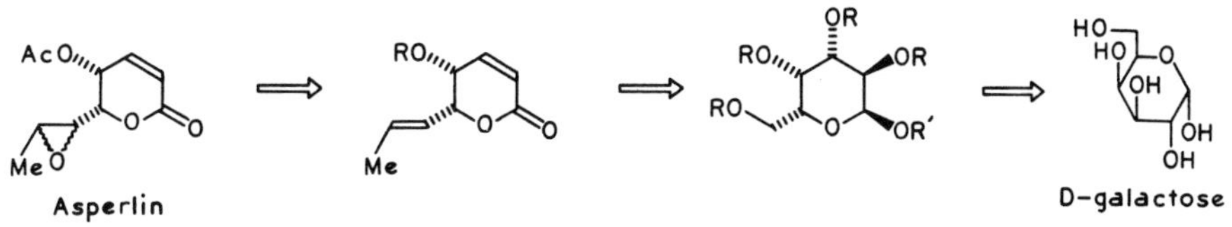

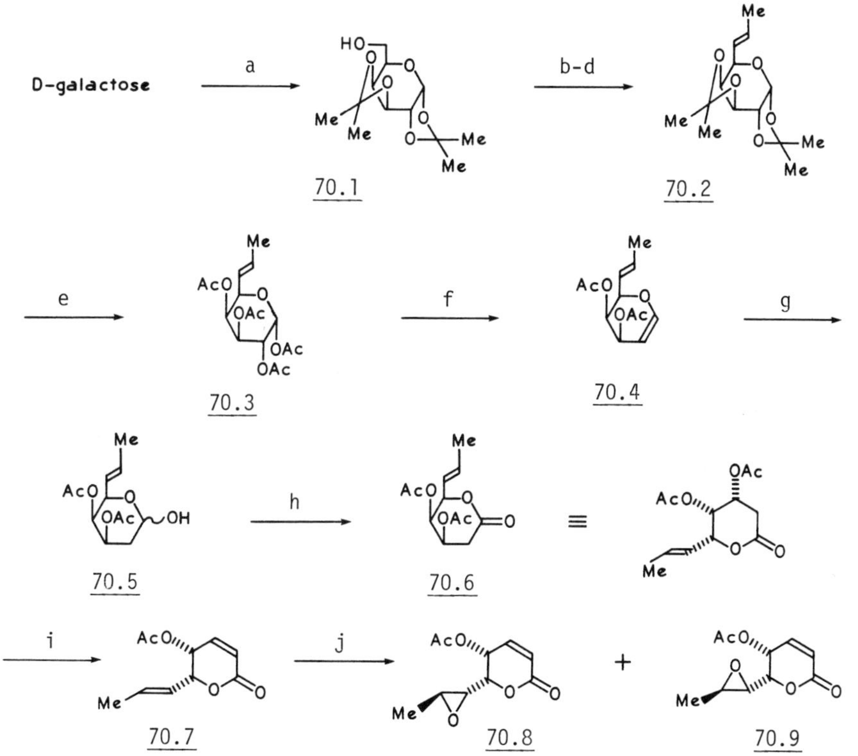

Asperlin   D-galactose

D-galactose →a→ 70.1 →b-d→ 70.2

70.3 →f→ 70.4

70.5 →h→ 70.6 ≡

70.7 →j→ 70.8 + 70.9

a. acetone, H^+, b. Ac_2O, DMSO; c. $Ph_3P=CHMe$; d. hν; e. Ac_2O, BF_3, Et_2O, 0^0;
f. HBr, AcOH, then Zn, NaOAc, $CuSO_4$; g. HCl, ether; h. PCC, CH_2Cl_2; i. Et_3N,
CH_2Cl_2; j. m-CPBA

structure of the lactone 71.6 reveals a substitution pattern consisting
of alternating C-methyl groups having a *cis*-orientation and a "*trans*"
side-chain with an additional chiral center. The orientation of the

Scheme 71

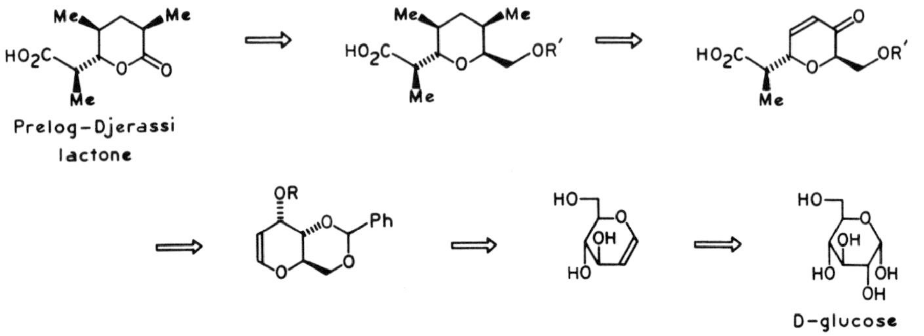

Prelog-Djerassi
lactone

D-glucose

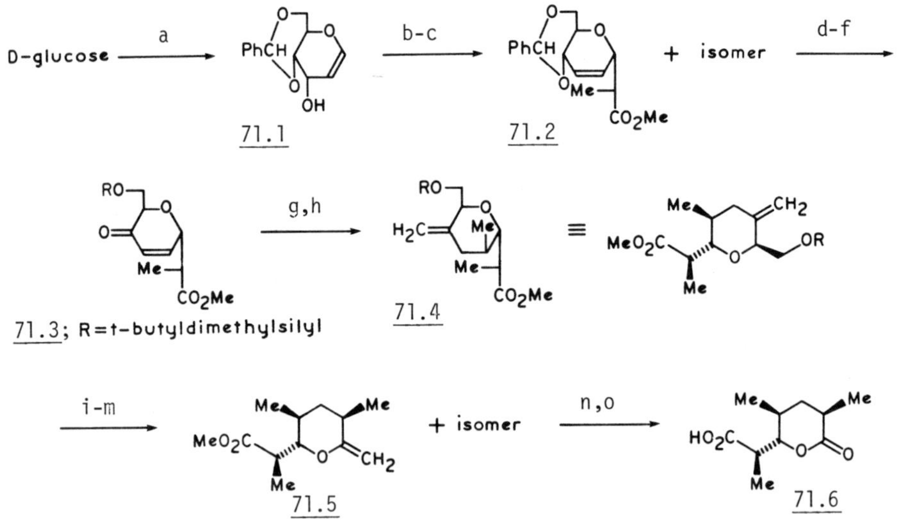

D-glucose $\xrightarrow{a}$ **71.1** $\xrightarrow{b-c}$ **71.2** + isomer $\xrightarrow{d-f}$

71.3; R=t-butyldimethylsilyl $\xrightarrow{g,h}$ 71.4 $\equiv$

$\xrightarrow{i-m}$ 71.5 + isomer $\xrightarrow{n,o}$ 71.6

a. several steps; b. propionic anhydride, DMAP, CH_2Cl_2; c. Li, HMDS, THF,
-100^0; then t-BuMe$_2$SiCl, HMPA, -100^0 25^0, then benzene, reflux; then CH_2N_2;
d. H$^+$, THF, 60^0; e. t-Bu Me$_2$SiCl, pyr., 0^0; f. PDC, CH_2Cl_2; g. Me$_2$CuLi,
ether, 0^0; h. Ph$_3$P=CH$_2$, THF; i. PtO$_2$, H$_2$, pentane; j. Bu$_4$NF, THF; k. TsCl,
pyr.; l. NaI, 2-butanone, reflux; m. AgF, pyr.; n. O$_3$, CH_2Cl_2, pyr.;
o. LiOH, MeOH

side-chain corresponds to that of a D-configuration at C-5. Thus a
viable strategy could be based on regio and stereocontrolled
introduction of C-methyl groups in a D-hexopyranose derivative and chain-
extension at C-6. The route devised by Ireland creates the intended
carboxylic acid appendage at C-1 (and not at C-6) via a stereocontrolled
Claisen enolate rearrangement using the glycal derivative 71.1 as a
template. The 9:1 mixture of isomers in favor of the desired one could
be separated by chromatography. Through a series of manipulations, the
C-glycoside 71.2 was then converted to the enone 71.3 and the latter
treated with lithium dimethylcuprate to introduce the first C-methyl
group with regio and stereocontrol as in 71.4. The second C-methyl
group was introduced by a Wittig reaction followed by hydrogenation.
The major isomer was that corresponding to the desired orientation as in
expression 71.5. There remained to transform the hydroxymethyl group
into the lactonic carbonyl function which was accomplished via ozonolysis of the exocyclic enol ether derivative 71.5. The Prelog-Djerassi
lactone thus obtained showed the highest yet recorded value for optical
rotation, $[\alpha]_D$ + 47.7°(CHCl$_3$). The strategy was therefore based
on an interchange between C-1 and C-5 in D-glucose. The former was used
as a branch point for the two-carbon acid chain, by "transfer of
chirality" from C-3. Excision of the original C-6 in D-glucose on the
other hand allowed conversion of C-5 into a lactone function. These
tactics further demonstrate the uniqueness of sugar derivatives in terms
of symmetry relationships (see Scheme 4)and their versatility in the
control of regio- and stereochemistry.

     A synthesis of the Prelog-Djerassi lactone and its 2-epimer from
D-glucose was reported by the Fraser-Reid group[6] (Scheme 72). The
orientation of substitution in the Prelog-Djerassi lactone can be
related to a 2,6-di-C-methyl-D-hexopyranose. Scheme 72 shows this
feature and puts the structure in a carbohydrate-type perspective. The
key steps in creating the chiral ring portion would then consist in the
stereocontrolled branching at positions 2- and 4- in a D-hexopyranose
derivative . There are several other syntheses of the Prelog-Djerassi
lactone utilizing other approaches.[7]

     c. Prelog-Djerassi lactone - Access to the chiral ring system -
Scheme 73 shows another route based on conformational bias and Wittig
methodology.[8] This approach is based on previous experience in the
construction of the carbon backbone of macrolide antibiotics such as

Scheme 72

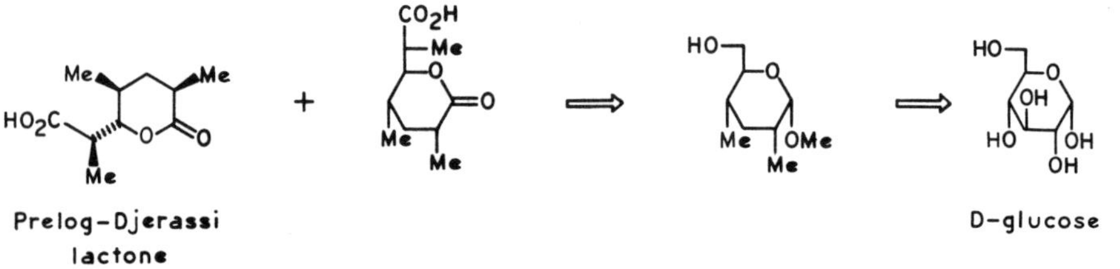

Prelog-Djerassi
lactone

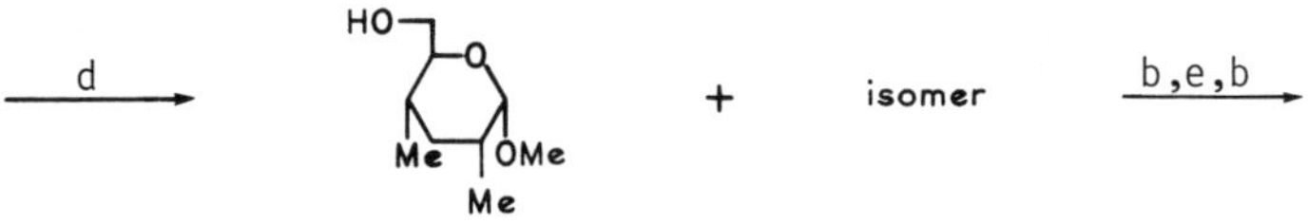

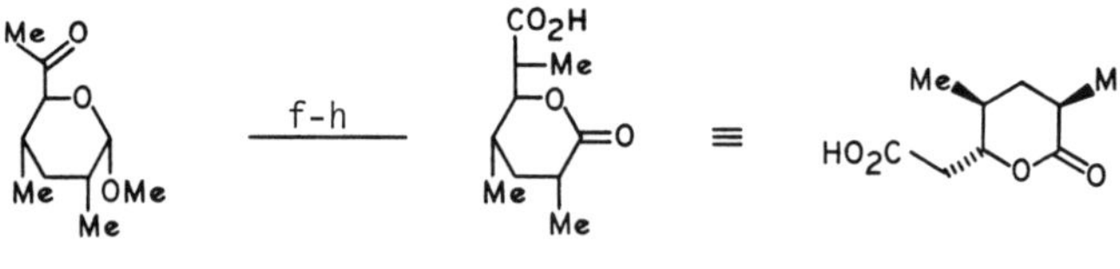

D-glucose

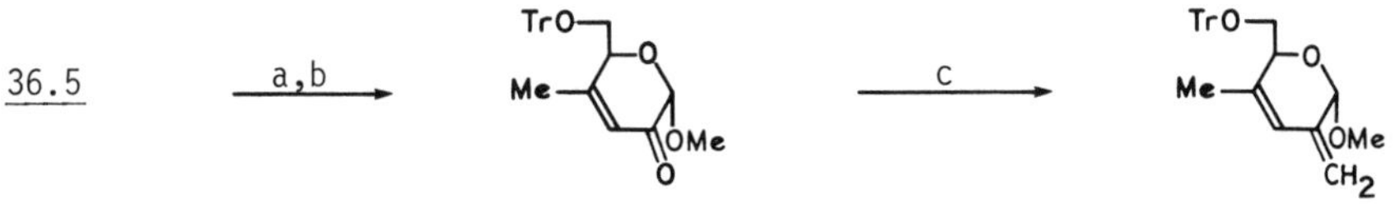

36.5   a,b   72.1   c   72.2

d   72.3   +   isomer   b,e,b

72.4   f-h   71.6   ≡

a. MeLi; b. PCC; c. Ph$_3$=CH$_2$; d. Pd/C,H$_2$; e. MeLi; f. Ph$_3$P=CHOMe;
g. aq. HC1; h. CrO$_3$

Scheme 73

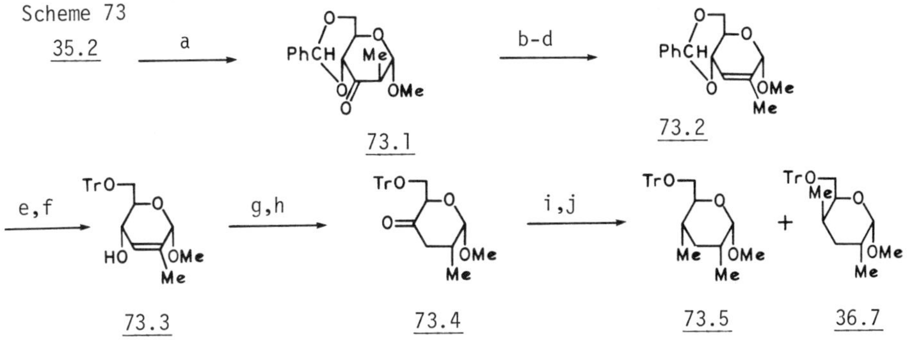

35.2 $\xrightarrow{a}$ 73.1 $\xrightarrow{b-d}$ 73.2

$\xrightarrow{e,f}$ 73.3 $\xrightarrow{g,h}$ 73.4 $\xrightarrow{i,j}$ 73.5 + 36.7

a. DMSO, Ac_2O; b. NaOMe, MeOH; c. $NaBH_4$, MeOH; d. Ph_3P, DEAD, MeI;
e. PPTS; f. TrCl, pyr.; g. Pt black, toluene; h. PCC; i. $Ph_3P=CH_2$;
j. $Pd(OH)_2$, H_2, THF

erythronolide A, etc, (see page 239). Note that in spite of the
relative lack of steric factors in intermediates such as 72.2 and 73.3,
hydrogenation from the "β"-side was not an exclusive process. Highly
regio- and stereoselective introduction of a C-methyl group at C-4 in a
D-glucopyranose system can be achieved by conjugate addition of an
organocuprate to 1,7-anhydro-3,4-didehydro-α-D-glycero-hexopyranos -2
ulose (levoglucosenone).[9]

d. 2(R)-Methyl-5(R)-hydroxyhexanoic acid lactone - An isomer of
the Carpenter bee pheromone - The major volatile component of the carpenter bee sex attractant consists of a 2-methyl-5-hydroxyhexanoic acid
lactone having a cis-configuration.[10] It is possible that other
isomers may also be present . Scheme 74 illustrates an easy entry into
this class of optically active α-substituted-γ-lactones by utilizing
carbohydrate precursors.[11] The readily available intermediate 58.1 (see
Scheme 58) can be reduced to provide a 9:1 mixture of 74.1 and its C-2
epimer 74.2 which can be conveniently separated by chromatography of
their t-butyldiphenylsilyl ether derivatives. The mixture can however
be converted to a mixture of lactones from which the desired 2(R), 5(R)
isomer 74.4 can be obtained crystalline. This sequence is very
practical since it provides access to deoxygenated derivatives at
C-3/C-4 essentially in one step from the readily available 2,3-unsatu-
rated glycosides (ex.48.2). Moreover, by the proper choice of aglycon
(ex. methyl rather than t-butyl), the stereochemistry of the reduction
of the diene derivative can be changed to provide a preponderance of the

Scheme 74

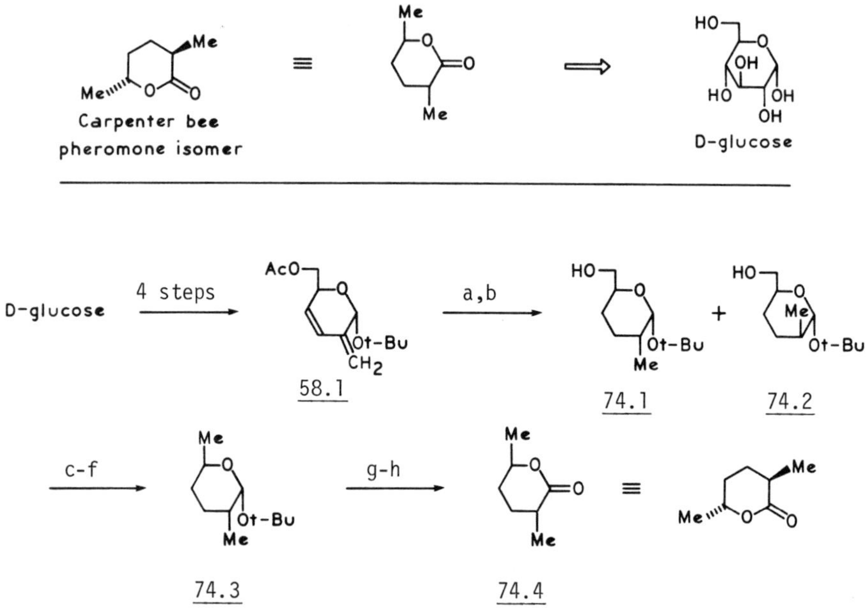

Carpenter bee
pheromone isomer

D-glucose

D-glucose $\xrightarrow{\text{4 steps}}$ 58.1 $\xrightarrow{\text{a,b}}$ 74.1 + 74.2

$\xrightarrow{\text{c-f}}$ 74.3 $\xrightarrow{\text{g-h}}$ 74.4

a. Pd/C, H_2, EtOAc; b. NaOMe, MeOH; c. Ph_2 t-BuSiCl, imidazole, DMF,
separate isomers; d. Bu_4NF; e. NBS, Ph_3P, DMF; f. $LiEt_3$BH, THF;
g. H_3O^+; h. PCC, NaOAc, CH_2Cl_2

C-2 epimer 74.2 (from a methyl glycoside), hence the 2(S), 5(R) isomer
of the lactone. By going through a 5,6-unsaturated intermediate (ex.
via dehydrobromination of the 6-bromo derivative, see Scheme 74),
followed by catalytic reduction it should be possible to invert the
configuration at C-5, hence have access to all four possible
diastereomeric lactones. These have been synthesized by alkylation of
δ-methylvalerolactone.[12] A synthesis of the *cis*-isomer has also been
described.[12]

    e. Malyngolide – A synthesis of (−)-malyngolide, a marine
antiobitic,[13] has been reported by Sinaÿ and coworkers,[14] starting from
D-glucose. An asymmetric synthesis[15] starting with (S)-2-hydroxy-2-

nonyl-6-heptenal and other routes[16,17] to racemic malyngolide have also reported.

REFERENCES

1. A.D. Argoudelis and J.F. Zeiserl, Tetrahedron Lett., 1969 (1966).

2. S. Lesage and A.S. Perlin, Can. J. Chem., 56, 2889 (1978).

3. D.G. Lance, W.A. Szarek, J.K.N. Jones and G.B. Howarth, Can. J.Chem., 47, 2871 (1969).

4. C. Djerassi and J.A. Zderic, J. Am. Chem. Soc., 78, 2907, 6390 (1956); R. Anliker, D. Dvornik, K. Gubler, H. Heusser and V. Prelog, Helv. Chim. Acta, 39, 1785 (1956).

5. R.E. Ireland and J.P. Daub, J. Org. Chem., 46, 479 (1981).

6. S. Jarosz and B. Fraser-Reid, Tetrahedron Lett., 22, 2533 (1981).

7. See for example, D.A. Evans and J. Bartroli, Tetrahedron Lett., 23, 807 (1982) and references cited therein, see also, R.E. Ireland and J.P. Daub, J. Org. Chem., 46, 479 (1981); K. Maruyama, Y. Ishihara and Y. Yamamoto, Tetrahedron Lett., 22, 4235 (1981); P.A. Bartlett and J.L. Adams, J.Am. Chem. Soc., 102, 337 (1980); M. Hirama, D.S. Garvey, L.D.-L. Lu and S. Masamune, Tetrahedron Lett., 3937 (1979); J.D.White and Y. Fukuyama, J.Am.Chem. Soc., 101, 228 (1979); G. Stork and V. Nair, J. Am. Chem. Soc., 101, 1315 (1979); P.A. Grieco Y. Ohfune, Y. Yokoyama and W. Owens, J. Am. Chem. Soc., 101, 4749 (1979), A. Nakano, S. Takimoto, J. Inanaga, T. Katsuki, S. Ouchida, K. Inoue, M. Aiga, N. Okukado and M. Yamaguchi, Chem. Lett., 1019 (1979).

8. S. Hanessian and I. Boessenkool, unpublished results.

9. F. Shafizadeh and P.P.S. Chin, Carbohydr. Res., 58, 79 (1977).

10. J.M. Wheeler, S.L. Evans, M.S. Blum, H.H. Velthius and J. M. F. de Camargo, Tetrahedron Lett., 4029 (1976).

11. S. Hanessian, G. Demailly, Y. Chapleur and S. Léger, J.C.S. Chem. Comm., 1125 (1981).

12. W.H. Pirkle and P.E. Adams, J. Org. Chem., 43, 378 (1978); 44, 2169 (1979); R. Bacardit and M. Moreno-Manas, Tetrahedron Lett., 21, 551 (1980).

13. J.H. Cardllina II, R.E. Moore, E.V. Arnold and J. Clardy, J. Org. Chem., 44, 4039 (1979).

14. J.-R. Pougny, P. Rollin and P. Sinaÿ, Tetrahedron Lett., 23, 4929 (1980).

15. Y. Sakito, S. Tanaka, M. Asami and T. Mukaiyama, Chem. Lett., 1223
    (1980).

16. G. Cardillo, M. Orena, G. Porzi and S. Sandri, J. Org. Chem., 46,
    2439 (1981).

17. S. Torii, T. Inokuchi and K. Yoritaka, J. Org. Chem., 46, 5030
    (1981).

PART FIVE

EXECUTION

Molecules Containing Partially-hidden
Carbohydrate-type Symmetry

CHAPTER ELEVEN

ACYCLIC MOLECULES

a. <u>(S)-(+)-Ipsdienol – Synthesis of the enantiomer</u>   (<u>S</u>)-(+)-
Ipsdienol (2-methyl-6-methylene-2,7-octadien-4-ol) is one of the aggregation pheromones of a bark beetle.[1] The synthesis of the enantiomer
<u>75.7</u> (Scheme 75) from (<u>R</u>)-glyceraldehyde confirmed the absolute configu-
ration of the natural component.[2] Examination of the structure in
question reveals that the asymmetric center containing a hydroxyl group
can be related to (<u>R</u>)-glyceraldehyde, which can be derived from D-mannitol as the 2,3-0-isopropylidene derivative <u>12.2</u> . Elaboration of the
aldehydic end with the appropriate Wittig reagent gave <u>75.1</u>. The
required two-carbon chain corresponding to the diene unit was incorporated by a malonate extension of the epoxide <u>75.3</u>. Note that the double
bond in <u>75.1</u> was "protected" by oxymercuration– demercuration, to render
the ensuing intermediates compatible with the reaction used. α-Methylenation to the lactone <u>75.4</u>, protection of the methylene group via the
phenylselenyl adduct and dehydration to <u>75.5</u> followed by extension of
the lactone carbonyl gave the desired target. Note that treatment of
<u>75.6</u> with the Wittig reagent led directly to <u>75.7</u> with concomitant
removal of the phenylselenelyl group by a <u>retro</u>-Michael-type reaction.
In passing, it should also be remarked that preliminary modification of
<u>12.2</u> by interchanging the aldehyde and hydroxymethyl groups via a series
of protection– deprotection steps could lead to the natural (<u>S</u>)-enantio-
mer by application of the same synthetic scheme. Racemic ipsdienol has
been synthesized by other routes[3].

Scheme 75

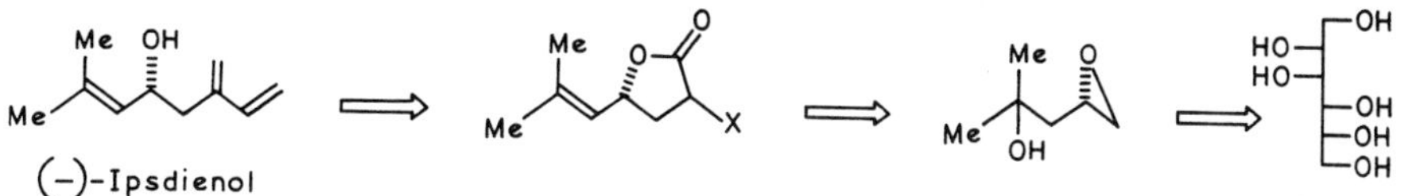

Me OH
Me⟍⟍⟍⟍⟍
(−)-Ipsdienol
⟹
[lactone structure] 75 with X
⟹
Me O (epoxide) Me OH
⟹
HO—OH / HO— / —OH / —OH / —OH / —OH
D-mannitol

D-mannitol —a,b→ **12.2** —c→ **75.1** —d-f→

75.2 (Me OH / Me OH / OTs) —g→ **75.3** (Me O epoxide / Me OH) —h-j→

75.4 (lactone) —k,l→ **75.5** (CH₂SePh) —m→

75.6 (CH₂SePh) —n→ **75.7**

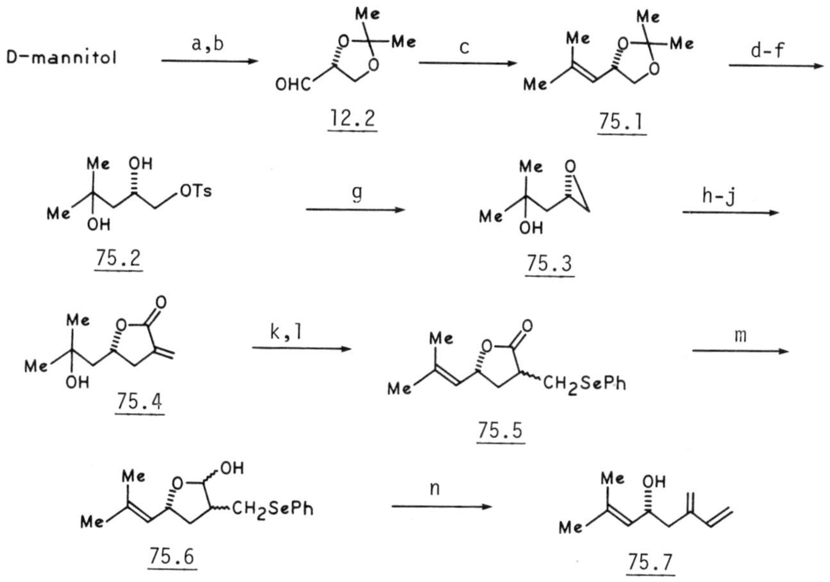

a. acetone, H⁺; b. Pb(OAc)₄; c. Ph₃P=CMe₂, DMSO; d. Hg(OAc)₂, THF,
aq. NaBH₄; e. aq. HCl; f. TsCl, pyr.; g. aq. KOH; h. CH₂(CO₂Et)₂,
NaOEt, EtOH; i. aq. KOH, then H⁺; j. CH₂O, Et₂NH; k. PhSeH; l. POCl₃,
pyr.; m. DIBAL, THF; n. Ph₃P=CH₂, DMSO

    b. (-)- and (+)-Sulcatol - Sulcatol is an aggregation pheromone
produced by the ambrosid beetle, a widely occurring timber pest in
western North America.[4] It has been shown to consist of a 65:35 mixture
of (S)-(+) and (R)-(-)-6-methyl-hept-5-en-2-ol. To investigate the
question whether both enantiomers were responsible for the aggregation
response, Schuler and Slessor[5] reported the synthesis of both components
from L-fucose and 2-deoxy-D-erythro-pentose (2-deoxy-D-ribose). Scheme
76 shows that the target structure consists of a seven-carbon backbone

Scheme 76

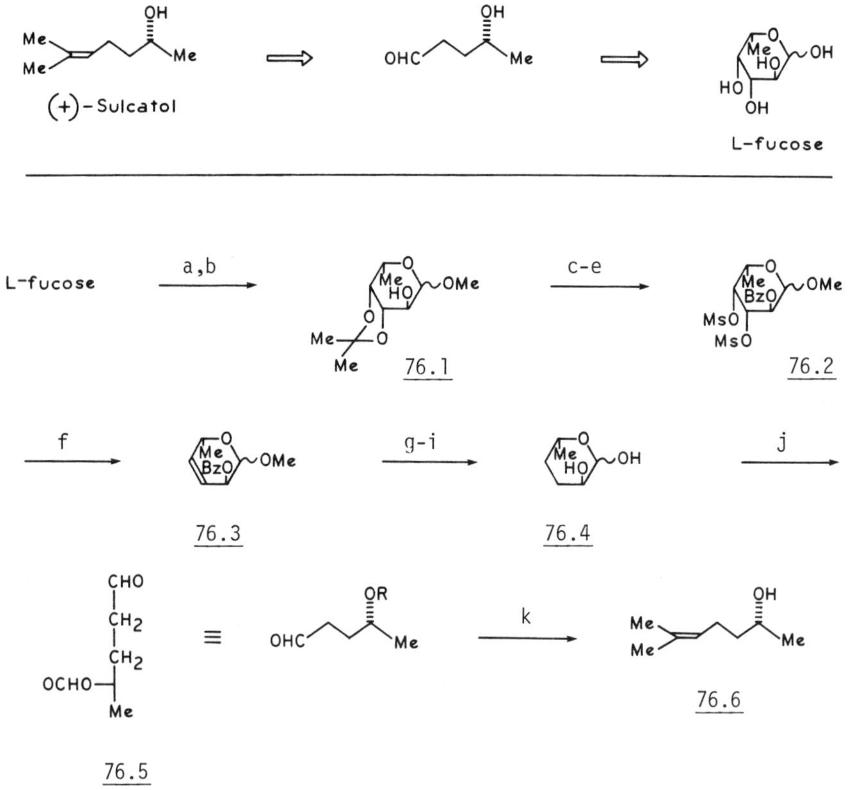

a. MeOH, HCl; b. acetone, H$^+$; c. BzCl, pyr.; d. aq. AcOH; e. MsCl, pyr.;
f. Zn, NaI, DMF, reflux; g. Pd/C, H$_2$; h. NaOMe; i. Dowex-50 (H$^+$);
j. aq. NaIO$_4$; k. Ph$_3$P=CMe$_2$, THF.

with one asymmetric center as exemplified by (+)-sulcatol, 76.6. The
absolute configuration of the hydroxyl-bearing carbon corresponds to
that in (S)-glyceraldehyde or an appropriate L-sugar. Indeed, the
target can be synthesized by chain-extension of a 3-deoxy-(S)-glycer-
aldehyde. Alternatively a higher deoxy sugar can be deoxygenated to the
desired level and further manipulated to arrive at the target. Note the
location of the hydroxyl group and the presence of sp^2 center ("rule of
five, or four"). The reported synthesis of (-)-sulcatol utilized such
an approach. The starting sugar was the commercially available L-fucose
(6-deoxy-L-galactose) which possesses the required stereochemistry at
C-5. Through a series of well known transformations intermediate 76.1
was converted into the 3,4-dideoxy derivative 76.4. Note the formation
of the olefinic intermediate 76.3 by the Tipson-Cohen method.[6] Periodate cleavage of 76.4 gave the desired five-carbon aldehyde 76.5 which
was chain-extended to (+)-sulcatol. The enantiomeric (-)-alcohol was
prepared from a 2-deoxy-D-erythro-pentose derivative by deoxygenation at
C-3 and C-5 and chain-extension. Considering the location of the
hydroxyl group and the double bond in the target 76.6 or its enantiomer,
one can think in terms of applying the "rule of four" using a pentose as
starting material precursor. The synthesis of the enantiomeric
sulcatols from (R)- and (S)-glutamic acids has been reported by Mori.[7]

   c. 3(S), 4(S)-4-methyl-3-heptanol - This aggregation pheromone[8]
has been shown to possess the threo 3(S), 4(S)configuration.[9]
Examination of the structure in question (Scheme 77) reveals the
presence of vicinal hydroxyl and C-methyl substitution on a seven-carbon
chain. Consideration of a carbohydrate precursor approach can lead to
several possibilities depending on the site of C-methylation on a cyclic
carbohydrate derivative. Following such a reaction, extensive
deoxygenation and a chain-extension must be envisaged. Note that the
threo relationship of the substituents leads one to consider opening of
an epoxide (see Scheme 3) as a viable route since the trans-product
would correspond to the desired relative stereochemistry. Scheme 77
shows how D-glucose was used as a precursor to the target 77.4.[10] The
readily available epoxide derivative 43.6[11] was converted into the
C-methyl derivative 43.7 with methylmagnesium chloride. The threo
configuration was thus secured in the sequence, except that as
anticipated the sense of chirality at the two centers is opposite to
that found in the target. The relatively rigid structure was then

manipulated via a oxidation-epimerization-reduction sequence (see also Scheme 129, page 240), to lead to intermediate 77.1 with the correct sense of chirality. This "template" was then further manipulated by transformation to an acyclic derivative 77.2 which was chain extended at both extremities to give the product in excellent overall yield. Thus, the principles of conformational bias were put to good advantage in controlling regio- and stereochemistry as had been demonstrated in other syntheses(see page 239).Several syntheses of the pheromone by other routes are also available.[12]

Scheme 77

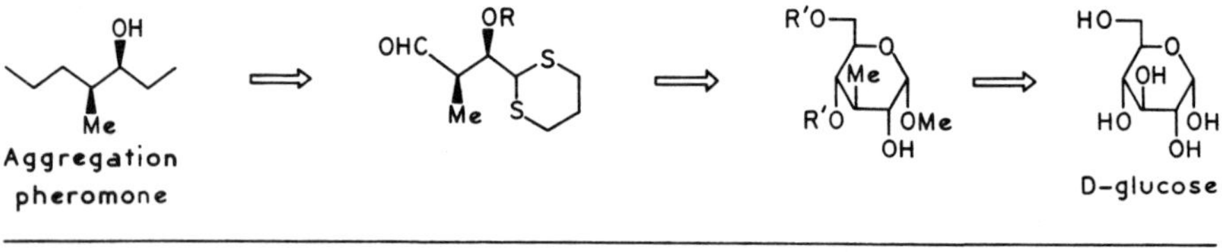

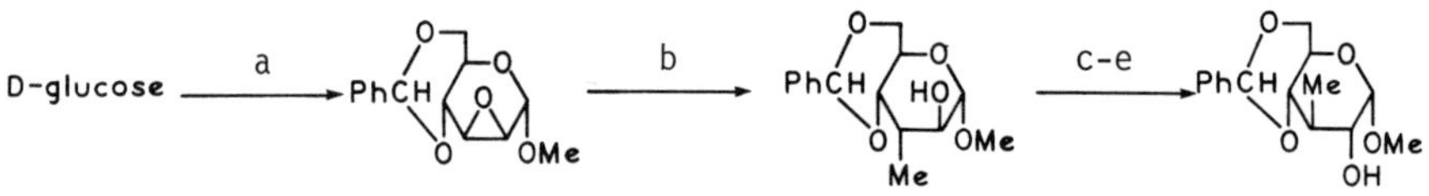

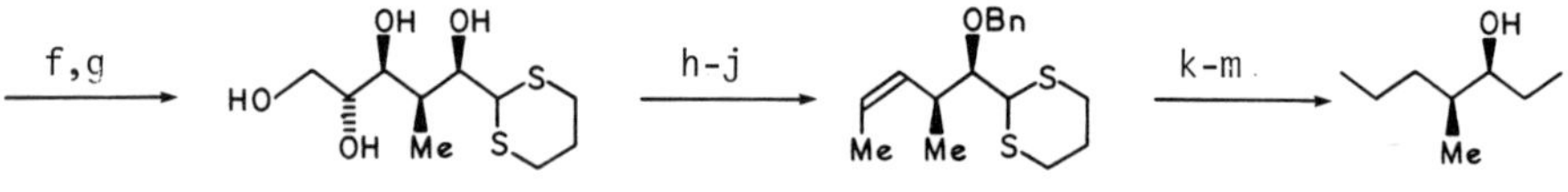

a. see Ref. 11; b. MeMgCl, ether; c. DMSO, TFAA, CH_2Cl_2, -70^0; d. DMF, Et_3N; e. LAH, ether; f. aq. H_2SO_4, reflux; g. $HSCH_2CH_2CH_2SH$, H^+, 0^0; h. $Pb(OAc)_4$, MeCN, -45^0; i. $Ph_3P=CHMe$, toluene; j. BnBr, DMF, BaO, $Ba(OH)_2$; k. MeI, aq. acetone, 55^0; l. $Ph_3P=CH_2$; m. Pd/C, H_2, ether.

d. Octadienic esters of the Tricothecenes - The trichothecene
family of sesquiterpenes feature components such as the trichodermadine-
diols which contain octadienoic acid esters[13] (Scheme 78). Access to
these pendant ester units can be gained from chain-extension of 6-deoxy-
hex-2-ene-al derivatives 78.3 and 78.6, available from the treatment of
6-deoxy-tri-O-acetyl-D-glucal 78.1 and 6-deoxy-tri-O-acetyl-D-galactal
78.4 respectively with aq. dioxane at reflux.[14] The best ratio of the
desired *cis-trans* diene was obtained by the Petersen procedure.[15] In
this manner the D-erythro and threo octadienoic ester diols 78.3 and
78.6 respectively were obtained.

e. 3(S), 4(S)-4-Amino-3-hydroxy-6-methylheptanoic acid - This
unusual amino acid has been isolated from hydrolyzates of several pep-
statins[16] which are specific inhibitors of acid proteases. Degradative
and X-ray crystallographic studies[17] established the structure of the
amino acid and its synthesis from a carbohydrate precursor corroborated
the absolute stereochemistry. Scheme 79 illustrates the synthetic route
starting from D-glucose.[18] It can be seen that the hydroxyl group in
the target 79.6 and that at C-4 of D-glucose are of the same absolute
configuration. Consideration of the pattern of deoxygenation and the
relative positions of the vicinal amino and hydroxyl group lead to the
selection of a D-glucofuranose derivative as starting material, as indicated in the retrosynthetic analysis. Thus, the critical C-4 oxygen
is protected in ring formation and positions 3 and 5 can be deoxygenated
and "aminated" respectively. There would remain peripheral transformations to attain the desired chain length and oxidation state. Kinoshita
and coworkers[18] transformed D-glucose to the aldehydo intermediate
79.1[19] and proceeded to introduce the 2-butyl side-chain in a stereoselective Grignard reaction. The newly formed center was subjected to a
double-inversion procedure to give the azido derivative 79.4 in which
both chiral centers of the target are secured. Completion of the synthesis consisted of an oxidative cleavage of the diol resulting from
79.4 to give the azido acid 79.5 and the amino acid 79.6. It is thus of
interest how D-glucose was first chemically manipulated and chain-
extended to achieve the desired level of chiral and functional overlap
with the target, then it was excised to reach the required number of
carbon atoms. The synthesis of the 3(S),4(R) diastereoisomer was also
achieved by following the same method. A synthesis of both
diastereoisomers from L-leucine has also been reported.[20]

Scheme 78

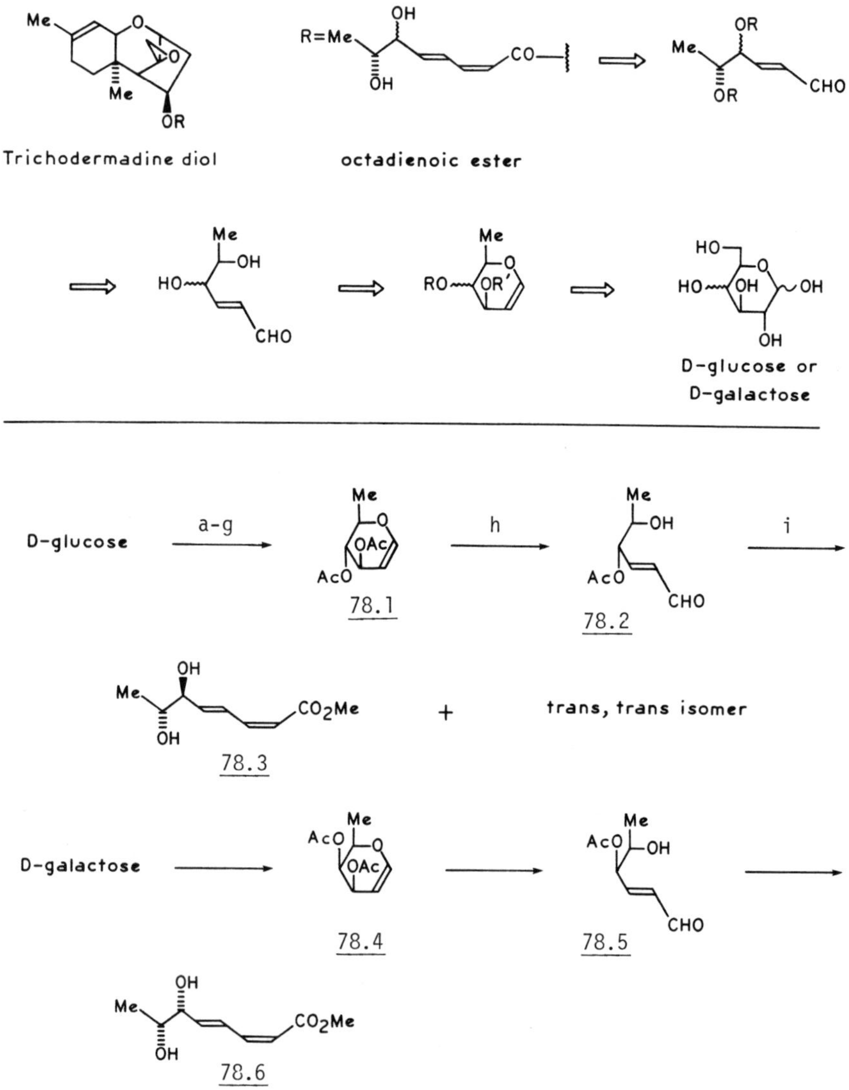
Trichodermadine diol
octadienoic ester
R=Me
D-glucose or
D-galactose
D-glucose
a-g
h
i
78.1
78.2
78.3
trans, trans isomer
D-galactose
78.4
78.5
78.6

a. Ac_2O, pyr.; b. HBr, AcOH; c. Zn, AcOH; d. NaOMe, MeOH; e. TsCl, pyr., then Ac_2O; f. NaI; g. Bu_3SnH; h. H_2SO_4, aq. diox; i. $Me_3SiCH_2CO_2Me$, BuLi, -78^0

Scheme 79

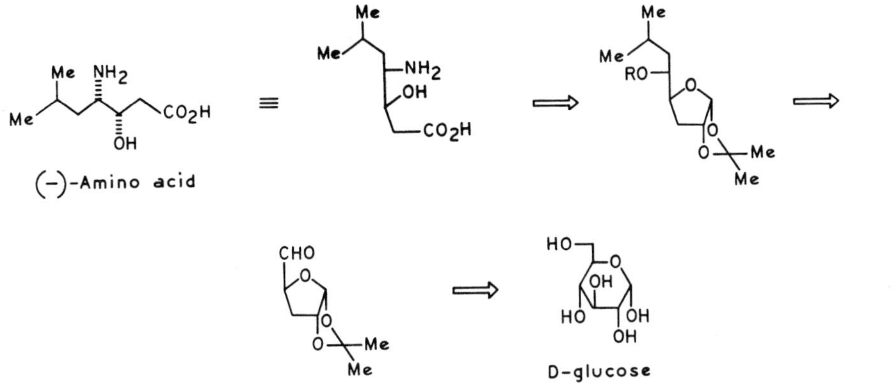

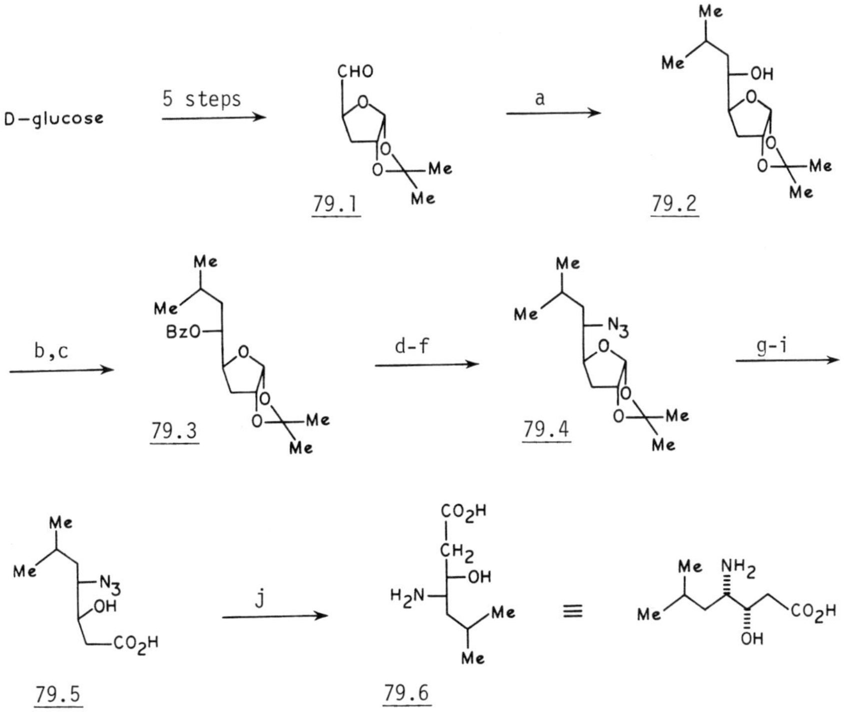

(−)-Amino acid

D-glucose

D-glucose $\xrightarrow{\text{5 steps}}$ 79.1 $\xrightarrow{a}$ 79.2

$\xrightarrow{b,c}$ 79.3 $\xrightarrow{d-f}$ 79.4 $\xrightarrow{g-i}$

79.5 $\xrightarrow{j}$ 79.6

a. Me$_2$CHCH$_2$MgBr; b. MsCl, pyr.; c. NaOBz, DMF; d. NaOMe, MeOH; e. TsCl, pyr.;
f. NaN$_3$, DMF; g. aq. AcOH; h. NaIO$_4$; i. CrO$_3$, AcOH, pyr.; j. Pd black, H$_2$

f. 2(R), 3(S), 4(S)-4-Amino-3-hydroxy-2-methyl-6-(3-pyridyl)-
pentanoic acid - The title amino acid was isolated as a degradation
product of the antibiotic pyridomycin[21] and its absolute configuration
was established based on X-ray crystallographic studies of the
antibiotic itself. A regio- and stereospecific synthesis of the amino
acid by Kinoshita and Mariyama[22] utilized D-glucose as a chiral starting
material (Scheme 80). Examination of the target structure 80.6 shows
the presence of four contiguous and different substituents on a
five-carbon acid chain. The presence of a pyridyl moiety, C-methyl and
amino substituents make the relationship with a carbohydrate somewhat
remote if not unnoticeable. However, the vicinal relationship of the
amino alcohol unit and the congruence of the hydroxyl-bearing carbon
atom with C-4 of D-glucose are good starting points for a
carbohydrate-based synthesis. With the configuration of the amino
alcohol unit predetermined, it becomes clear that a D-glucofuranose
derivative is an ideal precursor since the oxygen is "protected" in ring
formation and the secondary amine as well as the C-pyridyl moiety can be
introduced by elaboration of the side-chain.There remains to C-methylate
regio and stereoselectively at C-3 and to sacrifice C-1. Scheme 80
shows how these operations were performed in a sequence that introduces
the C-3 methyl (carbohydrate numbering) first by Wittig methodology.
Note that in the readily available intermediate 80.1 reduction is
virtually stereospecific. Introduction of the C-pyridyl unit was done
by opening of a terminal epoxide (80.2 → 80.3) and the resulting
secondary hydroxyl group was displaced via the mesylate with inversion
to provide 80.4. This intermediate 'chiron' now contains all four
crucial substituents. The carboxyl group was then produced by excision
of the C-1 — C-2 bond and the azido acid 80.5 reduced to the target
80.6. This synthesis nicely illustrates how a variety of contiguous
substituents can be incorporated in a carbon chain by relying on the
predictable regio and stereochemistry provided by a carbohydrate carbon
framework. The facile incorporation of a pyridyl unit should extend
this reasoning to the construction of a number of other unrelated
heterocyclic compounds containing multifunctional appendages.

f. (+)-Cerulenin - Cerulenin is an antibacterial and antifungal
substance[23] which has attracted recent attention because of its inhibi-
tory action in the biosynthesis of lipids and steroids. Its revised
constitutional structure was assigned as 2(R), 3(S)-epoxy-4-oxo-7,10-

Scheme 80

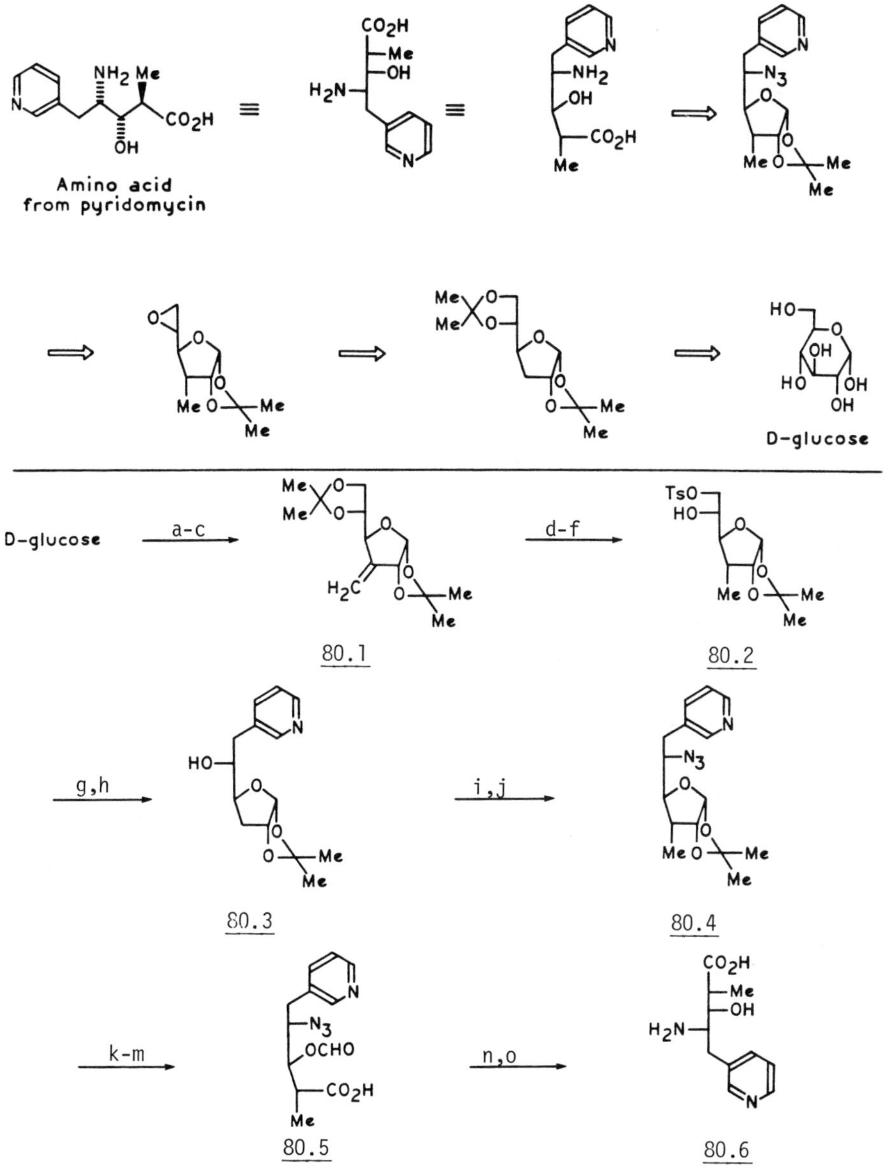

Amino acid
from pyridomycin

D-glucose

D-glucose $\xrightarrow{a-c}$

80.1

$\xrightarrow{d-f}$

80.2

$\xrightarrow{g,h}$

80.3

$\xrightarrow{i,j}$

80.4

$\xrightarrow{k-m}$

80.5

$\xrightarrow{n,o}$

80.6

a. acetone, H^+; b. RuO_4, CCl_4; c. $Ph_3P=CH_2$; d. Pd/C, H_2; e. aq. AcOH;
f. TsCl, pyr.; g. NaOMe, MeOH; h. 3-pyridyllithium, ether; i. MsCl,
pyr.; j. NaN_3, DMSO; k. aq. AcOH, reflux; l. $NaIO_4$; m. Br_2; n. aq. HCl;
o. Pd black, H_2

trans, trans-dodecadienoic acid amide,[24],[25] (Scheme 81). Two independent
stereoselective syntheses from D-glucose have been reported,[26],[27] which
rely on the same symmetry recognition criteria. Inspection of the
target structure reveals the presence of an α,β-epoxy-γ-oxo-
functionality on a carbon acid chain. Considering the two asymmetric
centers, the presence of an sp^2 center ($-CONH_2$) and an oxygen function
in the γ-position leads one to consider a carbohydrate precursor,
preferably in the furanose form in which the anomeric and C-4 carbon
atoms would correspond to C-1 and C-4 of the target respectively (note
oxidation states; "rule of four"). In a furanose derivative, the
side-chain can be extended to introduce the required crucial unsaturated
appendage. The introduction of the epoxide would be facilitated by the
proper choice of substituents and relative stereochemistry at C-2 and
C-3 of the sugar derivative. Consideration of these parameters leads to
a 1,2-O-isopropylidene-α-D-glucofuranose structure as an ideal substrate
which was used in both syntheses. Thus, in the Japanese approach[26]
(Scheme 81), D-glucose was converted to the readily available
3-protected derivative 81.1,[28] which was oxidized to 81.2 and chain
extended to 81.4. A key step was the introduction of the *trans, trans*
diene system which was made by generation of the acetylenic derivative
81.6, chain-extension with crotyl bromide in the presence of
ethylmagnesium bromide and copper chloride,[29] followed by debenzylation
in liq. ammonia, protection of the hydroxyl group and reduction of the
triple bond (liq. ammonia). This stepwise debenzylation-reduction
sequence was found to be necessary for efficient conversion into the
diene system. Subsequent operations were concerned with the generation
of the epoxy lactol 81.9, oxidation to the lactone 81.10, and
elaboration to the target 81.12. The independent French approach[27] was
essentially similar in that it too relied on acetylene chemistry for
chain-extension and made efficient use of the O-(2-methoxyisopropylidene
group). Thus, the syntheses of cerulenin are stereoselective by virtue
of the possible formation of isomeric dienes in the reduction of the
acetylenic intermediates and some isomerization. With regard to the
introduction of the epoxy function however, the operations are regio-
and stereospecific. Several syntheses of racemic cerulenin are
available.[30] Tetrahydrocerulenin has also been synthesized from
D-glucose[24] and D-xylose[25] respectively, essentially by the same
procedures as for cerulenin.

Scheme 81

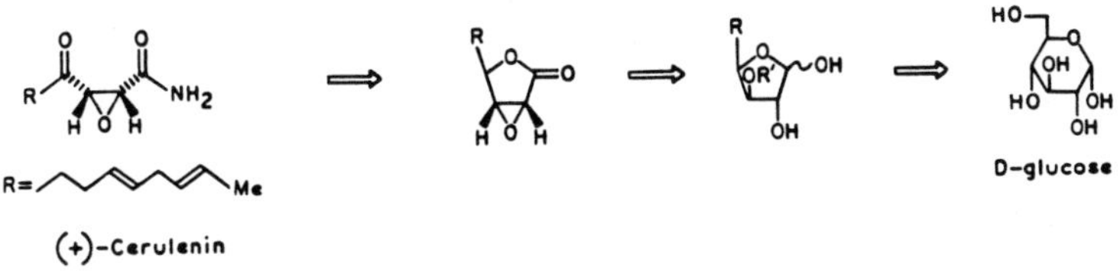

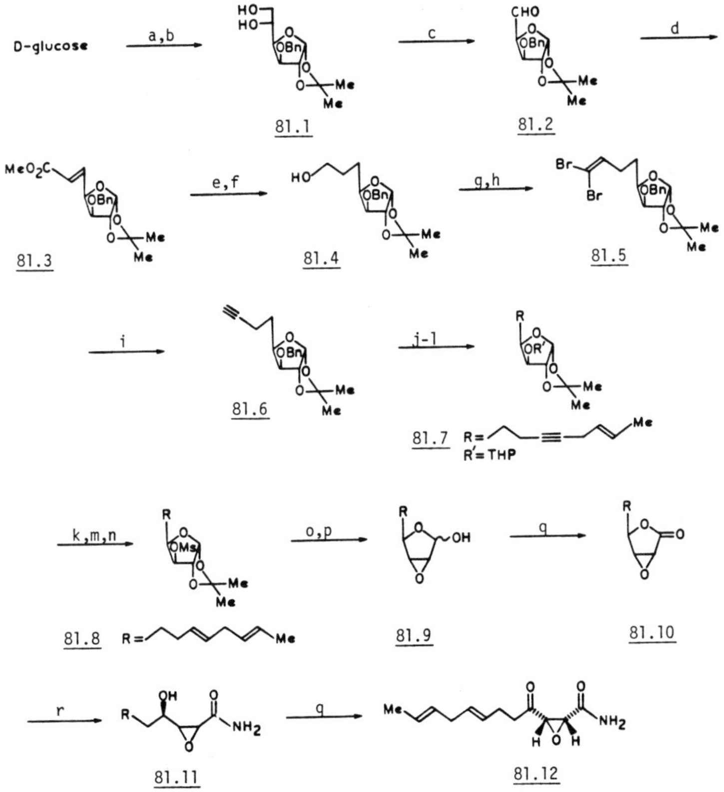

(+)-Cerulenin

---

a. acetone, H$^+$; b. BnBr, NaH, DMF then H$_3$O$^+$; c. aq. AcOH, Pb(OAc)$_4$,
benzene; d. (EtO)$_2$P(O)CH$_2$CO$_2$Me, NaH, ether; e. Pd/C, H$_2$, EtOH;
f. LAH, ether; g. DMSO, (COCl)$_2$, -60°; h. Ph$_3$P, CBr$_4$, Zn, CH$_2$Cl$_2$;
i. n-BuLi, THF; j. EtMgBr, crotyl bromide, Cu$_2$Cl$_2$; k. Li, NH$_3$, THF,
((NH)$_4$)$_2$SO$_4$, t-BuOH, -70°; 1. dihydropyran, PPTS, CH$_2$Cl$_2$; m. PPTS,
EtOH; n. MsCl, Et$_3$N, CH$_2$Cl$_2$; o. aq. TFA; p. NaOMe, THF, 0°; q. PCC,
CH$_2$Cl$_2$; r. NH$_3$, MeOH

   h. Methyl 8(R), 9(S), 11(R)-trihydroxy-(5Z), (12E), (14Z)-eicosa-
trienoate - The title compound has been suggested[31] to arise from arachidonic acid, although it has not as yet been isolated from the cascade
of biosynthetic products.[32] According to Just's hypothesis[31], it could
be derived from a sequence involving peroxidation, hydroxyperoxidation
and reduction of arachidonic acid. Biosynthetic arguments and convenience of synthetic manipulations have led to the consideration of D-
glucose as a starting material for the synthesis of the title compound.
The partially hidden carbohydrate-type symmetry in the target molecule
(Scheme 82) can be seen by focusing on the segment which contains contiguous and alternating hydroxyl groups. The presence of olefinic
linkages and a α-hydroxyl group leads one to consider the "rule of
five". Thus the 8(R),9(S) and 11(R) hydroxyl groups could, in
principle, individually correspond to a C-5 hydroxyl group in an
appropriate hexose derivative. In each case, the anomeric carbon would
be the future site of double bond formation and chain-extension.
Consideration of the stereochemical features present in the target led
Just and Luthe to a D-ribo-hexose configuration as in
3-deoxy-D-ribo-hexose (3-deoxy-D-glucose).[33] Thus, the C-1 — C-6 portion
of this sugar can be found in the segment C-12 — C-7 of the target.
With this type of stereochemistry decoded, the main challenge resides in
generating the appropriately protected hexose derivative and effecting
chain-extensions at both extremities. The readily available 3-deoxy
intermediate 25.1 was transformed into the epoxide 82.1 which in turn
was chain-extended by treatment with the dilithium salt of 5-hexynoic
acid and the resulting the product reduced. Protection of the hydroxyl
group, mercaptolysis and acetonation (note compatibility of the silyl
protecting group to acid-catalyzed reactions), gave the acyclic
dithioacetal derivative 82.4 which contains all three hydroxyl groups
needed in the target. The remainder of the sequence was based on well
precedented chain-extension reactions using Wittig methodology.

   i. Leukotriene C-1 - The so-called ("slow-reacting substance")
(SRS) of anaphylaxis[34] or leukotriene C-1, is one of several such components isolated from biological sources. Its total synthesis from a carbohydrate precursor by Corey and coworkers[35] provided unambiguous proof
for the previously proposed molecular structure. Examination of the
structure in question (Scheme 83) reveals that the carbon backbone of
the C-20 acid contains two contiguous asymmetric centers at C-5 and C-6

Scheme 82

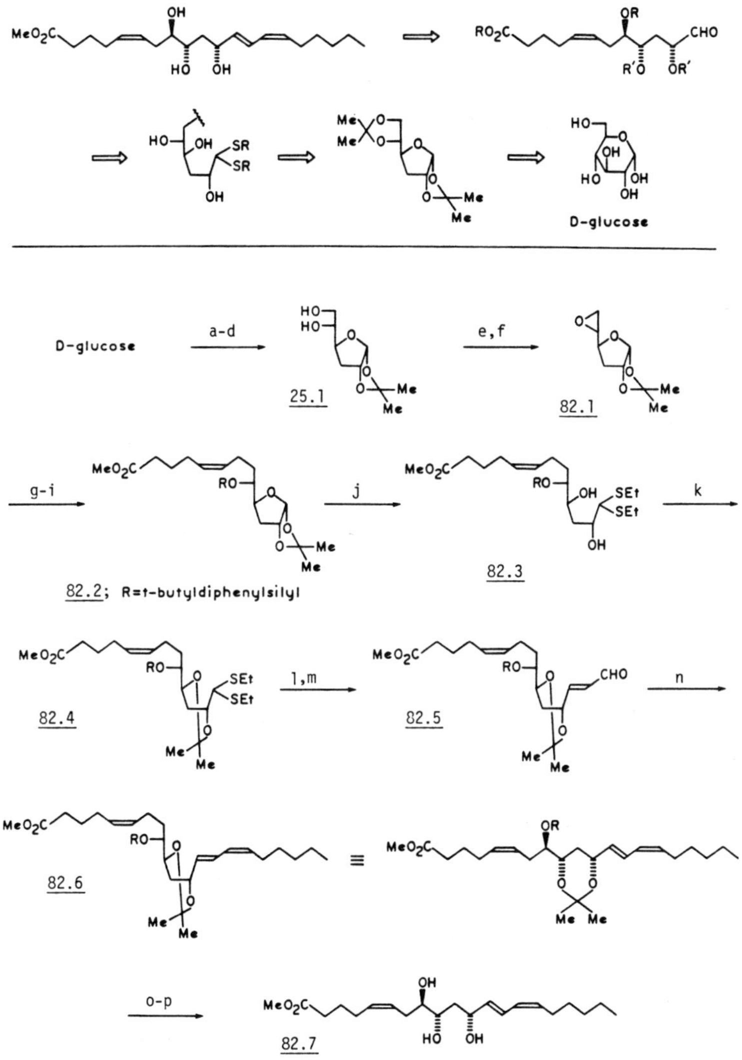

a. acetone, H$^+$; b. CS$_2$, MeI, NaH; c. Bu$_3$SnH, AIBN; d. aq. AcOH;
e. TsCl, pyr.; f. aq. KOH; g. 5-hexynoic acid, BuLi, HMPA 5^0,
then CH$_2$N$_2$; h. Ni(OAc)$_2$, NaBH$_4$, H$_2$; i. Ph$_2$t-BuSiCl, imidazole,
DMF, 78^0; j. EtSH, ZnCl$_2$, -10^0; k. Me$_2$C(OMe)$_2$, p-TsOH; l. HgO,
HgCl$_2$, aq. acetone; m. Ph$_3$P=CHCHO, DMF; n. Ph$_3$P=CHC$_5$H$_{11}$, THF;
o. Bu$_4$NF, THF; p. aq. HCl

consisting of hydroxyl and thiol functionality. A priori, while the
relationship of C-1 and C-5 in the target may converge with the same in
a hexose derivative ("rule of five"), the presence of an asymmetric
center at C-6 precludes utilization of a hexose as such. Since one can
assume that the hydroxyl group will be inherent to the structure of the
starting carbohydrate, the main challenge resides in the vicinal introduction of a thiol group with the desired regio and stereochemistry and
in subsequent chain-elongation. This translates itself into manipulating two secondary sites in a sugar derivative via an epoxide intermediate of predictable stereochemistry. Regioselective opening of the
epoxide by a thiol-containing moiety may be envisaged by virtue of
activation through its allylic disposition, as shown in the retrosynthetic analysis. The route pursued by Corey and coworkers[35] shows
how the readily available D-ribose was used in achieving complete regio
and stereocontrol in the introduction of the vicinal substitution in
question. The Wittig product 83.2[36] available from the reaction of
2,3,5-tri-O-benzoyl-D-ribose with methoxycarbonylmethylenetriphenyl-
phosphorane was transformed into a mixture of β,γ-unsaturated esters
83.4 by an interesting allylic elimination in the presence of zinc
amalgam and hydrogen chloride. Catalytic reduction then gave 84.3 which
encompasses the C-1 — C-7 portion of the target including the correct
stereochemistry at C-5. The crucial epoxide 83.5 was then obtained by
tosylation and base treatment. Note that possible benzoyl migration
(C-5 — C-6) or epoxide transposition to the terminal C-6 — C-7 position
were not observed. Oxidation of 83.5 gave the corresponding aldehyde
which was chain-extended by known methodology to give the critical epoxy
tetraene biosynthetic intermediate 83.7. Reaction with glutathione
under carefully controlled conditions led to the target leukotriene C-1,
83.8, characterized as its N-trifluoroacetyl trimethyl ester. Thus, this
scarcely available, biologically important substance can be obtained in
significant quantities by this route. It should be noted finally, that
the original five-carbon framework of D-ribose was incorporated as the
C-3 — C-7 segment of the target, hence partially hidden in the
structure. The synthesis of optically active leukotriene intermediates
starting with D-araboascorbic acid has been reported.[37]

        j. Leukotriene A4 - An efficient and versatile synthesis of leu-
kotriene A4 has been reported by the Merck-Frosst Group[38] which relies
on the utilization of (R)- or (S)-glyceraldehyde as a chiral precursor

Scheme 83

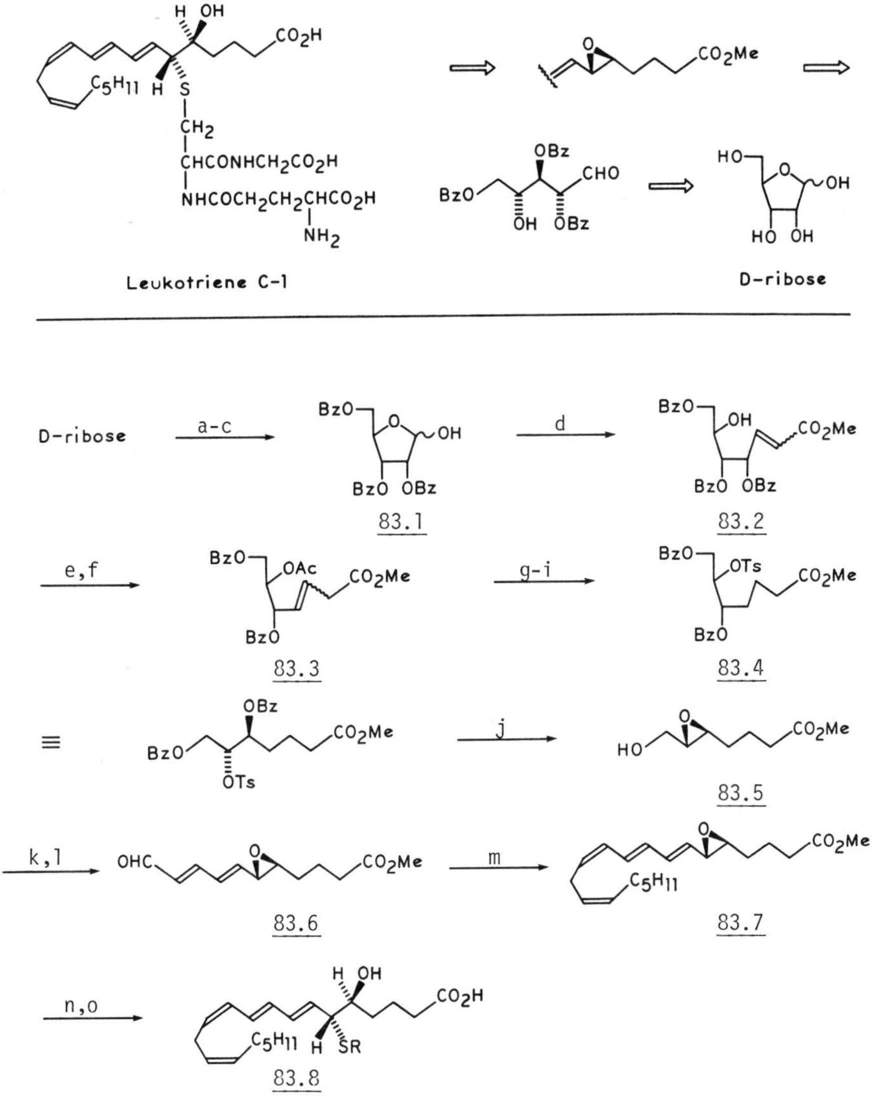

Leukotriene C-1

D-ribose

83.1

83.2

83.3

83.4

83.5

83.6

83.7

83.8

a. MeOH, HCl; b. BzCl, pyr.; c. H_3O^+; d. $Ph_3P=CHCO_2Me$, DME, $PhCO_2H$, reflux; e. Ac_2O, pyr.; f. Zn amalgam, ether, HCl; g. Pd/C, H_2; h. HCl, MeOH; i. TsCl, pyr.; j. K_2CO_3, MeOH; k. Collins; l. 1-lithio-4-ethoxy-butadiene, THF, -78°, then MsCl, Et_3N; m. Wittig; n. glutathione derivative, Et_3N, MeOH; o. aq. K_2CO_3, MeOH

(Scheme 84). Interestingly the chiral center in this 'chiron' is not
found in the target, rather it is cleverly used to achieve asymmetric
induction in an epoxidation step as well as to provide a two-carbon
segment of the target. Thus, (S)-glyceraldehyde acetonide 84.1, available from L-arabinose was transformed into the *cis*-olefin 84.2 which
was isomerized to the *trans*-isomer 84.3 (95% conversion). Epoxidation
of this olefinic derivative led to a 2:1 mixture of the desired epoxide
84.4 and the unnatural isomer. Oxidative cleavage to 84.5 and a double
homologation gave the dieneal 84.6 which was converted to leukotriene A_4
by another Wittig reaction. Initially, the use of (R)-glyceraldehyde
acetonide had led to a 2:1 mixture of epoxides in favor of the unnatural
isomer. A clever appreciation of the factors involved in asymmetric induction by the presence of chirality as well as functionality (dioxolane
ring), led the Merck group to using the enantiomeric (S)-aldehyde. It
should also be mentioned that other leukotrienes (C_4, D_4, E_4) can be
obtained from the A_4 component by reaction with the appropriate thiol
amino acid. Other syntheses of leukotrienes starting from carbohydrate
precursors have been reported by the Merck-Frosst group.[39]

   k. <u>6-Epileukotrienes and Leukotriene B</u> - The 6-epileukotrienes
are not found in substantial quantities compared to the leukotrienes.
Corey and Goto[40] as well as other groups have reported a synthesis of
6-epileukotrienes C and D in order to allow rigorous comparison with
material isolated from natural sources   (Scheme 85). The basic approach
is the same as in Scheme 83, except that the inverted stereochemistry at
C-6 of the target necessitates the creation of a different epoxide.
Retrosynthetic analysis shows that a D-mannofuranose derivative provides
an ideal source for a 2-deoxy-D-<u>arabino</u>-hexose intermediate, in which
C-3 and C-4 have the prerequisite stereochemistry for epoxide
formation. The readily available and crystalline 2,3:5,6-di-0-iso-
propylidene-D-mannofuranose[41] 85.1 was transformed into the glycal
derivative 85.2 by the Ireland procedure.[42] Hydration then led to the
monoacetonide derivative of the desired 2-deoxyhexose, which was
homologated via a Wittig reaction and the hydroxyl groups manipulated so
as to provide the proper substitution pattern as in 85.5. The remainder
of the sequence follows the same methodology used in the synthesis of
leukotriene A_4 (Scheme 84). Note the transformations 85.3 → 85.4 → 85.5
and the voluntary manipulation of functional groups. The total synthesis
and assignment of stereochemistry of leukotriene B has been accomplished

Scheme 84

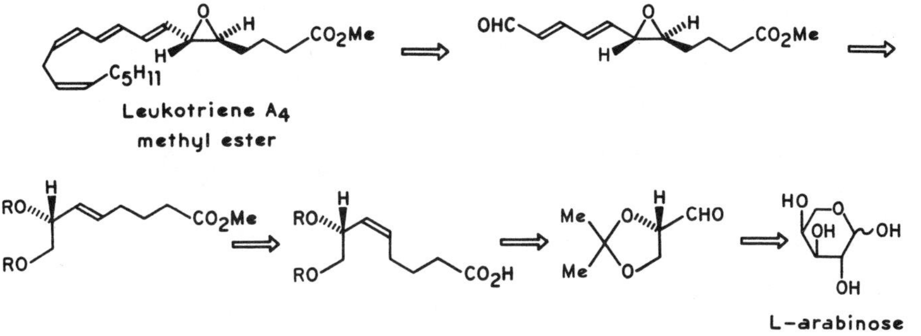

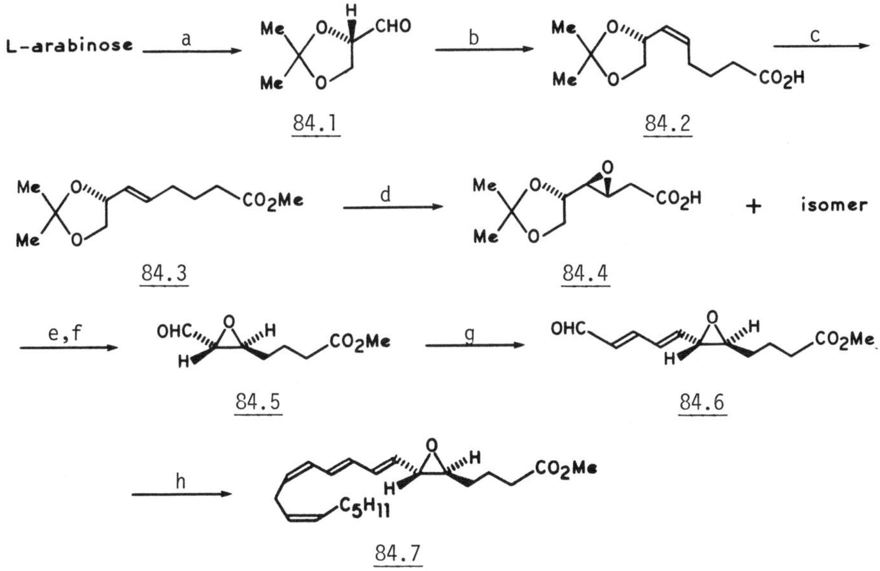

Leukotriene A₄
methyl ester

L-arabinose

L-arabinose $\xrightarrow{a}$ $\xrightarrow{b}$ $\xrightarrow{c}$

84.1   84.2

$\xrightarrow{d}$ + isomer

84.3   84.4

$\xrightarrow{e,f}$ $\xrightarrow{g}$

84.5   84.6

$\xrightarrow{h}$

84.7

a. 3 steps; b. $Ph_3P^+(CH_2)_4CO_2H\ Br^-$; c. $h\nu$, PhSSPh, then CH_2N_2;
d. m-CPBA; e. aq. AcOH; f. $NaIO_4$; g. $Ph_3P=CHCHO$; h. Wittig

Scheme 85

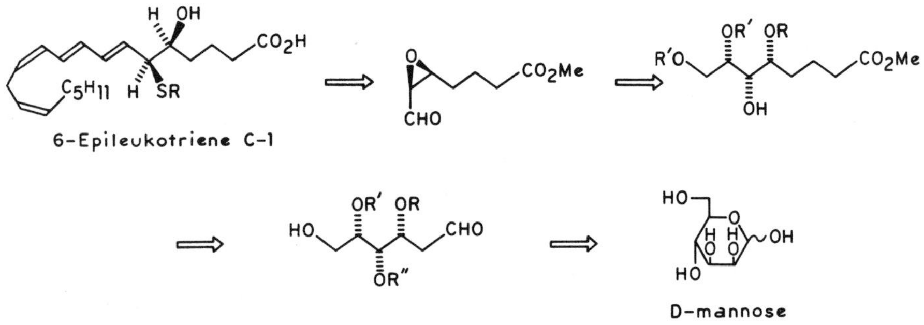

6-Epileukotriene C-1
D-mannose

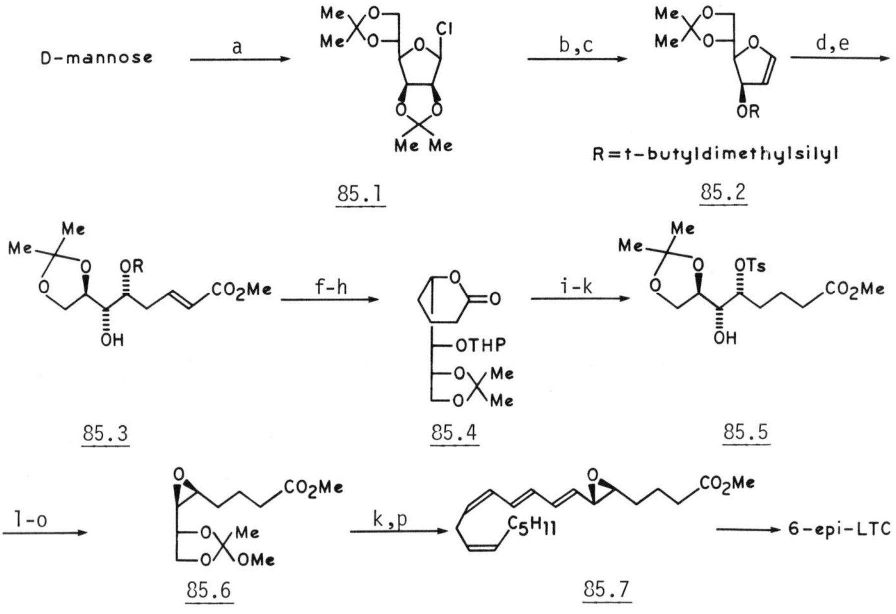

D-mannose
a
b,c
d,e
R=t-butyldimethylsilyl
85.1
85.2
f-h
i-k
85.3
85.4
85.5
l-o
k,p
6-epi-LTC
85.6
85.7

a. acetone, H^+, then Ph_3P, CCl_4; b. Na, liq. NH_3; c. t-$BuMe_2SiCl$, DMF, imidazole; d. $Hg(OAc)_2$, aq. THF, 0°, then NaI, 0° and $NaBH_4$, -10°; e. $Ph_3P{=}CHCO_2Me$, DME, benzoic acid, 70°; f. Pt/C, H_2; g. dihydropyran, PPTS, CH_2Cl_2; h. Bu_4NF, THF; i. MeOH, Et_3N; j. TsCl, pyr.; k. PPTS, MeOH; l. BzCl, pyr.; m. HCl, MeOH; n. $MeC(OMe)_3$, CH_2Cl_2, TsOH; o. K_2CO_3, MeOH; p. $Pb(OAc)_4$, CH_2Cl_2, Na_2CO_3, -45°; then Wittig.

starting with 2-deoxy-D-<u>erythro</u>-pentose.[43,44]

1. <u>Miscellaneous</u> - The acyclic segments of (-)-anamarin,[45] an
unsaturated δ-lactone with a tetraacetoxylated side-chain has been
synthesized from D-glucose as a chiral building block.[46]

The optically active C-1 — C-18 segment of carzinophilin A, a
natural intercalative bisalkylator with notable antitumor activity,[47]
was synthesized from D-glucose.[48] The lactone portion of the
hypocholesterolemic agent compactin has been synthesized from a
carbohydrate precursor.[49]

REFERENCES

1. R.M. Silverstein, J.O. Rodin and D.L. Wood, Science, <u>154</u>, 509
   (1966); J. Econ. Entomol, <u>60</u>, 944 (1967).

2. K. Mori, Tetrahedron Lett., 1609 (1976).

3. C.A. Reece, J.O. Rodin, R.G. Brownlee, W.G. Duncan and R.M.
   Silverstein, Tetrahedron, <u>24</u>, 4249 (1968); R.G. Riley, R.M.
   Silverstein, J.A. Katzenellenbogen and R.S. Lenox, J. Org. Chem.,
   <u>39</u>, 1957 (1974); K. Mori, Agr. Biol. Chem., <u>38</u>, 2045 (1974).

4. J.H. Borden and E. Stokkink, Can. J. Zool., <u>51</u>, 469 (1973).

5. H.R. Schuler and K. Slessor, Can. J. Chem., <u>55</u>, 3280 (1977).

6. R.S. Tipson and A. Cohen, Carbohydr. Res., <u>1</u>, 338 (1965).

7. K. Mori, Tetrahedron, <u>31</u>, 3011 (1975).

8. G.T. Pearce, W.E. Gore, R.M. Silverstein, J.W. Peacock, R.A.
   Cuthbert, G.N. Lanier and J.B. Simeone, J.Chem. Ecol., <u>1</u>, 115
   (1975).

9. K. Mori, Tetrahedron, <u>33</u>, 289 (1977). For a determination of the
   absolute configuration and synthesis of the enantiomer of
   semicornin, the sex pheromone of the female cigarette beetle, from
   D-glucose, see M. Mori, T. Chuman, M. Kohno, K. Kato, M. Noguchi,
   H. Nomi and K. Mori, Tetrahedron Lett., <u>23</u>, 667 (1982). The
   structure is that of 4<u>S</u>,6<u>S</u>,7<u>S</u>, 4,6-dimethyl-7-hydroxynonan-3-one.

10. J.-R. Pougny and P. Sinaÿ, J. Chem. Res., <u>5</u>, 1, 1982.

11. L.F. Wiggins, Methods Carbohydr. Chem., <u>2</u>, 188 (1963).

12. K. Mori and H. Iwasawa, Tetrahedron, <u>36</u>, 2209 (1980); see also,
    G. Fráter, Helv. Chim. Acta, <u>62</u>, 2829 (1979); J.P. Vigneron, R.
    Meric and M. Dhaenens, Tetrahedron Lett., <u>21</u>, 2057 (1980).

13. B.B. Jarvis, G. Pavanasasivam, C.E. Holmlund, T. DeSilva, G. P.

Stahly and E.P. Mazzola, J. Am. Chem. Soc., 103, 472 (1981).

14. D.B. Tulshian and B. Fraser-Reid, J.Am. Chem. Soc., 103 474 (1981); B.Fraser-Reid and B. Radatus, J. Am. Chem. Soc., 92, 5288 (1970).

15. D.J. Petersen, J. Org. Chem., 33, 780 (1968); K. Shimoji, H. Taguchi, K. Oshima, H. Yamamoto and H. Nozaki, J. Am. Chem. Soc., 96, 1620 (1974).

16. H. Morishima, T. Takita, T. Aoyagi, T. Takeuchi and H. Umezawa, J. Antibiotics, 23, 263 (1970).

17. H. Nakamura, H. Morishima, T. Takita, H. Umezawa and Y. Iitaka, J. Antibiotics, 26, 255 (1973).

18. M. Kinoshita, A. Hagiwara and S. Aburaki, Bull. Chem. Soc. Japan, 48, 570 (1975).

19. D.H. Murray and J. Prokop, J. Pharm. Sci, 54, 1468 (1965).

20. H. Morishima, T. Takita and H. Umezawa, J. Antibiotics, 26, 115 (1973).

21. G. Koyama, Y. Iitaka, K. Maeda and H. Uwezawa, Tetrahedron Lett., 3587 (1967).

22. M. Kinoshita and S. Mariyama, Bull. Chem. Soc. Japan, 48, 2081 (1975).

23. Y. Sano, S. Nomura, Y. Kamio, S. Omura and T. Hata, J. Antibiotics, 20, 344 (1967); B.H. Arison and S. Omura, J. Antibiotics, 27, 28 (1974) and references cited therein 193; S. Omura, Bacteriol Rev., 40, 681 (1976).

24. H. Ohrui and S. Emoto, Tetrahedron Lett., 2095 (1978); and references cited therein.

25. J.-R. Pougny and P. Sinaÿ, Tetrahedron Lett., 3301 (1978).

26. N. Sueda, H. Ohrui and H. Kuzuhara, Tetrahedron Lett., 2039 (1979).

27. M. Pietraszkiewicz and P. Sinaÿ, Tetrahedron Lett., 4741 (1979).

28. M.L. Wolfrom and S. Hanessian, J. Org. Chem., 27, 1800 (1962).

29. See for example, G. Fouquet and M. Schlosser, Angew. Chem. Int. Ed. Engl., 13, 82 (1974).

30. A.A. Jakubowski, F.S. Guziec, Jr., M. Sugiura, C.C. Tam, M. Tishler and S. Omura, J. Org. Chem., 47, 1221 (1982); R.K. Boeckman, Jr., and E.W. Thomas, J. Am. Chem. Soc., 101, 987 (1979); A.A. Jakubowski, F.S. Guziec, Jr., and M. Tishler, Tetrahedron Lett., 2399 (1977); E.J. Corey and D.R. Williams, Tetrahedron Lett., 3847 (1977).

31. G. Just and C. Luthe, Can. J. Chem., 58, 1799 (1980).

32. C. Pace-Asciak and L.S. Wolfe, Chem. Commun., 1234 (1970).

33. P. Szabo and L. Szabo, J. Chem. Soc., 5139 (1964).

34. See for example, W. Feldberg and C.H. Kellaway, J. Physiol. (London), 94, 187 (1938); R.C. Murphy, S. Hammarström and B. Samuelsson, Proc. Natl. Acad. Sci., U.S.A., 76, 4275 (1979); and references cited therein.

35. E.J. Corey, D.A. Clark, G. Goto, A. Marfat, C. Mioskowski, B. Samuelsson and S. Hammarström, J. Am. Chem. Soc., 102, 1436 (1980).

36. J.G. Buchanan, A.R. Edgar, M. J. Power and P. D. Theaker, Carbohydr. Res., 38, C22 (1974); see also S. Hanessian, T. Ogawa and Y. Guindon, Carbohydr. Res., 38, C12 (1974); H. Ohrui, G.H. Jones, J.G. Moffatt, M.L. Maddox, A.T. Christensen and S.K. Byram, J. Am. Chem. Soc., 97, 4602 (1975).

37. N. Cohen, B.L. Banner, R.J. Lopresti, F. Wong, M. Rosenberger, Y. Liu, E. Thom and A.A. Liebman, J. Am. Chem. Soc., 105, 3661 (1983), N. Cohen, B.L. Banner and R.J. Lopresti, Tetrahedron Lett., 21, 4163 (1980).

38. J. Rokach, R.N. Young, M. Kakushima, C.-K. Lau, R. Seguin, R. Frenette and Y. Guindon, Tetrahedron Lett., 22, 979 (1981).

39. See for example Y. Guindon, R. Zamboni, C.-K. Lau and J. Rokach, Tetrahedron Lett., 23, 739 (1982); J. Rokach, C.-K. Lau, R. Zamboni and Y. Guindon, Tetrahedron Lett., 22, 2763 (1981); J. Rokach, R. Zamboni, C.-K. Lau and Y. Guindon, Tetrahedron Lett., 22, 2759 (1981).

40. E.J. Corey and G. Goto, Tetrahedron Lett., 21, 3463 (1980);

41. J.B. Lee and T.J. Nolan, Tetrahedron, 23, 2789 (1967).

42. R.E. Ireland, S. Thaisrivongs, N. Vanier and C.S. Wilcox, J. Org. Chem., 45, 48 (1980).

43. E.J. Corey, A. Marfat, G. Goto and F. Brion, J. Am. Chem. Soc., 102, 7984 (1980).

44. D.P. Marriott and J.R. Bantick, Tetrahedron Lett., 22, 3657 (1981).

45. A. Alemany, C. Marquez, C. Pasqual, S. Valverde, A. Perales, J. Fayos, and M. Martinez-Ripoll, Tetrahedron Lett., 3579 (1979).

46. F. Gillard and J.-J. Riehl, Tetrahedron Lett., 24, 587 (1983).

47. N. Shimada, M. Uekusa, T. Denda, Y. Ishii, T. Iizuka, Y. Sato, T. Hatori, M. Fukui and M. Sudo, J. Antiobitics, 8, 67 (1955); A.

Terawaki and J. Greenberg, Nature, 209, 481 (1966); J.W. Lown and
K.C. Majumdar, Can. J. Biochem., 55, 630 (1977).

48. M. Shibuya, Tetrahedron Lett., 24, 1175 (1983).

49. See for example, P.A. Grieco, R.E. Zelle, R. Lis and J. Finn, J.
Am. Chem. Soc., 105, 1403 (1983); Y.-L. Yang and J.R. Falck,
Tetrahedron Lett., 23, 4305 (1982); J.D. Prugh and A.A. Deana,
Tetrahedron Lett., 23, 281 (1982); S. Hanessian and A. Ugolini,
unpublished results.

CHAPTER TWELVE

CYCLIC MOLECULES

a. <u>(-)-Pestalotin - Synthesis of the enantiomer</u> - Pestalotin
(LL.P8802) is a gibberellin synergist whose structure and stereochemistry have been elucidated.[1],[2] The synthesis[3] of 6(<u>R</u>):1'(<u>R</u>)-pestalotin
from (<u>R</u>)-glyceraldehyde confirmed the assignment of a 6(<u>S</u>):1'(<u>S</u>)-stereo-
chemistry to the natural product. Examination of the structure of this
fungal lactone (Scheme 86) reveals the presence of two asymmetric carbon
atoms comprising a diol unit as part of a 4-methoxy-5,6-dihydro-α-pyrone
moiety. One of the oxygen atoms is involved in the lactone ring. In
their synthetic scheme Mori, Matsui and coworkers[3] envisaged the C-6
carbon atom bearing the lactone ring oxygen as originating from (<u>R</u>)-
glyceraldehyde. It follows that the C-1' center could arise from a
chain-extension from the aldehyde end, creating a new asymmetric center,
and the remaining four carbon "acid" portion can be generated by elonga-
tion of the hydroxymethyl end. Thus, the readily available (<u>R</u>)-glyce-
raldehyde derivative <u>12.2</u> was treated with n-butylmagnesium bromide to
give a mixture of epimeric alcohols <u>86.1</u> which was not separated, but
converted to the epoxides <u>86.2</u>. Subsequent steps leading to the product
were carried out according to Carlson's general method of α-pyrone synthesis.[4] Note the transformation of the acetylenic alcohol <u>86.3</u> to the
α-pyrone directly in the presence of sodium methoxide. The final
product consisted of a mixture of 6(<u>R</u>):1'(<u>R</u>)-pestalotin and its
1'-epimer which were separated by chromatography. There are several
other syntheses of pestalotin including an "asymmetric synthesis" in an
optical yield of 10%.[5]

Scheme 86

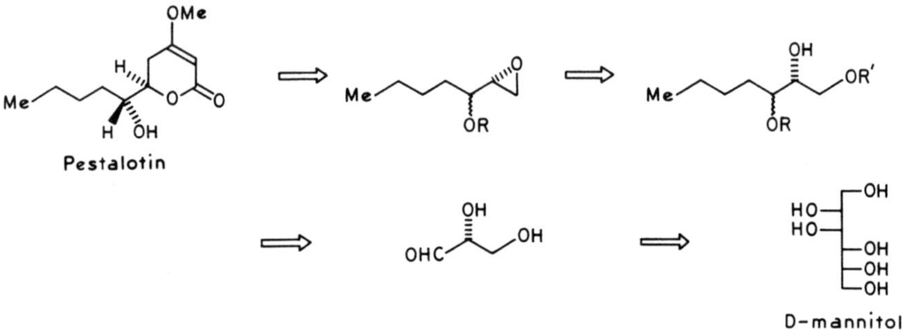
OMe
Me
Pestalotin
OR
OH
OR'
OHC
D-mannitol

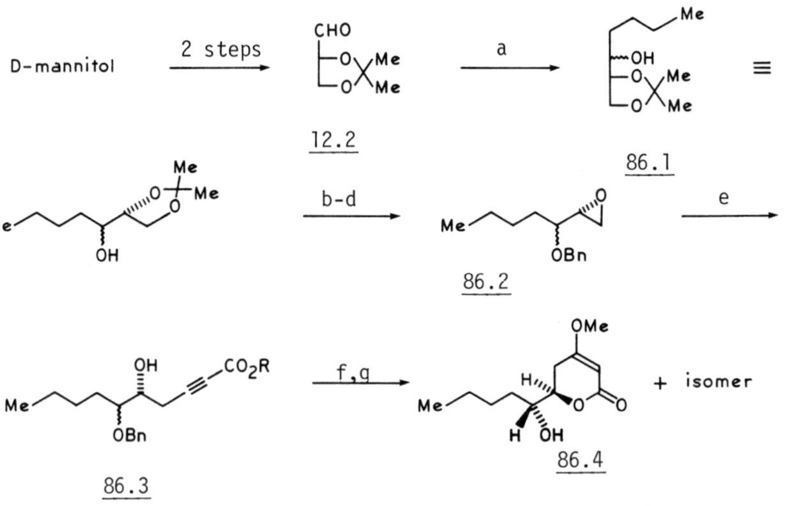
D-mannitol
2 steps
CHO
12.2
a
86.1
b-d
OBn
86.2
e
CO2R
86.3
f,g
86.4
+ isomer

a. BuMgBr, ether; b. BnCl, NaOH, DMSO; c. dil. HCl, MeOH, 60°; d. TsCl, pyr, then NaOMe; e. aq. KOH, propiolic acid, LDA, THF, HMPA, then esterification; f. NaOMe, MeOH; g. Pd/C, EtOH

b. (-)-Multistriatin - The synthesis of (-)-multistriatin from D-glucose[6] was discussed previously (see page 77). Here we discuss another approach to this molecule which utilizes (R)-glyceraldehyde as the starting material and leads to the formation of isomeric multistriatins.[7] The Mori approach (Scheme 87) consists of securing the stereo-

Scheme 87

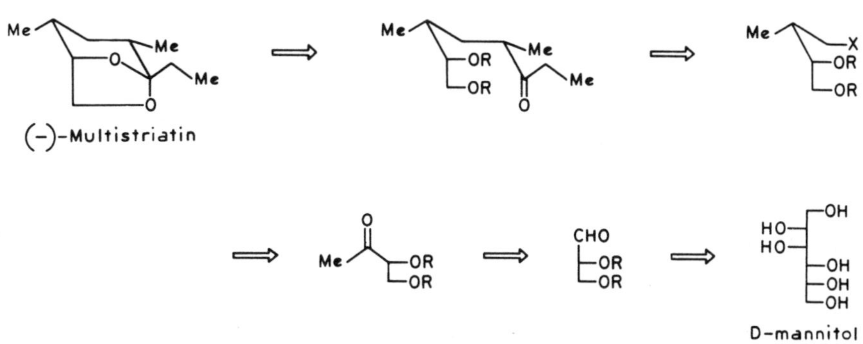

(-)-Multistriatin

D-mannitol

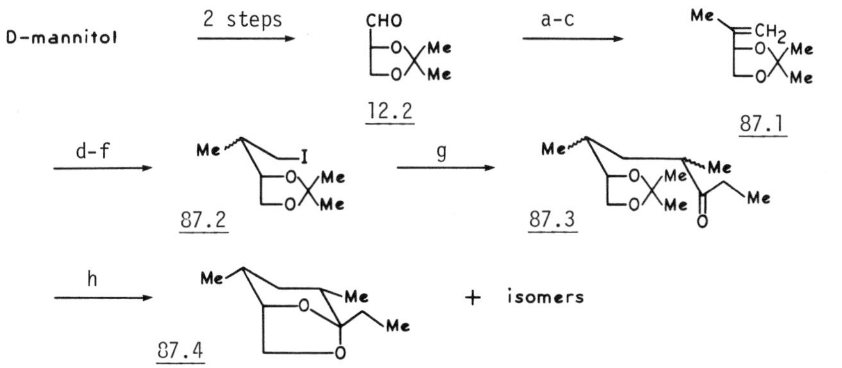

a. MeMgI, ether; b. CrO$_3$, acetone; c. Ph$_3$P=CH$_2$, DMSO; d. B$_2$H$_6$, NaOH, H$_2$O$_2$; e. TsCl, pyr.; f. LiI, acetone, reflux; g. 3-pentanone enamine derivative, EtMgBr, THF; h. H$_3$O$^+$

chemistry of C-1 bearing the 6-membered ring oxygen, and building up the remaining carbon backbone of the molecule by chain-extension from the aldehyde end of (R)-glyceraldehyde. No consideration was given to controlling the stereochemistry of methyl-bearing carbon atoms as in the previous synthesis from D-glucose.[6] Thus, (R)-glyceraldehyde was chain-extended to give the isopropenyl derivative 87.1 which was hydroborated and chain-extended again via the iodide 87.2 to give a mixture of ketones 87.3. Acid treatment gave a mixture of diastereoisomeric multistriatins from which the desired levorotatory 1(S), 2(R),4(S),5(R) isomer was isolated by chromatography. For other syntheses see Chapter seven.

c. 1(S),3(R),5(R)-1,3-Dimethyl-2,9-dioxabicyclo[3:3:1]nonane - The title compound has been isolated from a fermentation source[8] and identified as endo-1,3-dimethyl-2,9-dioxabicyclo [3:3:1] nonane.[9] Redlich and coworkers[10] have synthesized the title compound and its diastereoisomers from D-glucose (Scheme 88). Analysis of the stereochemistry shows the presence of two secondary alcohols bearing a 1,3 relationship, which are involved in intramolecular acetal formation. Thus a 2,4,6-trideoxy-hexose with predisposed asymmetric centers secures the crucial stereo-centers in the molecule. The retrosynthetic analysis shows that a three-carbon extension from the aldehyde provides the entire carbon framework of the target. Thus, methyl α-D-glucopyranoside was converted to the 4,6-dichloro derivative 88.1 and the dideoxy analog 88.2 in high yield by treatment with sulfuryl chloride in pyridine followed by reduc-tion. Note that in this efficient and useful transformation, the primary and one of three secondary equatorially disposed hydroxyl groups are converted into chlorides.[11] The reaction proceeds via critical formation of chlorosulfate esters which are subject to S_N2 displacement by chloride ions. The selectivity of attack at C-4 in this system may be attributed to stereoelectronic factors (note α-glycoside, see Scheme 3). With the desired level of deoxygenation attained 88.2 was then converted to an acyclic analog 88.3 which was ideally set up for β-elimination and subsequent reduction of the resulting ketene dithioacetal derivative 88.4. By this expedient methodology, the required 2,4,6-trideoxy-D-threo-hexose was obtained in the form of the dithian-acetal structure 88.5. Alkylation of the corresponding carbanion with propylene oxide followed by oxidation and acid treatment led to the desired target 88.8, isolated by preparative gas chromatography. By inverting the hydroxyl groups in 88.5 individually, it was also possible to prepare the other isomeric systems.[12]

Scheme 88

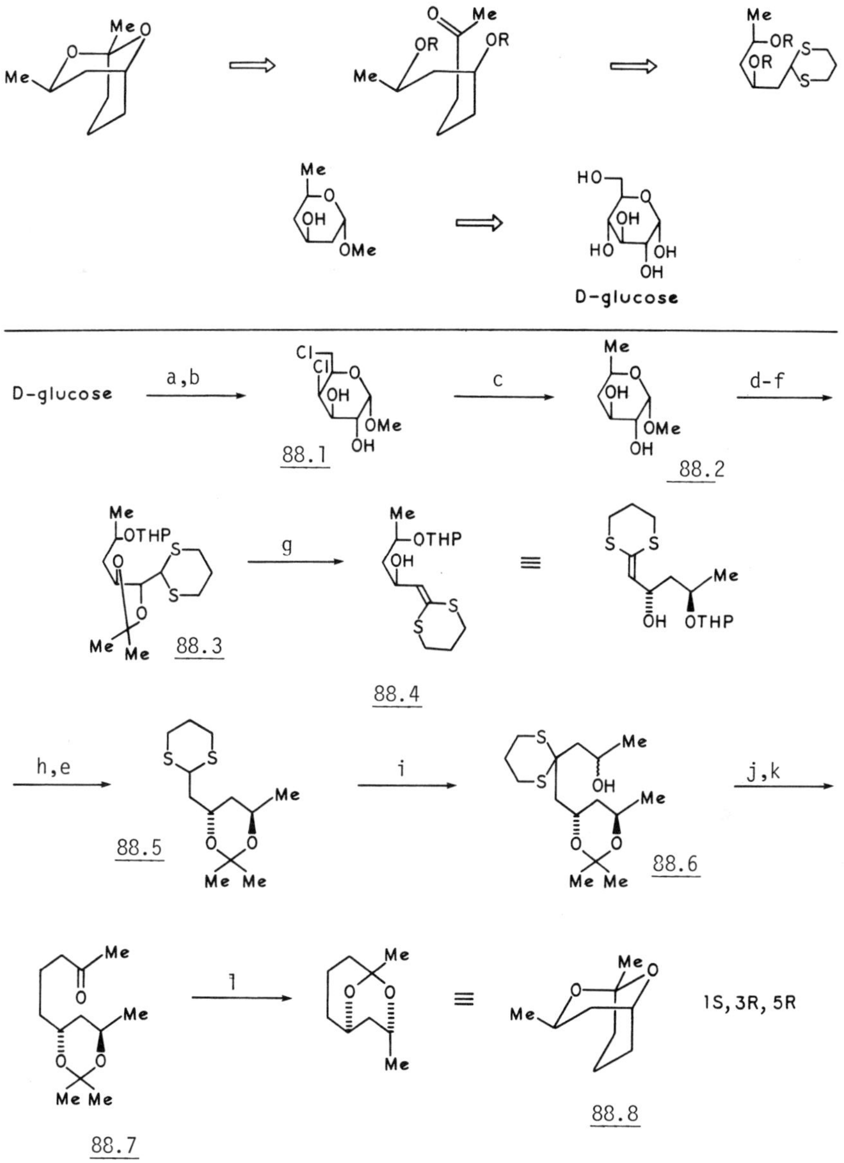

a. MeOH, HCl; b. SO_2Cl_2, pyr.; c. n-Bu_3SnH, AIBN; d. $HSCH_2CH_2CH_2SH$, H^+; e. acetone, H^+; f. dihydropyran, H^+; g. BuLi, THF; h. LAH; i. propylene oxide, BuLi, THF; j. Ra-Ni, EtOH; k. PCC; l. H^+, ether, 5Å sieves.

d. Chalcogran - Chalcogran is the principal aggregation pheromone of a species of beetles considered to be a pest of Norway spruce.
The natural product has been shown to be a 1:1 mixture of diastereoisomers[13] at C-2 (Scheme 89). Consideration of a carbohydrate-derived
synthesis would logically build the unsubstituted tetrahydrofuran ring
by chain-extension of an existing ring having the desired C-ethyl appendage. Thus, Redlich and Franck[14] utilized D-glucose as the source of
the more substituted ring recognizing a six-carbon framework ending with
the acetal junction ("rule of four"). A series of deoxygenations of the
D-glucofuranose structure led to the 2,3,5,6-tetradeoxy derivative 89.2
which was extended at C-1 via the corresponding dithian analog. Deprotection of the resulting 89.4 then gave a mixture of chalcograns diastereoisomeric at the acetal ring juncture. The enantiomeric 2($\underline{S}$), 5($\underline{S}$)
and 2($\underline{S}$), 5($\underline{R}$) chalcograns were obtained by inversion of configuration
at the C-4 position and subsequent manipulation as in Scheme 89. Syn-
theses of optically active[14] and racemic[15,16] chalcogran are also available.

e. Streptolidine - Streptolidine is a unique amino acid isolated from the hydrolyzate of the antibiotic streptothricin.[17] Its structure was elucidated by chemical degradation[18] and later by X-ray crystallographic studies.[19] Syntheses of streptolidine have been
accomplished from D-ribose[20] and D-xylose[21]. The latter more expedient
route will be discussed here (Scheme 90). The target amino acid
possesses three chiral centers with a D-arabino stereochemistry and
access can be seen through the stereocontrolled introduction of amino
functions at C-2 and C-3 of an appropriate sugar derivative. The
trimesylate derivative 90.1, readily available from D-xylose was
subjected to a double-inversion process by treatment with sodium azide
in refluxing DMF. This reaction led to four products resulting from
selective displacements at C-3 and C-5, in addition to a low yield of
the desired 90.2. This is in major part due to the difficulty in
effecting S_N2-type displacements in furanoside derivatives, which is
compounded by the unfavorable disposition of a β-glycosidic substituent.
The α-anomer appears to be more favorable to such a displacement at C-2.
Having introduced the amino groups regio- and stereospecifically, there
remained to transform the 'chiron', 90.2, into the target acid by
routine manipulation. Thus, by focusing on the carbon chain rather than
the heterocyclic guanidino unit, the carbohydrate-type symmetry becomes
apparent.

Scheme 89

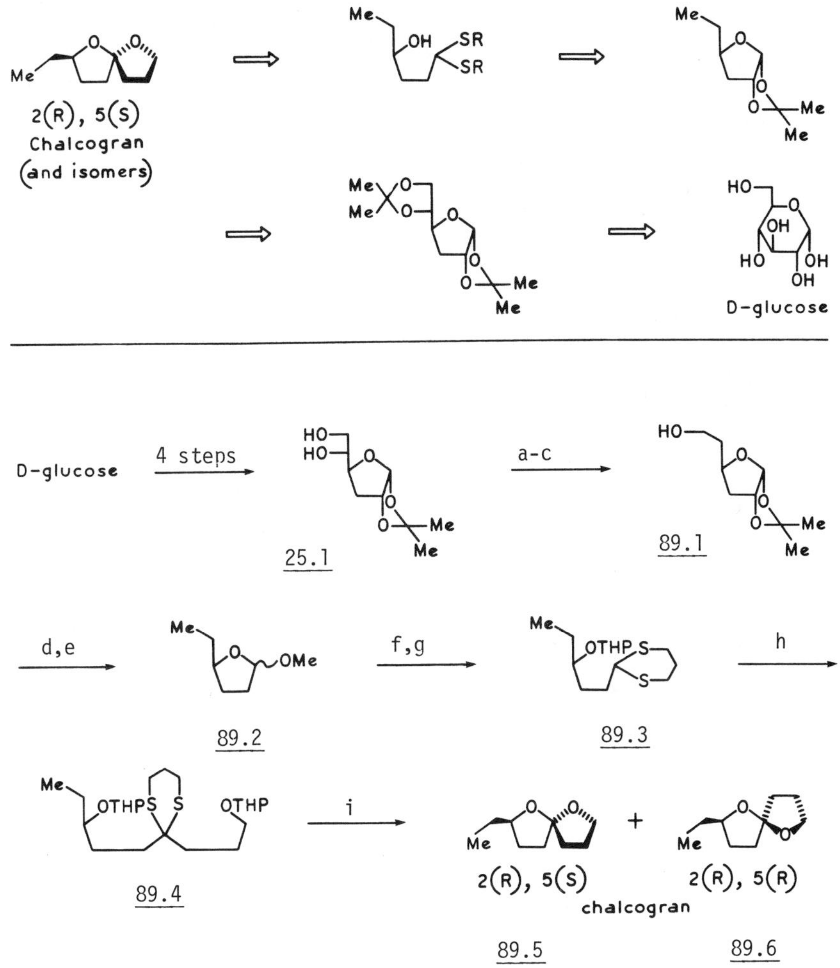

a. BzCl, pyr.; b. TsCl, pyr.; c. LAH, ether; d. MeOH, HCl; e. NaH, CS$_2$, MeI; then Bu$_3$SnH; f. HSCH$_2$CH$_2$CH$_2$SH, H$^+$; g. dihydropyran, H$^+$; h. Br(CH$_2$)$_3$OTHP, BuLi; i. HgO, collidine, HCl

Scheme 90

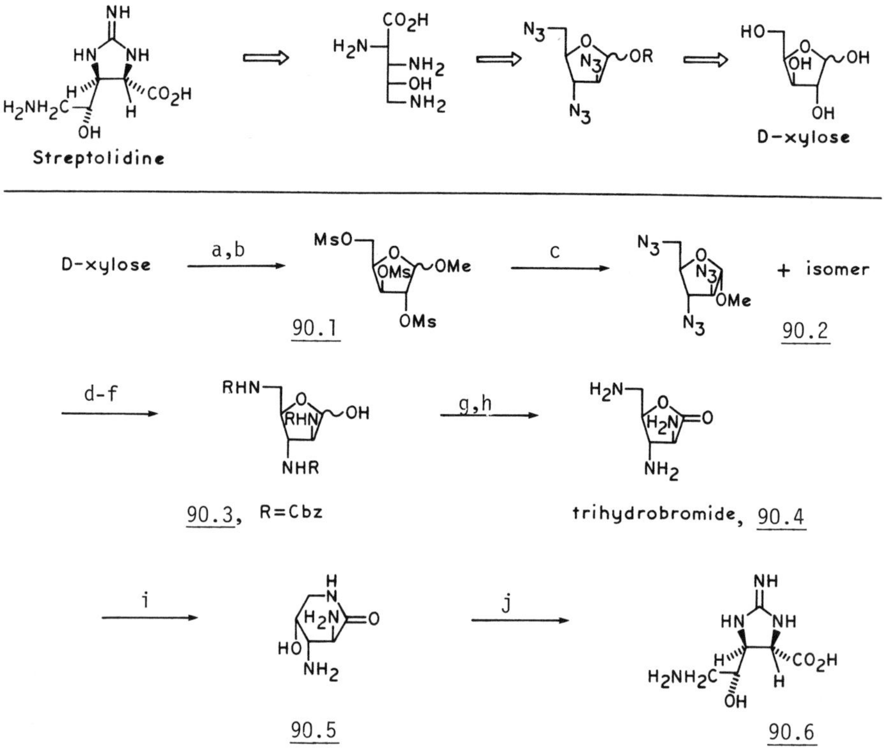

Streptolidine   ⟸   [structure]   ⟸   [structure]   ⟸   D-xylose

---

D-xylose $\xrightarrow{a,b}$ [structure] 90.1 $\xrightarrow{c}$ [structure] 90.2 + isomer

$\xrightarrow{d-f}$ [structure] 90.3, R=Cbz $\xrightarrow{g,h}$ [structure] trihydrobromide, 90.4

$\xrightarrow{i}$ [structure] 90.5 $\xrightarrow{j}$ [structure] 90.6

a. MeOH, HCl; b. MsCl, pyr.; c. NaN$_3$, DMF, reflux; d. Pd black, H$_2$, EtOH; e. CbzCl, Et$_3$N; f. aq. TsOH, diox., reflux; g. CrO$_3$, AcOH; h. HBr, AcOH, anisole; i. IR-45(OH$^-$); j. BrCN, aq. NaOH; then HCl, reflux.

   f. L-erythro-β-Hydroxyhistidine – This amino acid is a consti-
tuent of bleomycin[22,23]. A recent synthesis by Hecht and coworkers[24]
utilizes an amino sugar as starting material (Scheme 91). Examination
of the structure reveals that the stereochemistry at C-2 and C-3 is
related to 2-amino-2-deoxy-D-mannose[25] 91.1 which in turn is available
from the D-gluco analog by epimerization or from D-arabinose by chain
extension. Alternatively 2-amino-2-deoxy-D-mannono-1,4-lactone can be
used which is prepared from an oxidation-epimerization sequence from 2-
amino-2-deoxy-D-glucose. The synthesis of the target amino acid relied

Scheme 91

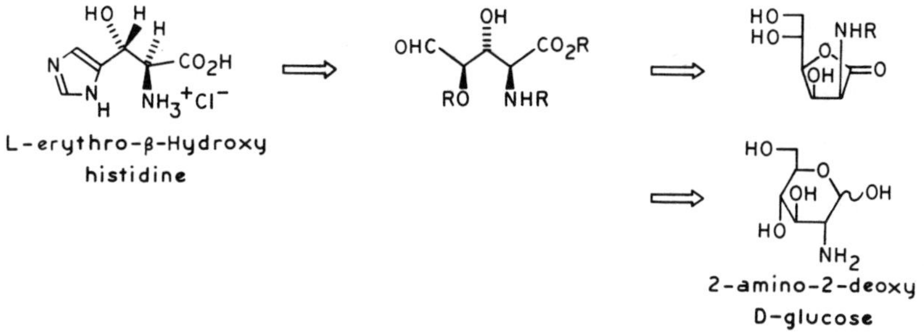
L-erythro-β-Hydroxy
histidine
2-amino-2-deoxy
D-glucose

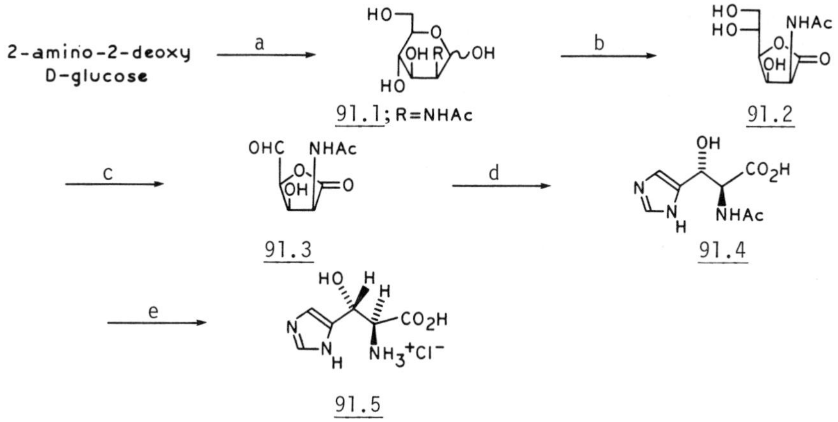
2-amino-2-deoxy
D-glucose
a
b
91.1; R=NHAc
91.2
c
d
91.3
91.4
e
91.5

a. see ref. 25 ; b. aq. Br_2; c. aq. $NaIO_4$, 0^0; d. aq. NH_4OAc, $Cu(OAc)_2$, CH_2O, 110^0; e. aq. HCl, reflux

on the generation of a five-carbon aldehydo derivative 91.3 which was condensed with formaldehyde in the presence of copper acetate and ammonium acetate to give the imidazole derivative 91.4 by a well-known precedent.[26] Acid-catalyzed hydrolysis of the N-acetyl function led to the desired amino acid 91.5 without noticeable epimerization. Another synthesis of the amino acid is based on the treatment of imidazole-4-carboxaldehyde with copper glycinate and provides a 2.5:1 mixture of the racemic <u>erythro</u> and <u>threo</u> species[27]. In the Hecht synthesis two asymmetric centers are destroyed and one carbon excised to provide the entire carbon framework of the target except for C-2 of the imidazole ring which originates from formaldehyde.

g. <u>L-erythro-Biopterin</u> - This ubiquitous, naturally occurring pteridine is involved in a variety of biological functions.[28] Scheme 92 shows that the molecule consists of a pteridine with a three carbon appendage at C-6, in which an L-<u>erythro</u> diol is situated. Taylor and Jacobi[29] described a synthesis of L-<u>erythro</u>-biopterin <u>92.6</u> starting with 5-deoxy-L-arabinose <u>92.2</u> and relying on a unique hetero-cyclization. Since the starting sugar was not available except via modification of L-arabinose, it was obtained by degradation of L-rhamnose by a well-known sequence.[30] The α-ketoaldoxime <u>92.3</u> was then condensed with benzyl α-aminocyanoacetate to give the pyrazine 1-oxide derivative <u>92.4</u>, which was transformed into the pteridone 8-oxide <u>92.5</u> and lastly to L-<u>erythro</u>-biopterin <u>92.6</u>. Thus, by virtue of the C-1/C-2 dicarbonyl system, the original 5-deoxy-L-arabinose was incorporated in the pyrazine, hence the partially-hidden carbohydrate-type symmetry of the target.

h. <u>Aglycones of the aureolic acids (Olivin, Chromomycinone, etc.)</u> - <u>Syntheses of models of the chiral side-chain</u> -   A group of cytostatic antibiotic agents known as the aureolic acids[31] are composed of closely related tetrahydroanthracenone derivatives. Representative members, olivomycin A[32] and Chromomycin A$_3$[33] are oligosaccharides of olivin and chromomycinone, the respective aglycone portions of the antibiotics (Scheme 93). Thiem and Wessel[34] have described a synthesis of a model representing the chiral C-1' — C-5' portion of chromomycinone (and olivin). Inspection of the structure in question reveals the presence of a five-carbon acyclic appendage containing three chiral centers which can be related to an ω-deoxy sugar with a D-<u>xylo</u> configuration. In one approach D-galactose was used as the chiral sugar precursor. Through a series of routine transformations a 6-deoxy derivative <u>93.3</u> was obtained which was in turn degraded to a 4-deoxy-D-threose propylene dithioacetal derivative <u>93.4</u>. This intermediate has two of the three required chiral centers. The remaining center was introduced by reaction of the dianion with benzaldehyde and separation of two epimers.

Another model in this series was described by Franck and John[35] (Scheme 94), in which the sugar portion could provide three asymmetric

Scheme 92

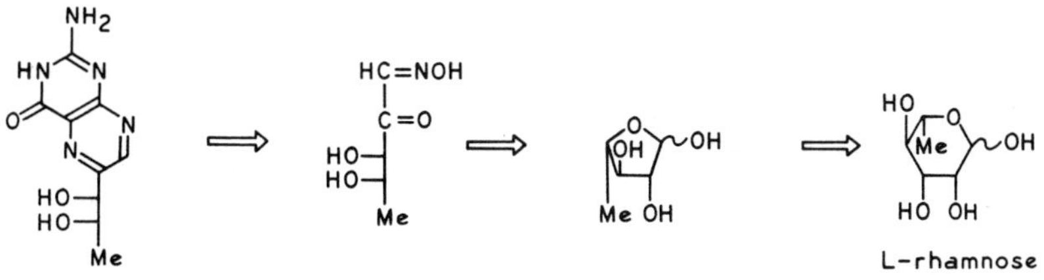

L-erythro-Biopterin

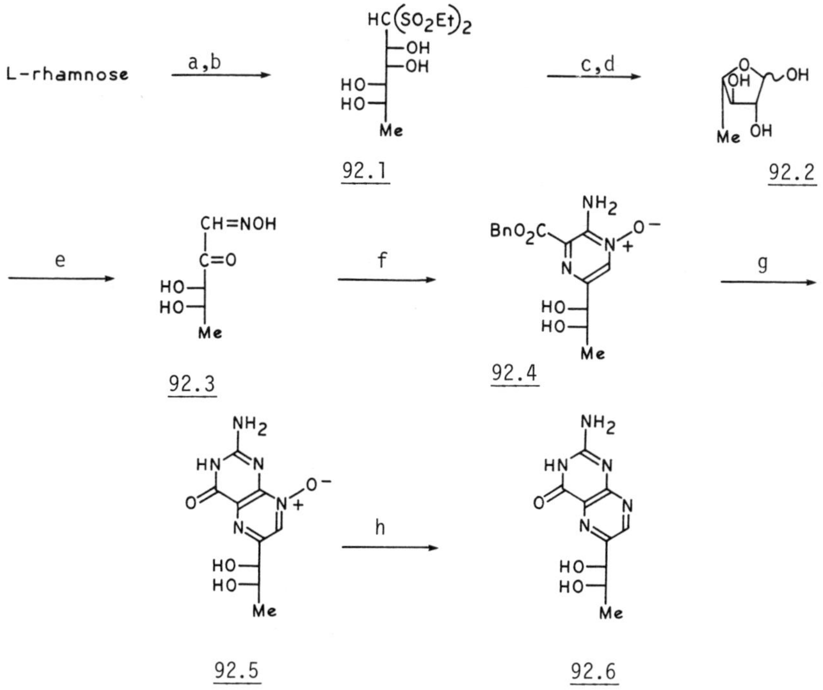

a. EtSH, HCl; b. m-CPBA, diox.; c. aq. NH$_4$OH; d. Cu(OAc)$_2$, aq. EtOH, reflux; e. acetone-oxime; f. benzyl α-aminocyanoacetate, MsOH, EtOH; g. guanidine, HCl, NaOMe, DMF, reflux; h. aq. sodium dithionate, reflux

Scheme 93

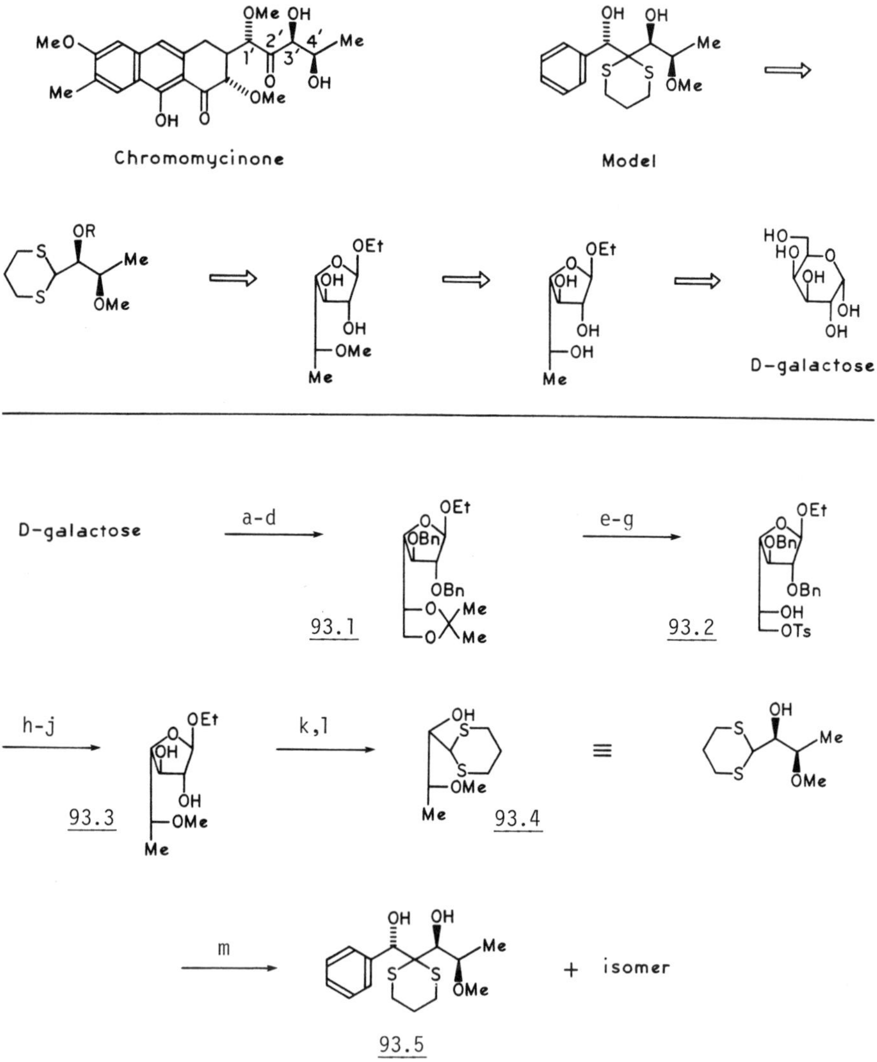

Chromomycinone

Model

D-galactose

D-galactose $\xrightarrow{a-d}$ 93.1 $\xrightarrow{e-g}$ 93.2

$\xrightarrow{h-j}$ 93.3 $\xrightarrow{k,l}$ 93.4 $\equiv$

$\xrightarrow{m}$ 93.5 + isomer

a. EtSH, HCl; b. acetone, H^+; c. BnCl, NaH, DMF; d. $HgCl_2$, HgO, EtOH;
e. H_3O^+; f. Bu_2SnO, MeOH; g. TsCl; h. $NaBH_4$, DMSO; i. MeI, Ag_2O, DMF;
j. Pd/C, H_2; k. H_5IO_6; l. $HS(CH_2)_3SH$, HCl; m. BuLi then PhCHO

Scheme 94

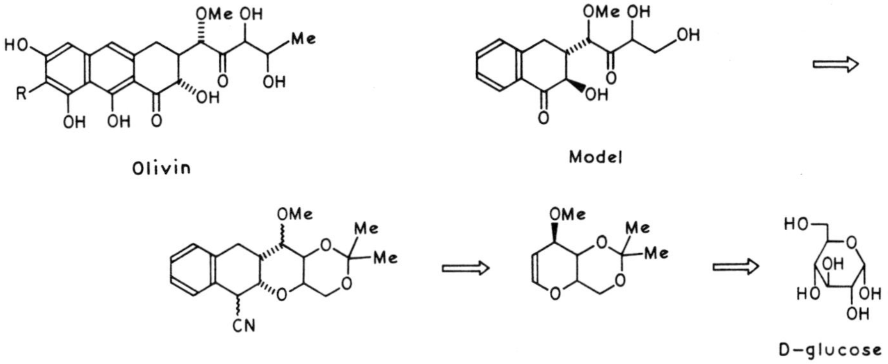

Olivin

Model

D-glucose

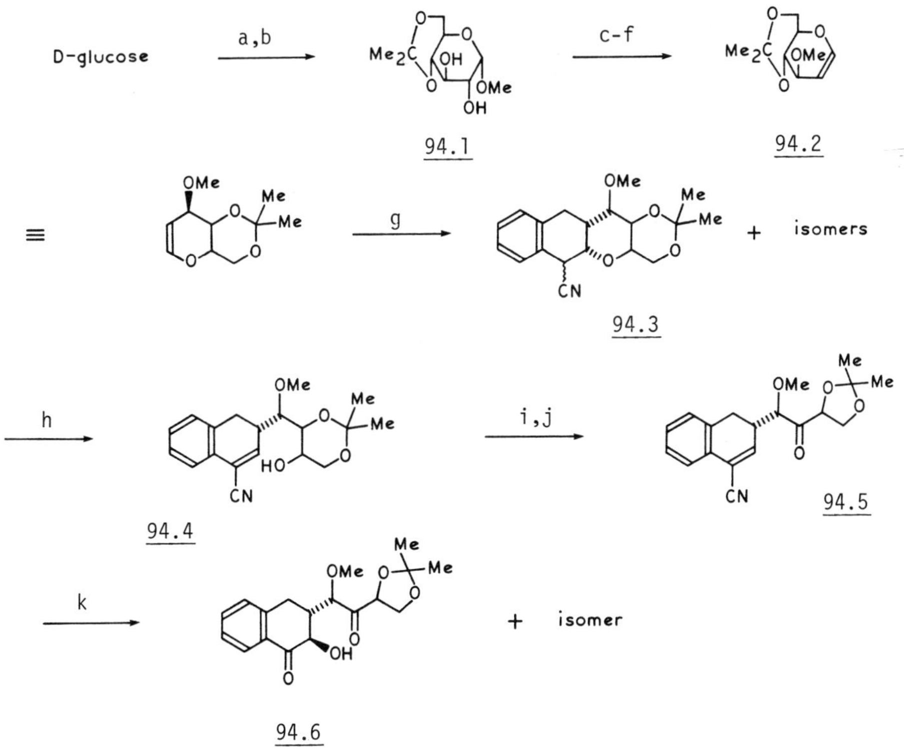

D-glucose $\xrightarrow{a,b}$ 94.1 $\xrightarrow{c-f}$ 94.2

94.1

94.2

$\equiv$ $\xrightarrow{g}$ 94.3 + isomers

94.3

$\xrightarrow{h}$ 94.4 $\xrightarrow{i,j}$ 94.5

94.4

94.5

$\xrightarrow{k}$ 94.6 + isomer

94.6

a. MeOH, H^+; b. $Me_2C(OMe)_2$, H^+; c. TsCl, pyr.; d. NaOMe; e. MeLi;
f. MeI, NaH, DMF; g. cyanobenzocyclobutene; h. MeOH, reflux;
i. acetone, H^+; j. DMSO, TFAA; k. $Ph_3P^+Me\ MnO_4^-$, $-78°$

centers (C-1', C-2' and C-3') in addition to two rings in the aglycone
portion. The approach depends on a Diels-Alder reaction using the readily available cyclic enol ether derivative 94.2. Reaction of cyanobenzocyclobutene (used as a quinone methide precursor) with 94.2 led to a
mixture of isomers, from which the desired isomer 94.3 could be isolated
by chromatography, in 14% yield. Elaboration of this intermediate as
shown in the scheme, gave the model target 94.6 in which C-2 and C-1 are
of the opposite configuration relative to the natural product. Using a
6-deoxy-D-galactose derivative corresponding to 94.2 would lead to the
correct absolute stereochemistry. In this case, two extra carbon atoms
(C-4' and C-5') would have to be added to the chain, including a new
asymmetric center. Synthetic studies toward the aureolic acids have
also been described by Weinreb and collaborators.[36]

i. The Ezomycins and Octosyl Acids - Synthesis of model ring
systems - The ezomycins are a group of antifungal antibiotics produced
by a strain of Streptomyces.[37] Their constitutional structures including
absolute stereochemistry have been established[38] and the structure of
the A_2 component is shown in Scheme 95. The octosyl acids isolated from
a strain of Streptomyces that produces the polyoxins[39] are structurally
similar to the ezomycins, except that they are constitutionally
simpler. Of interest is the presence of a bicyclic ring structure in
the ezomycins and octosyl acids which can be related to a *trans*-fused
perhydrofuropyran. Added challenges are the high level of functionalization including the presence of a pyrimidine base having a 1,2-*trans* -
relationship. Thus, these substances can be regarded as bicyclic fused
β-D-ribofuranosyl pyrimidine nucleosides derived from a 3,7-anhydro-
octose. The model ring system as well as a model bicyclic uracil nucleoside have been synthesized[40,41] starting with D-galactose, which provides the requisite substitution pattern for ultimate elaboration into
the ezomycin as well as octosyl acid series. Thus, if the ring system
containing the pyrimidine portion can be constructed, there would remain
to manipulate the hydroxyl groups in the pyranose portion to achieve a
synthesis of the targets in question. Scheme 96 shows how D-galactose
was transformed into two model derivatives 96.5 and 96.10. Preferential
substitution via the orthoester intermediate 96.1 gave the chloride 96.2
which after treatment with vinyl magnesium chloride led to the 1,2-
C-glycoside derivative 96.3. The critical perhydrofuropyran system was

Scheme 95

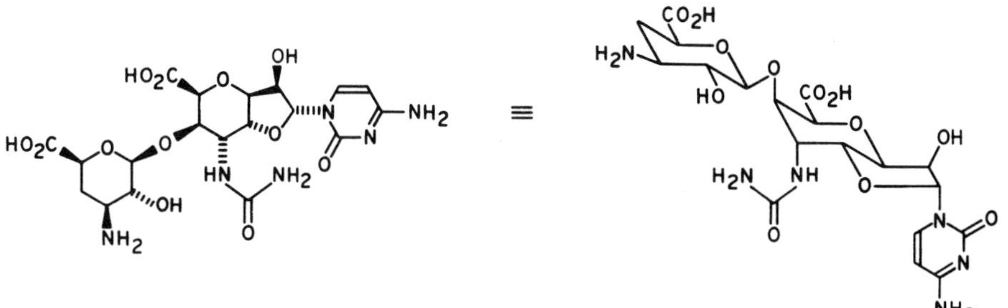

Ezomycin A$_2$

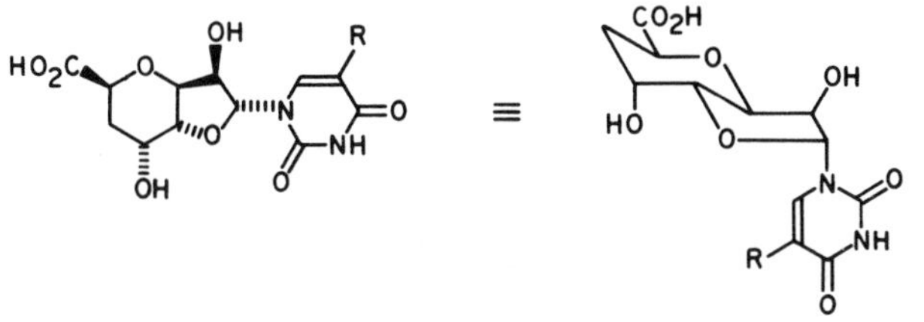

R = CO$_2$H, octosyl acid A
CH$_2$OH, octosyl acid B
H, octosyl acid C

constructed by stereoselective epoxidation and acid-catalyzed cyclization to give 96.5. For the synthesis of the 2'-deoxy nucleoside derivative, 96.10, the allyl derivative 96.6 was converted to the aldehyde
96.8, then to the dithioacetal 96.9, which was condensed with N-acetylcytosine mercury to give a 1:1 mixture of 96.10 and the epimer. The
anomeric configuration was established by spectroscopic studies and correlation with the corresponding 2'-deoxyadenosine analog previously prepared and of established structure.[41]   A bicyclic *trans*-fused structure,
derived from the intramolecular cyclization of a 3-malonyl ester in a
furanose derivative has been reported[42] in an attempt to construct the
octosyl acid skeleton. A recent report by Szarek and coworkers describes
another approach.[43]

In connection with the above studies (Scheme 96) it was also possible to synthesize a bicyclic *trans*-fused α-methylene lactone 96.7 from
the corresponding allyl C-glycoside by oxidative cleavage, lactonization

Scheme 96

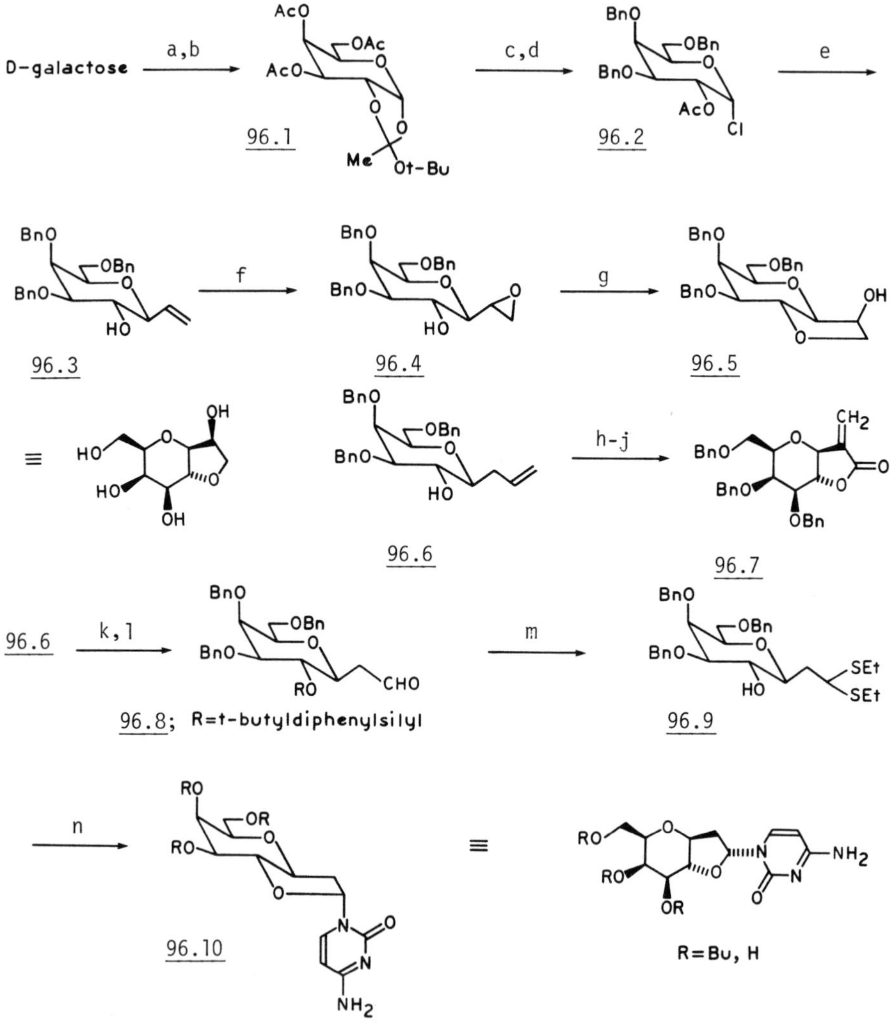

a. Ac$_2$O, pyr.; b. AlCl$_3$, t-BuOH; c. NaOMe, MeOH; d. BnBr, NaH, DMF;
e. H$_2$C=CHMgBr, CH$_2$Cl$_2$; f. m-CPBA, CH$_2$Cl$_2$; g. CSA, CH$_2$Cl$_2$, reflux;
h. KMnO$_4$, NaIO$_4$, then CH$_2$N$_2$; i. CSA, benzene reflux; j. Me$_2$N$^+$=CH$_2$I$^-$,
then base; k. OsO$_4$, NaIO$_4$ or O$_3$, CH$_2$Cl$_2$; l. Ph$_2$t-BuSiCl, imidazole,
DMF; m. EtSH, H$^+$; n. N-acetylcytosine mercury.

and α-methylenation.[40]   Access to such terpenoid-like structures from
carbohydrates via this route is therefore relatively easy.

j. (+)-Spectinomycin – This aminocyclitol antibiotic[44] is endowed
with a unique structure and an intriguing biological profile. It has
been isolated from various Streptomyces[45] and its structure has been
elucidated by chemical[46,47] and X-ray crystallographic studies.[48] It
forms a crystalline dihydrochloride hydrate which can be transformed to
the keto form by azeotropic dehydration.

Examination of the structure of spectinomycin (Scheme 97) reveals
the presence of a 2-hydroxy-1,4-dioxane central ring formed as a result
of the acetal type linkages between actinamine, the aminocyclitol
portion, and a highly oxidized 2-hydroxy dihydropyran derivative(spectinose). Shown in three perspectives, it becomes evident that the antibiotic could exist in equilibrium with the "open" hydroxy-ketone form.
Since the intramolecular acetalization is highly stereoselective involving one of two diastereotopic hydroxyl groups on the actinamine
ring,[49,50] the main challenge in the synthesis of spectinomycin centers
around the generation of a precursor to such a keto intermediate from an
appropriately masked form of spectinose and actinamine by a regio- and
stereocontrolled glycosylation reaction. These criteria were the basic
tenets in two independently conceived and operationally different syntheses of spectinomycin.[49,51]

Scheme 98 illustrates the basis of the approach by the Upjohn
group[51] which starts with L-glucose and generates a 2-oximino-α-L-glyco-
side of actinamine. This is further elaborated to spectinomycin via a
unique set of transformations. The actual synthesis is shown in Scheme
99. The choice of L-glucose ensured the formation of the desired α-L-
linkage (corresponds to β-D in the target) by taking advantage of the
nitrosyl chloride glycosylation method of Lemieux.[52]   Intramolecular
cyclization of the product 99.3 was both efficient and stereocontrolled
to give the bicyclic spectinomycin skeleton 99.4. At this point, all
that remained was to deoxygenate at C-4 and C-6 (sugar numbering), to
invert the stereochemistry at C-5 and to generate the keto function at
C-4 (spectinomycin numbering). These transformations were achieved in a
remarkable series of reactions. Thus, base treatment of 99.4 gave the
enone derivative 99.5, which upon hydrolysis and reduction gave
spectinomycin characterized as its crystalline dihydrochloride salt.

Scheme 97

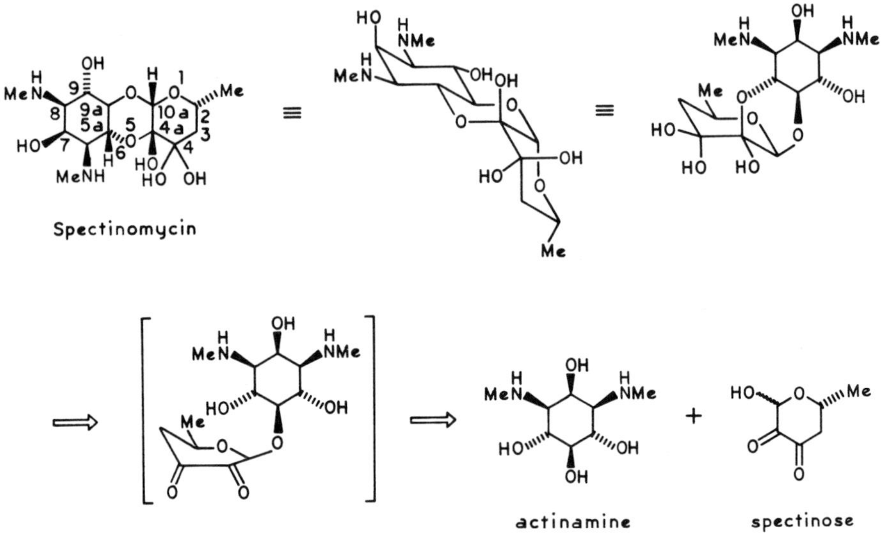

Spectinomycin

actinamine   spectinose

In another approach,[49] the more readily available D-glucose was
used as the chiral precursor (Scheme 100). Here, the idea was to
specifically generate 4a(R), 4(R) tetrahydrospectinomycin, 4(R)-dihydro-
spectinomycin, and ultimatly the target itself by systematic manipulation of hydroxyl groups in specific intermediates. Since there is
evidence that 4(R)-dihydrospectinomycin is a biosynthetic precursor of
spectinomycin,[53] the synthesis could also be considered to be partly a
biomimetic one. Details of the synthesis are given in Scheme 101, in
which D-glucose was transformed into the 4,6-dideoxy analog in an
efficient two-step process[11] (see also Scheme 88). Inversion of
configuration at C-3 in 88.2 was easily achieved by regioselective
O-benzoylation of the 2-O-tri-n-butyltin ether according to literature
procedures,[54] followed by oxidation and reduction. Note that reduction
is highly stereoselective because of the α-orientation of the anomeric
substituent. Very efficient regio- and stereocontrolled glycoside
synthesis was achieved using silver triflate in tetrahydrofuran, a
procedure[55] found useful in several other instances.[56] The 4a(R),
4(R)-tetrahydrospectinomycin derivative 101.5 thus obtained was

Scheme 98

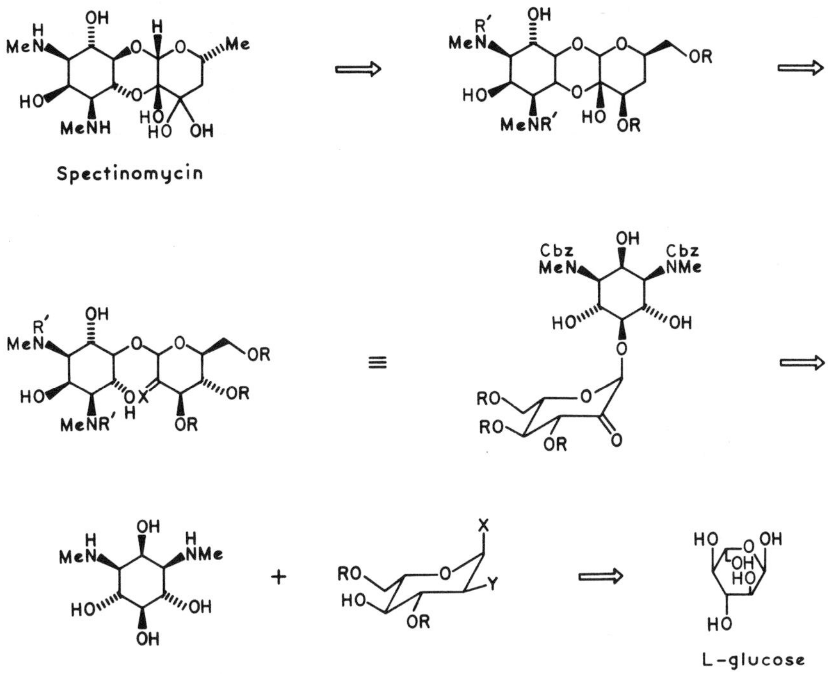

Spectinomycin

L-glucose

transformed into the orthoester (or orthoamide)derivative 101.6,
which, when subjected to mild acid treatment led to a preferentially
substituted derivative 101.7. It is therefore evident why the original
plan had called for the generation of the 4a(R), 4(R)-dihydro derivative
with a *cis*-orientation of the vicinal diol. Thus, the cleavage of the
orthoester to provide the axial-acetate/equatorial alcohol as in 101.7
was anticipated under kinetically controlled conditions. Also of
interest was the protection of the hydroxyl groups in the actinamine
portion as trichloroethylcarbonate esters (rather than acetates) as
originally reported.[49] Oxidation of 101.7 was effected with pyridinium
chlorochromate in refluxing benzene[57] to give the desired keto
derivative 101.8. Treatment with zinc and acetic acid led directly to
the 4(R)-dihydrospectinomycin derivative 101.9 demonstrating the highly
diastereoselective acetal formation. There remained to oxidize this

Scheme 99

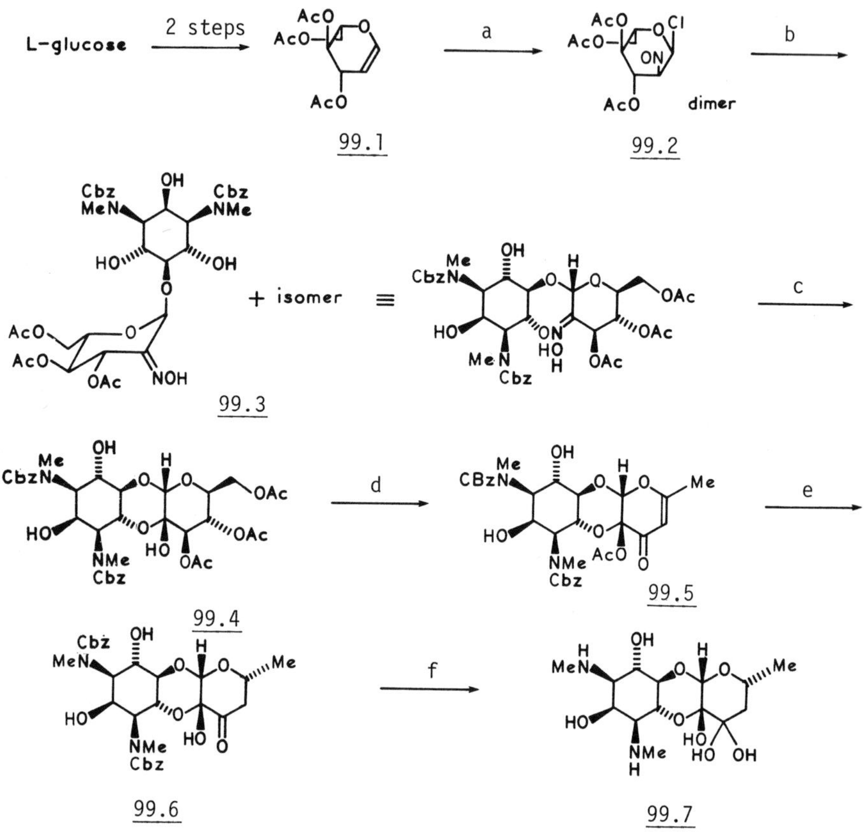

a. NOCl; b. Cbz-actinamine, DMF; c. MeCHO, MeCN, aq. HCl; d. KHCO$_3$, MeCN; e. K$_2$HPO$_4$, MeOH; f. Pd/BaSO$_4$, pyr., H$_2$

intermediate to N,N-dibenzyloxycarbonyl spectinomycin 101.11 and to remove the N-protecting group to complete the synthesis. Note that there are four hydroxyl groups and the choice of a selective oxidant becomes critical. Once again, organotin chemistry proved extremely useful, since oxidation[58] of the cyclic stannylidene derivative 101.10 with bromine under controlled conditions was predictably regiospecific and gave the desired spectinomycin derivative 101.11 in high yield. Again, the choice of the 4(R)-dihydro derivative as precursor was

evident since it readily formed the required organotin acetal.
Hydrogenolysis then led to crystalline (+)-spectinomycin
dihydrochloride. It should be pointed out that N,N-dibenzyloxycarbonyl
spectinomycin is susceptible to an acid-catalyzed tertiary ketol
rearrangement.[59] A similar rearrangement can be brought about in

Scheme 100

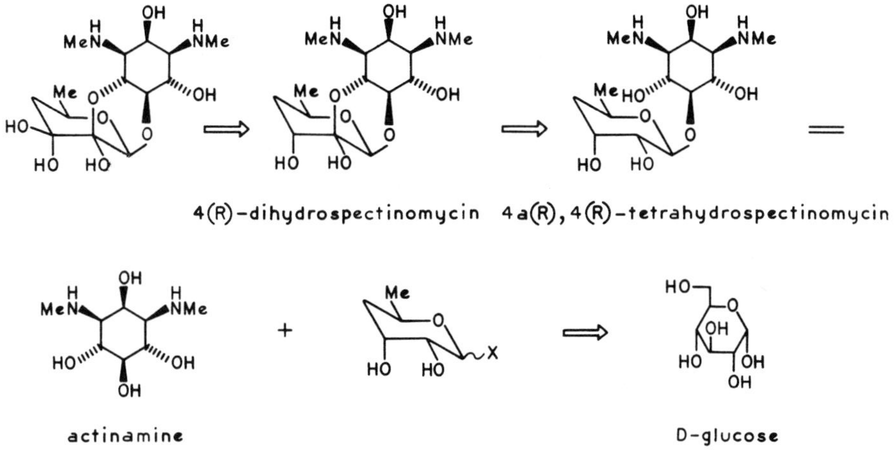

4(R)-dihydrospectinomycin   4a(R),4(R)-tetrahydrospectinomycin

actinamine   D-glucose

a mixture of bis-tributyltin oxide and bromine.[59] A synthesis of
4a(R),4(S)-tetrahydrospectinomycin in low yield has also been reported
by Suami and coworkers[60] by glycosylation of actinamine with an appropriate sugar precursor. Application of the silver triflate procedure
led to a 67% yield of this derivative.[49] It should be noted that these
glycosylations (see also Scheme 101) are highly regioselective and it
was not necessary to protect the hydroxyl groups in actinamine, the
desired hydroxyl group being the least hindered. Actinamine has been
previously synthesized by several groups,[61] and is readily available
from acid hydrolysis of the antibiotic.[47] With regard to other
synthetic approaches to spectinomycin and related compounds, the
extensive studies reported on carbohydrate enolones may be pertinent.[62]

   k. (-)-Sarracenin - A synthesis of the irridoid monoterpene (-)-
sarracenin[63] has been reported starting from D-glucose as a

Scheme 101

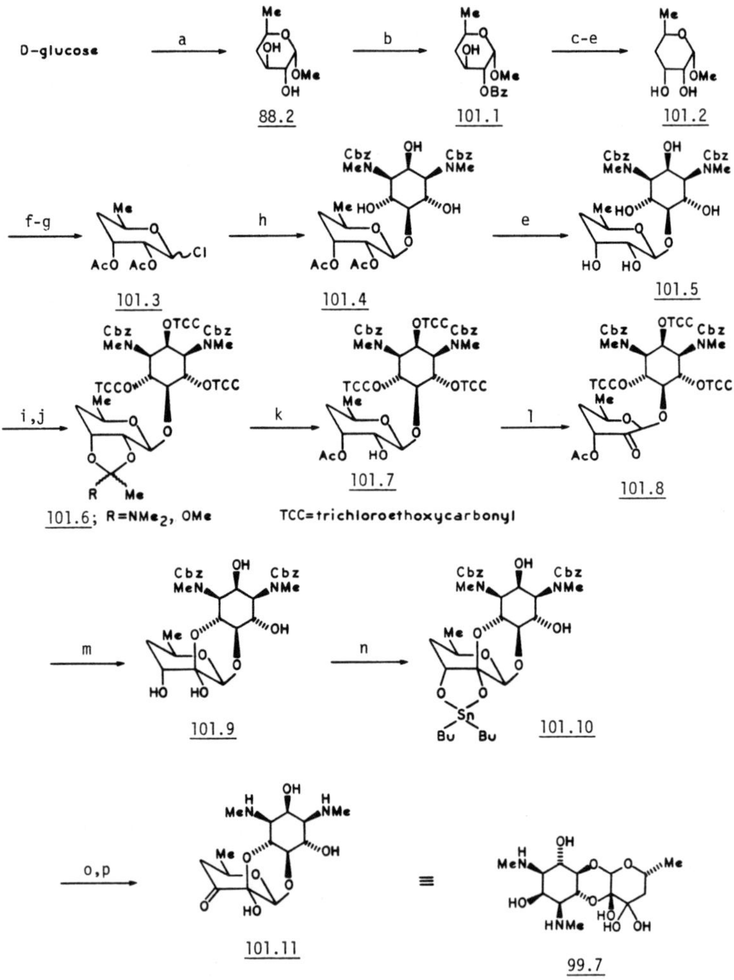

a. see Scheme 88; b. Bu₂SnO, MeOH, reflux; then BzCl, c. PCC; d. NaBH₄; e. NaOMe, MeOH; f. Ac₂O, BF₃.Et₂O; g. HCl, ether; h. Cbz-actinamine, Ag triflate, THF, -40°; i. MeC(OMe)₃, TsOH, benzene; j. ClCO₂CH₂CCl₃, pyr.; DMAP; k. aq. AcOH; l. PCC, benzene, reflux; m. Zn, MeOH; n. (Bu₃Sn)₂O, MeOH, reflux; o. NBS, benzene; p. Pd/C, H₂, 2-PrOH

"chiral template".[64]   Other syntheses from non-carbohydrate precursors
are also available.[65-67]

REFERENCES

1. Y. Kimura, K. Katagiri and S. Tamura, Tetrahedron Lett., 3137
    (1971); Agr. Biol. Chem., 36, 1925 (1972).

2. G. A. Ellestad, W.J. McGahren and M. P. Kunstmann, J. Org. Chem.,
    37, 2045 (1972).

3. K. Mori, M. Oda and M. Matsui, Tetrahedron Lett. 3173 (1976).

4. R.M. Carlson, A.R. Oyler and J.R. Peterson, J. Org. Chem., 40,
    1610 (1975).

5. R.M. Carlson and A.R. Oyler, Tetrahedron Lett., 2615 (1974);
    D.Seebach and H. Meyer, Angew. Chem., Int. Ed. Engl. 13, 77
    (1974);

6. P.E. Sum and L. Weiler, Can. J. Chem., 60, 327 (1982); D.E.
    Plaumann, B.J. Fitzsimmons, B.M. Ritchie and B. Fraser-Reid, J.
    Org. Chem., 47, 941 (1982).

7. K. Mori, Tetrahedron, 32, 1979 (1976).

8. J. Bauer and J.A. Vité, Naturwiss., 62, 539 (1975); J.P. Vité and
    A. Bakke, Naturwiss., 66, 528 (1979).

9. H. Gerlach and P. Künzler, Helv. Chim. Acta, 60, 638 (1977).; V.
    Heemann and W. Francke, Naturwiss., 63, 344 (1976).

10. H. Redlich, B. Schneider and W. Francke, Tetrahedron Lett., 21,
    3009 (1980).

11. H. Parolis, W.A. Szarek and J.K.N. Jones, Carbohydr. Res., 19, 97
    (1971) and references cited therein; see also H. Paulsen, B.
    Sumfleth and H. Redlich, Chem. Ber., 109, 1362 (1976).

12. H. Redlich, B. Schneider and W. Francke, Tetrahedron Lett., 21,
    3013 (1980).

13. W. Francke, V. Heemann, B. Gerken, J.A.A. Reuwick and J. P. Vité,
    Naturwiss., 64, 590 (1977).

14. H. Redlich and W. Francke, Angew. Chem. Int. Ed. Engl., 19, 630
    (1980).

15. L. R. Smith, H.J. Williams and R.M. Silverstein, Tetrahedron
    Lett., 323 (1978); K. Mori, M. Sasaki, S. Tamada, T. Suguro and
    S. Matsuda, Tetrahedron, 35, 1601 (1979).

16. R.E. Ireland and D. Häbich, Tetrahedron Lett., 21, 1389 (1980).

17. H.E. Carter, R.K. Clark, P. Kohn, J.M. Rothrock, W.R. Taylor,

C.A. West, G.B. Whitfield and W.G. Jackson, J. Am. Chem. Soc.,
76. 566 (1954); see also K. Nakanishi, T. Ito and Y. Hirata,
ibid., 76, 2845 (1954);   H. Brockmann and H. Musso, Chem. Ber.,
88, 648 (1955).

18. H.E. Carter, C.C. Sweeley, E.E. Daniels, J.E. McNary, C.P.
    Schaffner, C.A. West, E.E. van Tamelen, J.R. Dyer, and H. A. Whaley,
    J.Am. Chem. Soc., 83, 4296 (1961).

19. B.W. Bycroft and T.J. King, J. Chem. Soc., Chem. Commun., 652,
    (1972).

20. S. Kusumoto, S. Tsuji and T. Shiba, Bull. Chem. Soc. Japan, 47,
    2690 (1974).

21. S. Kusumoto, S. Tsuji, K. Shima and T. Shiba, Bull. Chem. Soc.
    Japan, 49, 3611 (1976).

22. T. Takita, Y. Muraoka, T. Nakatani, A. Fuji, Y. Umezawa, H.
    Nuganawa and H. Umezawa, J. Antibiotics, 31, 801 (1978) and
    references cited therein; for recent review, see H. Umezawa, in
    Anticancer Agents Based on Natural Product Models , J.M. Cassady
    and J.D. Douros, eds. Medicinal Chemistry Series Monographs, vol.16,
    1980, p.148, Academic Press, N.Y..

23. For recent total syntheses, see: T. Takita, Y. Umezawa, S. Saito,
    H. Morishima, H. Naganawa, H. Umezawa, T. Tsuchiya, T. Miyake, S.
    Kageyama, S. Umezawa, Y. Muraoka, M. Suzuki, M. Otsuka, M. Narika,
    S. Kobayashi and M. Ohno, Tetrahedron Lett., 23, 521 (1982); Y.
    Aoyagi, K. Katano, H. Suguna, I. Primeau, L. Chang and S.M. Hecht,
    J. Am. Chem. Soc., 104, 5537 (1982).

24. S.M. Hecht, K.M. Rupprecht and P.M. Jacobs, J. Am. Chem. Soc.,
    101, 3982 (1979); for the synthesis of another amino acid
    constituent of bleomycin from L-rhamnose, see T. Ohgi and S.M.
    Hecht, J. Org. Chem., 46, 1232 (1981).

25. J.C. Sowden and M.L. Oftedahl, J. Am. Chem. Soc., 82, 2303 (1960);
    see also C.T. Spivak and S. Roseman, ibid., 81, 2403 (1959).

26. M. Henze, Hoppe-Zeyler, Z. Physiol. Chem., 198, 82 (1931); M.
    Henze, ibid., 200, 232 (1931); J.K. Hamilton and F. Smith, J. Am.
    Chem. Soc., 74, 5162 (1952).

27. T. Takita, T. Yoshioka, Y. Muraoka, K. Maeda and H. Umezawa, J.
    Antibiotics, 24, 795 (1971).

28. See for example, H. Rembold and W.L. Gyure, Angew, Chem., Ind.
    Ed. Engl., 11, 1061 (1972); W. Pfleiderer, ibid., 3, 114 (1964).

29. E.C. Taylor and P.A. Jacobi, J. Am. Chem. Soc., 98, 2301 (1976).

For a synthesis of englenapterin, a pteridine derivative related
to biopterin, see P.A. Jakobi, M. Martinelli and E.C. Taylor, J.
Org. Chem., 46, 5416 (1981).

30. L. Hough and T.J. Taylor, J. Chem. Soc., 3544 (1955) and
    references cited therein.

31. See for example, W.A. Remers, in The Chemistry of Antitumor
    Antibiotics; Wiley & Son, New York, N.Y. 1979, p. 139.

32. Y.A. Berlin, S.F. Esipov, M.N. Kolosov and M.M. Shemyakin,
    Tetrahedron Lett., 1431, 1643 (1966).

33. N. Narada, K. Nakanishi and S. Tatsuoka, J. Am. Chem. Soc., 91,
    5896 (1969).

34. J. Thiem and H.P. Wessel, Ann., 2216 (1981); Tetrahedron Lett.,
    21, 3571 (1980).

35. R.W. Franck and T.V. John, J. Org. Chem., 45, 1170 (1980).

36. R.P. Hatch, J. Shringapure and S.M. Weinreb, J. Org. Chem., 43,
    4172 (1978); J.H. Dodd and S.M. Weinreb, Tetrahedron Lett., 3593
    1979).

37. K. Sakata, A. Sakurai and S. Tamura, Agr. Biol. Chem., 37, 697
    (1973).

38. K. Sakata, A. Sakurai and S. Tamura, Agr. Biol. Chem., 38, 1883
    (1974); 39, 885 (1975); S. Sakata, A. Sakurai and S. Tamura,
    Tetrahedron Lett., 4327 (1974); K. Sakata, A. Sakurai and S.
    Tamura, ibid., 3191 (1975).

39. K. Isono, P.F. Crain and J.A. McCloskey, J. Am. Chem. Soc., 97,
    943 (1975).

40. S. Hanessian, T.J. Liak and D.M. Dixit, Carbohydr. Res., 88, C14
    (1981).

41. S. Hanessian, D.M. Dixit and T.J. Liak, Pure & Appl. Chem., 53, 129
    (1981).

42. K. Anzai and T. Saita, Bull. Chem. Soc., Japan, 50, 169 (1977).

43. K.S. Kim and W.A. Szarek, Can. J. Chem., 59, 878 (1981).

44. For a recent review, see, W. Rosenbrook, Jr., J. Antibiotics, 32,
    Suppl., S211 (1979).

45. D.J. Mason, A. Dietz and R.M. Smith, Antibiot. Chem., 11, 118
    (1961).

46. P.F. Wiley, A.D. Argoudelis and H. Hoeksema, J. Am. Chem. Soc.,
    85, 2652 (1963).

47. A.L. Johnson, R.H. Gourlay, D.S. Tarbell and R.L. Autrey, J. Org.
    Chem., 28, 300 (1967).

48. T.G. Cochran, D.J. Abraham and L.L. Martin, J. Chem. Soc., Chem. Commun., 494 (1972).

49. S. Hanessian and R. Roy, J. Am. Chem. Soc., 101, 5839 (1979).

50. S. Hanessian and R. Roy, J. Antibiotics, 32, Suppl. S-73 (1980); see also L. Foley and M. Weigele, J. Org. Chem., 43, 4355 (1978).

51. D.R. White, R.D. Birkenmeyer, R.C. Thomas. S. L. Mizsak and V.H. Wiley, Tetrahedron Lett., 2737 (1979).

52. R.U. Lemieux, T.L. Nagabushan and I.K. O'Neill, Can. J. Chem., 43, 413 (1968).

53. H. Hoeksema and J.C. Knight, J. Antibiotics, 28, 240 (1975).

54. See for example, D. Wagner, J.P.H. Verheyden and J.G. Moffatt, J.Org. Chem., 39, 24 (1974); R.M. Munavu and H.H. Szmant, ibid., 41, 1832 (1976); T. Ogawa and M. Matsui, Carbohydr. Res., 56, C1 (1977); C. Augé, S. David and A. Veyrières, J. Chem. Soc. Chem. Commun., 375 (1976); M.A. Nashed and L. Anderson, Tetrahedron Lett., 3503 (1976).

55. S. Hanessian and J. Banoub, Carbohydr. Res., 53, C13 (1977); Adv. Chem. Ser., 39, 36 (1976); see also F.J. Kronzer and C. Schuerch, Carbohydr. Res., 27, 379 (1973); R.U. Lemieux and H. Driguez, J. Am. Chem. Soc., 97, 4069 (1975).

56. F. Arcamone, S. Penco, S. Redaelli and S. Hanessian, J. Med. Chem., 19, 1424 (1976).

57. D.H. Hollenberg, R. S. Klein and J.J. Fox, Carbohydr. Res., 67, 491 (1978).

58. For examples of oxidation of vicinal diols via stannylidene acetals, see, Y. Ueno and M. Okawasa, Tetrahedron Lett., 4597 (1976); S. David and A. Thieffrey, J. Chem. Soc., Perkin I., 1568 (1979) and previous papers.

59. S. Hanessian and R. Roy, Tetrahedron Lett., 22, 1005 (1981).

60. T. Suami, S. Nishiyama, H. Ishikawa, H. Okada and T. Kinoshita, Bull. Chem. Soc. Japan, 50, 2754 (1977).

61. See for example, M. Nakajima, N. Kurihara, A. Hasegawa, and T. Kurokawa, Ann., 689, 243 (1965); F. Lichtenthaler, H. Leinert and T. Suami, Chem. Ber., 100, 2383 (1967).

62. F. Lichtenthaler, Pure Appl. Chem., 50, 1343 (1978).

63. D.H. Miles, U. Kokpol, J. Bhattacharya, T.L. Atwood, K.E. Stone, T.A. Bryson and C. Wilson, J. Am. Chem. Soc., 98, 1569 (1976).

64. S. Takano, K. Morikawa and S. Hatakeyama, Tetrahedron Lett., 24, 401 (1983).

65. J.K. Whitesell, R.S. Matthews, M.A. Minton and A.M. Helbing, J.
    Org. Chem., 43, 784 (1973); J. Am. Chem. Soc., 103, 3468 (1981).

66. S.W. Baldwin and M.T. Crimmins, J. Am. Chem. Soc., 104, 1132 (1982).

67. L.-F. Tietze, K.-H. Glüsenkamp, M. Nakane and C.R. Hutchinson,
    Angew. Chem. Int. Ed. Engl., 21, 70 (1982).

PART SIX

EXECUTION

Molecules Containing Hidden Carbohydrate-type
Symmetry

CHAPTER THIRTEEN

CARBOCYCLIC MOLECULES

a. <u>Prostaglandin A$_2$ (PGA$_2$)</u> – The total synthesis of PGA$_2$ is
the first example of the synthesis of a naturally occurring prostaglandin[1] from a sugar,[2] which is a source of C-14 — C-16 in the target
(Scheme 102). The carbohydrate precursor in this case is L-erythrose
which was used to construct the C-14 — C-17 segment of the target harbo-
ring an asymmetric center at C-15. An added feature which is of pivotal
importance and reflects the elegance of the conception is use of the L-
erythrose framework as a template to "transfer" chirality, and create
the E-olefin. The required starting material, 2,3-0-isopropylidene-L-
erythrose <u>102.1</u>[3] was obtained from the readily available L-rhammose.
Homologation to the vinyl carbinol <u>102.2</u> and Claisen rearrangement gave
the *trans*-olefinic derivative <u>102.3</u>. Manipulation of the hydroxyl
groups gave <u>102.4</u> which was ideally suited for a second Claisen rearrangement involving a transfer of chirality. Indeed, the entire "acid"
chain was introduced in this manner to give <u>102.6</u>. Note that the
necessary framework for cyclopentane ring formation is present in this
intermediate. Completion of the synthesis involved selective reduction
of the triple bond, ring formation via a phenylselenyl derivative, and
chain-elongation to give the C$_5$H$_{11}$ appendage. Thus, the original
C-3 hydroxyl group in L-erythrose was maintained, but the C-2 hydroxyl
group was sacrificed in order to introduce the C-8 — C-12 bond stereospecifically via the Claisen rearrangement. All four carbon atoms of
the starting sugar were incorporated in the target and the need for
optical resolution was obviated. Other syntheses of PGA$_2$ are also
reported in the literature.[1,4]

Scheme 102

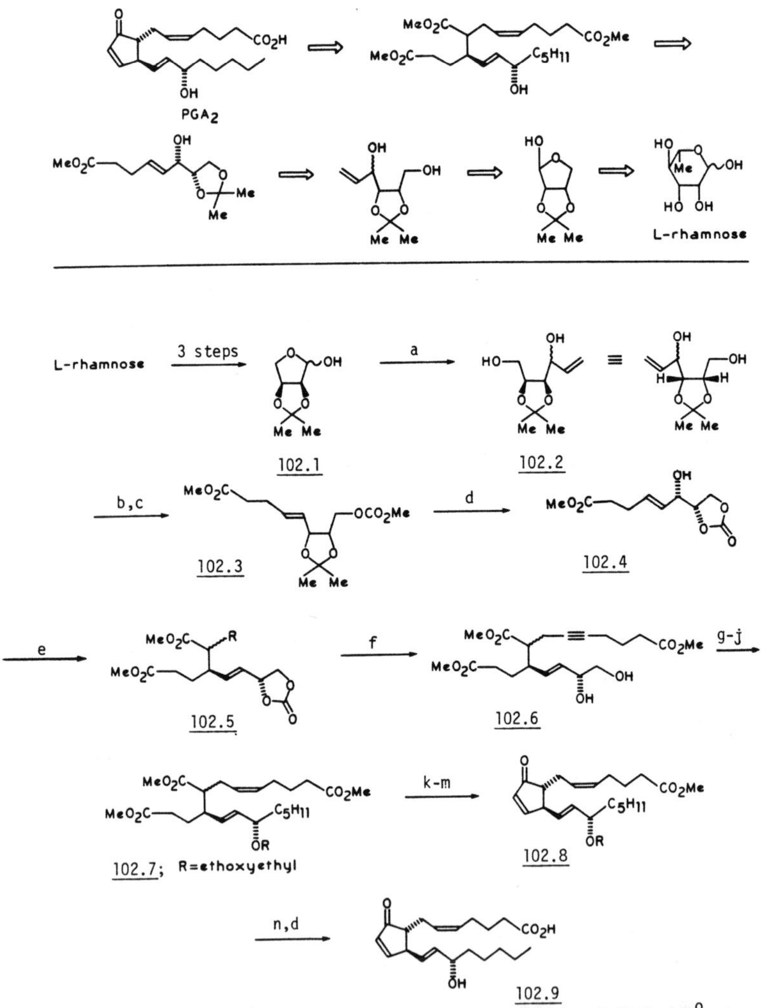

PGA$_2$

L-rhamnose

102.1

102.2

102.3

102.4

102.5

102.6

102.7; R=ethoxyethyl

102.8

102.9

a. CH$_2$=CHMgCl, THF, CH$_2$Cl$_2$; b. ClCO$_2$Me, pyr.; c. MeC(OMe)$_3$, PrCO$_2$H, 140°; d. aq. AcOH,
then Et$_3$N; e. Claisen with (MeO)$_3$C(CH$_2$)$_2$-C≡C-(CH$_2$)$_3$CO$_2$Me; xylene, reflux; f. K$_2$CO$_3$.
MeOH; g. Pd/BaSO$_4$, H$_2$; h. TsCl, pyr.; i. ethylvinyl ether; j. Bu$_2$CuLi, ether;
k.   t-BuOK, THF; then NaOH; l. LDA, THF, then PhSeCl; m. NaIO$_4$, then base; n. base

b. Prostaglandin E$_2$ (PGE$_2$) — The strategy of "chirality
transfer" from an optically active "template" was pursued by Stork and
Takahashi[5] in their synthesis of another naturally-occurring prostaglandin, namely, PGE$_2$ (Scheme 103, see also Scheme 9). Here, (R)-gly-
ceraldehyde was used as the source of C-10 — C-12 segment of the molecule which incorporates an asymmetric carbinol group at C-11. An aldol
condensation with methyl oleate provided 103.1 which was transformed
into the lactone 103.2. This allowed the elaboration of the existing
functionality to give the cyanohydrin 103.3 which could be cyclized into
the cyclopentane derivative 103.4. Thus, the synthesis was devised to
take advantage of the protected cyanohydrin as an acyl anion equivalent. This functionality was also found to be compatible with oxidative
cleavage of the double bond to generate the ester 103.5. Completion of
the synthesis was based on a conjugate addition of the appropriate racemic vinylcuprate derivative 103.5. Interestingly, this reaction took
place with remarkable enantiospecificity to give the desired diastereoisomer 103.6 with no detectable amounts of the unwanted isomer!
Finally, the 13-*cis*-15(R)-side-chain in 103.6 was isomerized to the
desired 13-*trans*-15(S) counterpart by literature procedures[6] to give the
target. PGE$_2$ has been synthesized by other routes also.[7]

c. Prostaglandin F$_2$α (PGF$_2$α) — Synthetic efforts in the
area of prostaglandins from carbohydrate precursors reached a high point
with the synthesis of PGF$_2$α from D-glucose by Stork and coworkers[8]
(Scheme 104). In this case D-glycero-D-guloheptose (available from
D-glucose and from the commercially available D-glycero-D-guloheptono-
1,4-lactone), was used as source for the C-10 — C-16 portion of the
target, encompassing two chiral centers (C-11,C-15) and a *trans*-double
bond. The key transformation involved a "transfer" of chirality via the
Claisen rearrangement to provide an acetic acid side chain (C-8 — C-12
bond) from which the remainder of the carbon framework was constructed.
Thus, the D-glycero-D-guloheptose derivative 104.2 was transformed into
the protected alditol 104.3, which when subjected to an elimination
reaction[9] via the corresponding orthoester derivative, led to the
desired *trans*-olefin 104.4. Manipulation of the hydroxyl groups
produced the allylic alcohol 104.5. Note that this is formed by first
exchanging the acetate ester in 104.4 for a methoxycarbonyl derivative,
hydrolysis of the acetal functions and cyclic carbonate formation.
Intermediate 104.5 now had the desired functionality for transfer of

Scheme 103

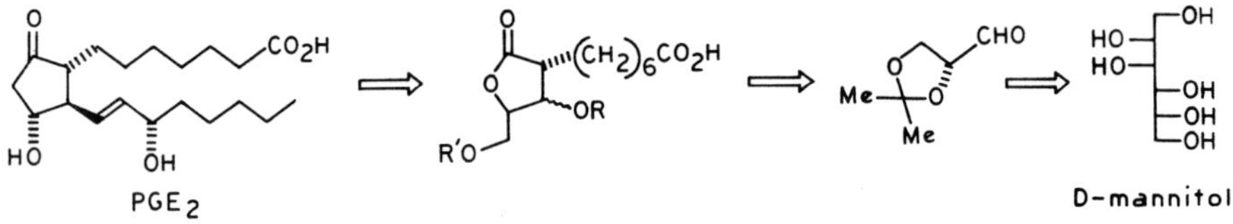

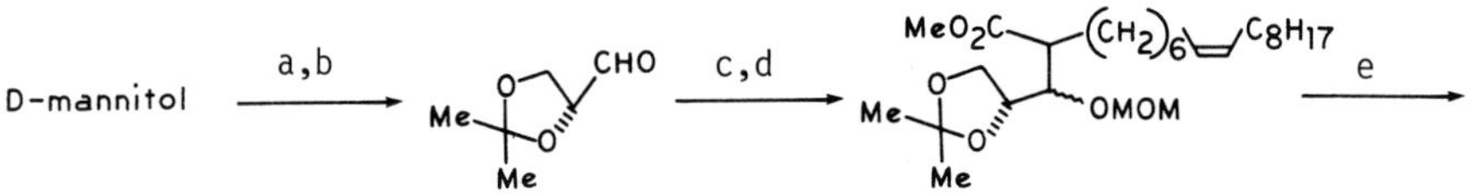

PGE$_2$ ⟹ (lactone intermediate) ⟹ (dioxolane CHO intermediate) ⟹ D-mannitol

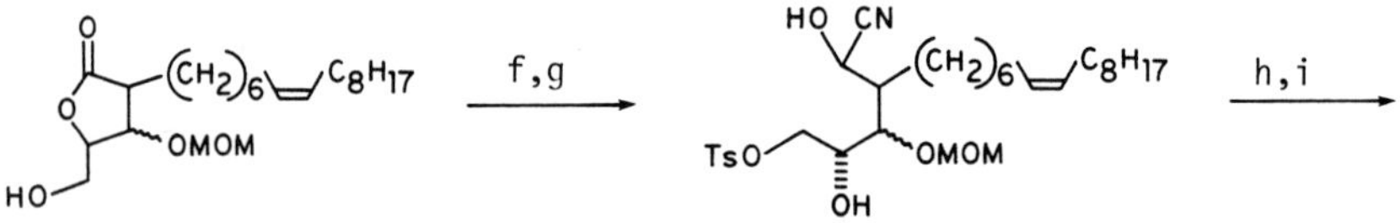

D-mannitol →(a,b)→ 12.2 →(c,d)→ 103.1 →(e)→

103.2 →(f,g)→ 103.3 →(h,i)→

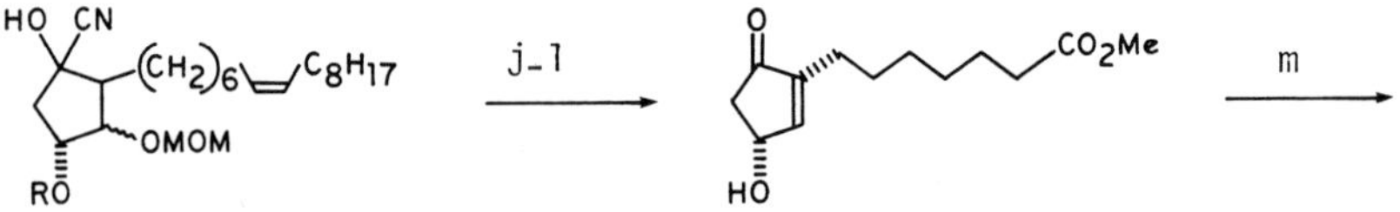

103.4 R=ethoxyethyl →(j-1)→ 103.5 →(m)→

103.6 R′=methoxyisopropyl → → 103.7

a. acetone, $\overset{+}{H}$; b. NaIO$_4$; c. methyl oleate, LDA, THF, HMPA, -78°;
d. ClCH$_2$OMe, R$_2$NH; e. aq. H$_2$SO$_4$, THF; f. TsCl, pyr.; g. HCN, EtOH;
h. EtOCH=CH$_2$, $\overset{+}{H}$; i. (Me$_3$Si)$_2$NH, benzene, reflux; j. NaIO$_4$, KMnO$_4$, then CH$_2$N$_2$;
k. aq. acid ; l. aq. NaOH, THF; m. 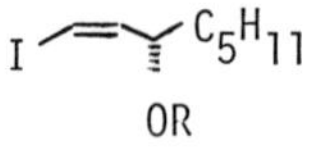 I⌒⌒$\overset{}{\diagup}$C$_5$H$_{11}$
                                        OR

Scheme 104

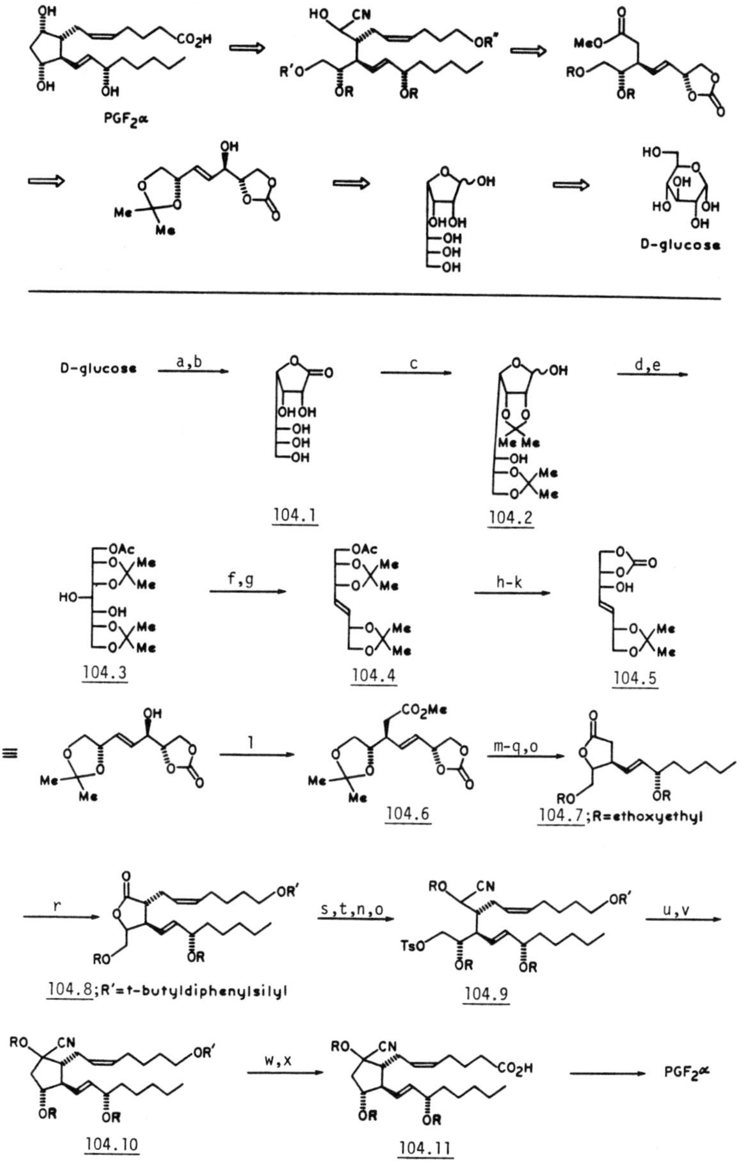

PGF$_{2\alpha}$

D-glucose

D-glucose $\xrightarrow{a,b}$ 104.1 $\xrightarrow{c}$ 104.2 $\xrightarrow{d,e}$

104.3 $\xrightarrow{f,g}$ 104.4 $\xrightarrow{h-k}$ 104.5

$\equiv$ $\xrightarrow{l}$ 104.6 $\xrightarrow{m-q,o}$ 104.7;R=ethoxyethyl

$\xrightarrow{r}$ 104.8;R'=t-butyldiphenylsilyl $\xrightarrow{s,t,n,o}$ 104.9 $\xrightarrow{u,v}$

104.10 $\xrightarrow{w,x}$ 104.11 $\longrightarrow$ PGF$_{2\alpha}$

a. NaCN, H$_3$O$^+$; b. NaBH$_4$, pH ~3.5; c. acetone, H$^+$; d. NaBH$_4$, MeOH; e. Ac$_2$O, pyr.;CHCl$_3$ -7^0; f. Me$_2$NCH(OMe)$_2$, CH$_2$Cl$_2$; g. Ac$_2$O, reflux; h. NaOMe, MeOH; i. MeOCOCl, pyr.,0^0; j. CuSO$_4$, aq. MeOH, reflux; k. acetone, H$^+$; l. MeC(OMe)$_3$, H$^+$, 140^0; m. K$_2$CO$_3$, MeOH; n. TsCl, pyr; o. EtOCH=CH$_2$, H$^+$; p. Bu$_2$CuLi, ether, -40^0; q. aq. H$_2$SO$_4$; r. 7-bromo-cis-5-hepten-1-ol t-butyldiphenylsilyl ether, Li N[SiMe$_3$]$_2$, THF, HMPA, -40^0; s. DIBAL THF; t. HCN, EtOH; u. aq. AcOH, THF; v. KN[(SiMe$_3$)]$_2$, benzene, reflux; w. Bu$_4$NF, THF; x. Collins, then AgNO$_3$, aq. KOH; y. L-selectride

chirality in a key Claisen rearrangement to give the chiral C-branched
alditol 104.6. From this point on the synthetic operations were similar
to the above described syntheses and relied on the formation of the
cyanohydrin derivative 104.9 which was cyclized to give the cyclopentane
ring system 104.10. Note that both asymmetric centers at C-11 and C-15
are intact. Completion of the synthesis followed established
methodology to give optically active $PGF_2\alpha$. Thus, the entire
seven-carbon framework of the starting carbohydrate was used, destroying
two asymmetric centers in order to generate the *trans*-double bond, and
setting the stage for the stereocontrolled Claisen rearrangement. This
is an example of the efficient use of carbohydrate alditols in complex
natural product synthesis, based on the recognition of symmetry and
chirality. While it was fortuitous that the sense of chirality at C-11
and C-15 in the target corresponded to those at C-6 and C-2 respectively
in the starting carbohydrate 104.2, other stereochemical situations
could have been easily accommodated by appropriate inversions of
configuration. Based on the elegant studies by Stork and his coworkers,
it is tempting to propose a synthesis for a $PGF_2\alpha$ in which, for example,
the entire C-8 — C-16 chiral segment is derived from a carbohydrate
precursor. An optically active cyclopentanoid intermediate for the
synthesis of prostaglandins has been prepared from D-glucose.[10]

      d. Brefeldin A and Prostaglandins – Synthesis of optically
active cyclopentane precursors – Brefeldin A is a fungal metabolite
whose constitutional structure and absolute configuration have been
determined.[11] As illustrated in Scheme 105, brefeldin A consists of a
bicyclic cyclopentane derivative fused to a 13-membered cyclic ether
with five asymmetric centers. Clearly, there are several challenges in
constructing the carbon framework of the trisubstituted cyclopentane
ring. Ohrui and Kuzuhara[12] have devised an approach from D-glucose
which unravels the presence of hidden carbohydrate-type symmetry in the
molecule. The retrosynthetic analysis of their plan shows that the
cyclopentane ring is assembled from C-2 — C-5 of a branched-chain
derivative of D-glucose. Thus the C-5 and C-2' (acetate) carbon atoms
are the extremities involved in ring formation. The furanoside
structure is merely a "template" onto which appropriate functionality is
introduced as required, and the cyclopentane ring is constructed around
the furanoid structure in a "scaffold process". With this analysis, it

Scheme 105

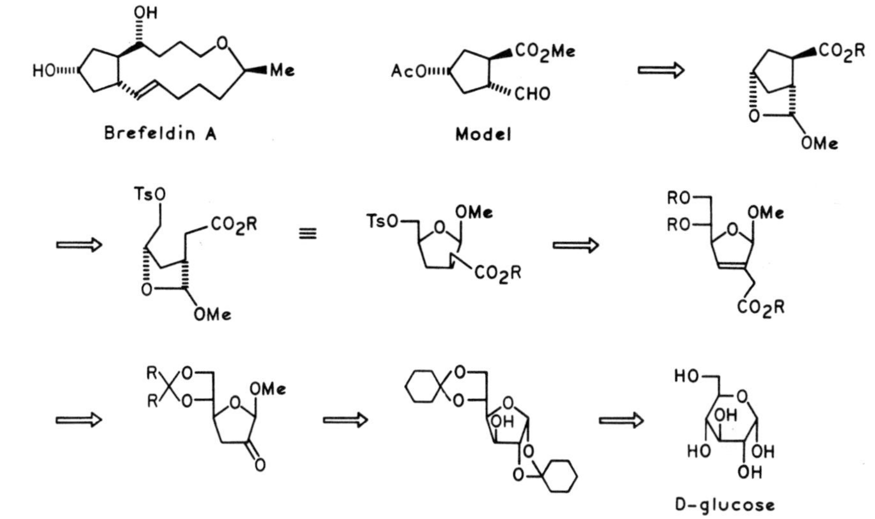

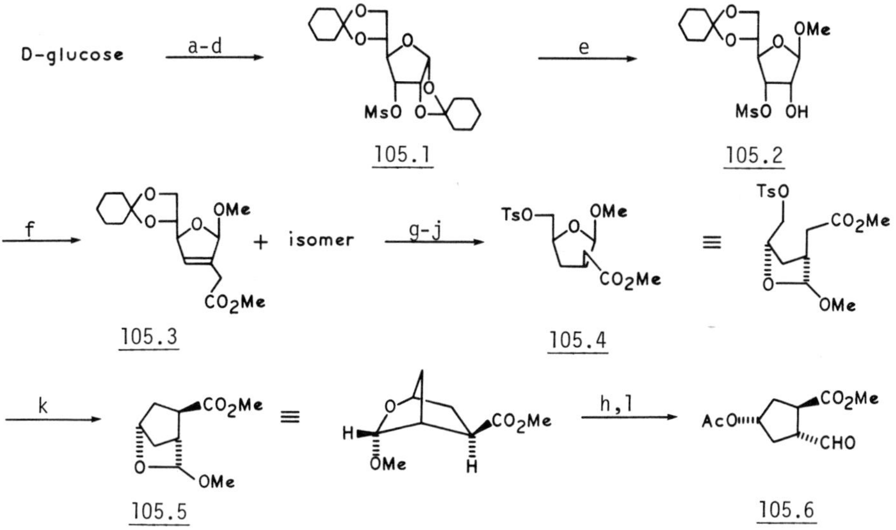

Brefeldin A   Model

D-glucose

D-glucose $\xrightarrow{\text{a-d}}$ 105.1 $\xrightarrow{\text{e}}$ 105.2

$\xrightarrow{\text{f}}$ 105.3 + isomer $\xrightarrow{\text{g-j}}$ 105.4

$\xrightarrow{\text{k}}$ 105.5 $\xrightarrow{\text{h,l}}$ 105.6

a. cyclohexanone, H^+; b. oxid.; c. $NaBH_4$; d. MsCl, pyr.; e. MeOH, H^+;
f. $(MeO)_2P(O)CH_2Li$, THF; g. Pd/C, H_2, MeOH; h. H_3O^+; i. aq. $NaIO_4$;
then $NaBH_4$; j. TsCl, pyr.; k. $(Me_3Si)_2NLi$, DME, HMPA; l. Ac_2O, pyr.

becomes clear that the key transformations reside in a regio and stereocontrolled branching at C-2 in a furanoid structure and in carbocyclic
ring formation. Once again D-glucose emerges as an ideal starting
material. Intermediate 105.1, readily available from D-glucose in four
steps was transformed into the methyl furanoside derivatives, 105.2, thus
exposing the C-2 hydroxyl group to chemical manipulation. Note the
oxidative elimination and branching reaction undergone by 105.2 to give a
9:1 mixture of 105.3 and the exocyclic isomer respectively. Catalytic
hydrogenation and subsequent manipulation of this mixture gave 105.4,
which, written in a different perspective, brings out the carbohydrate
precursor of the desired cyclopentane ring. Base treatment resulted in
intramolecular cyclization to give 105.5 which was transformed into the
chiral trisubstituted cyclopentane derivative 105.6 by standard methods.
Of interest is the formation of a single major isomer in the cyclization
process, corresponding to a thermodynamically stable orientation of the
carbomethoxy substituent. The corresponding nitrile was also prepared by
a comparatively easier process. The racemic derivative corresponding to
105.6 has been previously prepared[13] and converted into dl-brefeldin A.
Other syntheses of racemic brefeldin A have been reported.[14] The
successful intramolecular cyclization of conformationally biased furanoid
derivatives containing required appendages was extended to the synthesis
of a tetrasubstituted chiral cyclopentane 106.3, a derivative which can be
considered as being an immediate precursor to prostaglandins (Scheme
106).[12] Note that the relative as well as absolute stereochemistry at
C-8, C-9; C-11, C-12 (PG) can be secured by this route.

        e. (-) Pentenomycin - The antibiotic pentenomycin has been the
subject of structural and chemical studies[15,16] including X-ray
analysis.[17] Structurally remote from a carbohydrate, analysis of the
molecule reveals hidden cabohydrate-type symmetry which was uncovered and
cleverly exploited by Moffatt, Verheyden, and coworkers.[18] Analysis of
the structure of pentenomycin (Scheme 107) reveals the presence of a
functionalized cyclopentenone system, which in principle can be generated
from an intramolecular aldol condensation of a keto-aldehyde precursor.
The retrosynthetic analysis shows how such an acyclic precursor can be
generated form a carbohydrate. Essentially, if one ignores the branched
hydroxymethyl group, the five-carbon backbone can be related to a carbohydrate of known stereochemistry at C-2. Note that the aldehyde function
needed for eventual cyclization and the location of the ketone carbonyl

Scheme 106

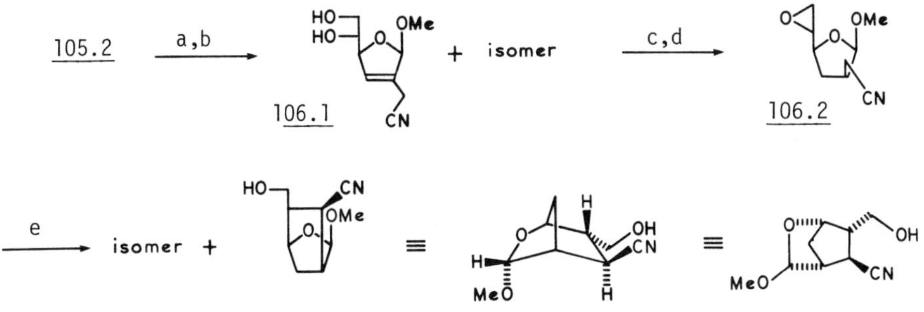

105.2 $\xrightarrow{a,b}$ [structure 106.1] + isomer $\xrightarrow{c,d}$ [structure 106.2]

106.1

106.2

$\xrightarrow{e}$ isomer + [structure 106.3] $\equiv$ [structure] $\equiv$ [structure]

106.3

a. $(MeO)_2P(O)CH_2CN$, BuLi; b. H_3O^+; c. TsCl, pyr.; d. NaOMe, MeOH;
e. $(Me_3Si)_2NLi$, THF, reflux.

function comply with the "rule of four", hence the possibility of a
'chiron' derived from a furanoid template.

   The configuration at the secondary hydroxyl group converges with
C-2 in 1,2:5,6-di-O-isopropylidene-α-D-glucofuranose and using such a
derivative also ensures stereospecific chain-branching. Thus, the two
chiral centers in the target can be secured early in a carbohydrate precursor. The main challenge resides in the cyclization of an appropriate
derivative to the cyclopentenone — an unprecedented transformation
from carbohydrate precursors. Scheme 107 shows that the hydroxymethyl
group was easily introduced via a nitromethane reaction to give 107.1.
Transformation of the nitromethyl group to a protected hydroxymethyl and
chain-shortening afforded derivative 107.3 with the two asymmetric
centers secured. Methanolysis followed by acetonation led to 107.4
which could be subjected to base-catalyzed elimination to provide the
enol ether 107.5 (a masked methyl ketone). Acid treatment afforded the
desired keto-aldehyde precursor 107.6, which after careful experimenta-
tion was subjected to intramolecular aldol condensation (note conditions) to give the projected pentenomycin derivative 107.7. Note how
debenzylation was done under acetolysis conditions to avoid hydrogenation. Finally, enzymatic deacetylation and acid hydrolysis gave (-)-
pentenomycin identical with authentic natural product. The total synthesis of pentenomycin by this route also confirmed its proposed

Scheme 107

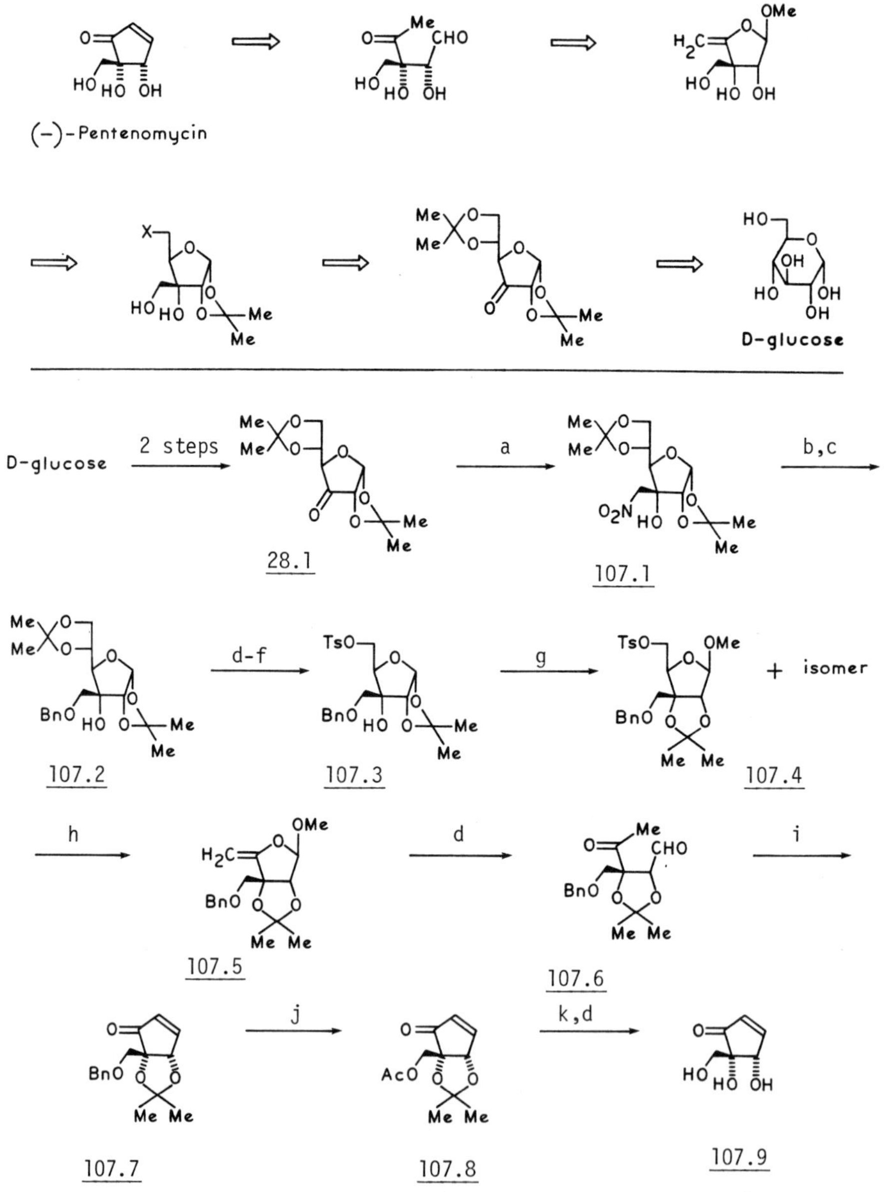

(−)-Pentenomycin

D-glucose

D-glucose $\xrightarrow{\text{2 steps}}$ 28.1 $\xrightarrow{a}$ 107.1 $\xrightarrow{b,c}$

28.1   107.1

107.2   107.3   107.4   + isomer

107.5   107.6

107.7   107.8   107.9

a. MeNO$_2$; b. alk. KMnO$_4$, then NaBH$_4$; c. BnBr, NaH, DMF; d. H$_3$O$^+$; e. NaIO$_4$, then NaBH$_4$; f. TsCl, pyr.; g. MeOH, Me$_2$C(OMe)$_2$, H$^+$; h. t-BuOK; i. alumina 100-180°, 30 mm Hg; j. Ac$_2$O, H$_2$SO$_4$, AcOH; k. lipase deacetylation

absolute configuration. The synthesis of 6-deoxypentenomycin followed a
similar route. Some key structures are shown in Scheme 108.

Scheme 108

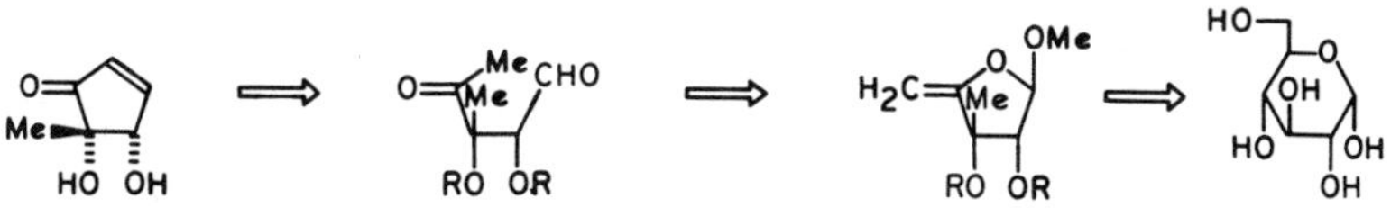

6-Deoxypentenomycin   D-glucose

     f. <u>Synthetic approaches to tetrodotoxin</u> - Several studies have
been made notably in the late R.B. Woodward's laboratory for the use of
carbohydrates as intermediates in the synthesis of tetrodotoxin, a
potent neurotoxin found in the Japanese puffer fish.[19] Unfortunately,
none of the approaches culminated in a synthesis of this complex molecule, but they were conceptually exciting and they are depicted in
Schemes 109 and 110. Only the retrosynthetic analysis is shown as the
approaches are reconstructed from the abstracts of several
theses.[20,21] The basic idea in the Woodward strategy relied on the manipulation of carbohydrate derivatives so as to create the highly functionalized cyclohexane ring of the target by an intramolecular cyclization of an ω-nitroaldehyde. Note that the stereochemistry at C-5 and
C-7 of tetrodotoxin is the same as C-2 and C-4 in D-glucose, hence the
possibility of using it as a chiral precursor. Thus, it can be seen in
Scheme 109 that if attainable, <u>109.2</u> can be a useful advanced intermediate since it can generate the C-8a—C-9 bond as well as the guanidine unit. The hydroxyl groups at C-4 and C-9 will have to be contended
with separately. The scheme is self-evident and shows how the critical
intermediate <u>109.2</u> was produced from D-glucose. However when R was H,
intramolecular cyclization did not give the desired <u>109.3</u>, rather an
isomer in which the nitro substituents were of opposite orientation and
the molecule in the other conformation.[20] An ingenious modification of
this scheme called for the cyclization of a glyoxylate ester produced
from a cinnamate ester <u>109.3b</u>. In this way, the bicyclic system of tetrodotoxin could be produced in one step, hopefully with the desired
stereochemistry. Indeed cyclization occurred as in the previous

Scheme 109

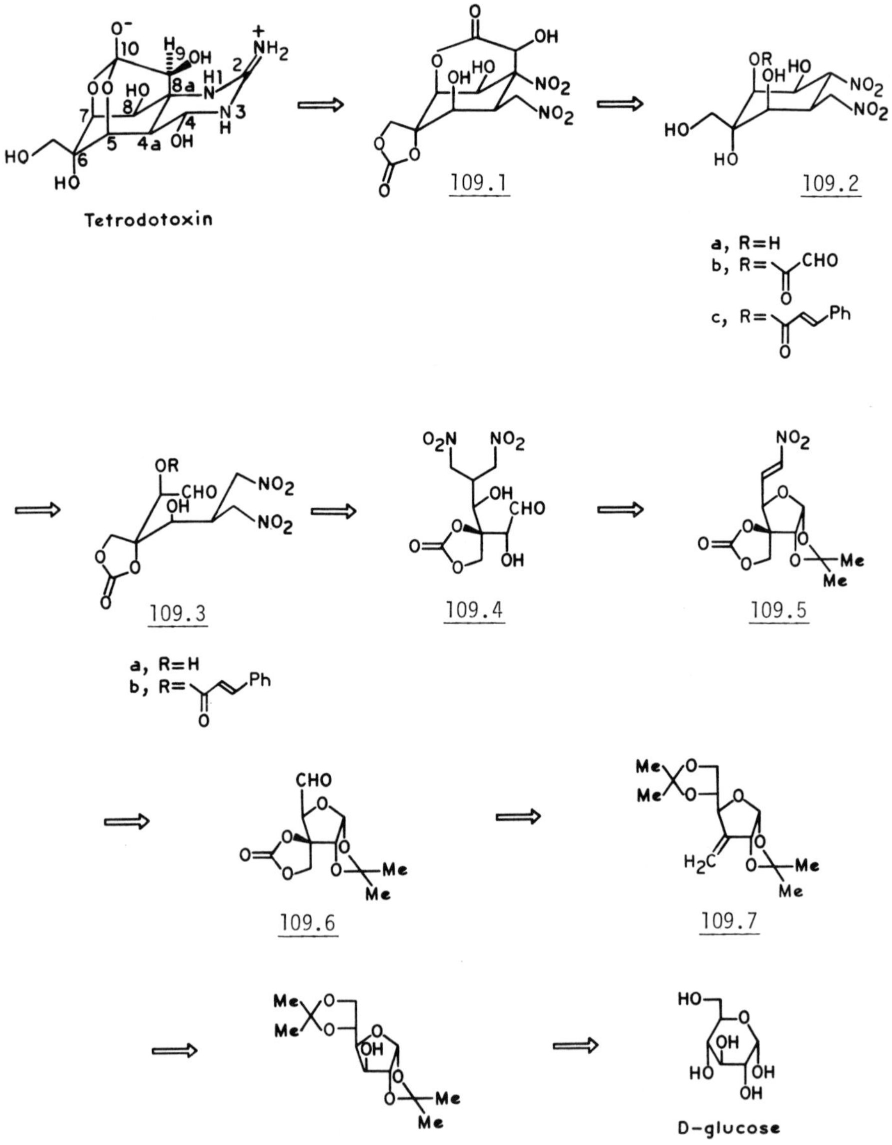

Tetrodotoxin

109.1

109.2

a, R=H
b, R= CHO
c, R= Ph

109.3

a, R=H
b, R= Ph

109.4

109.5

109.6

109.7

D-glucose

example, but again giving the wrong stereoisomer. The idea however was
to oxidize 109.3 to the glyoxylate ester, hoping that a cycloreversion
would generate 109.2, (R=∿CHO), which in turn would cyclize to
the desired 109.1.

     In another approach[21] (Scheme 110) the nitroolefin intermediate

Scheme 110

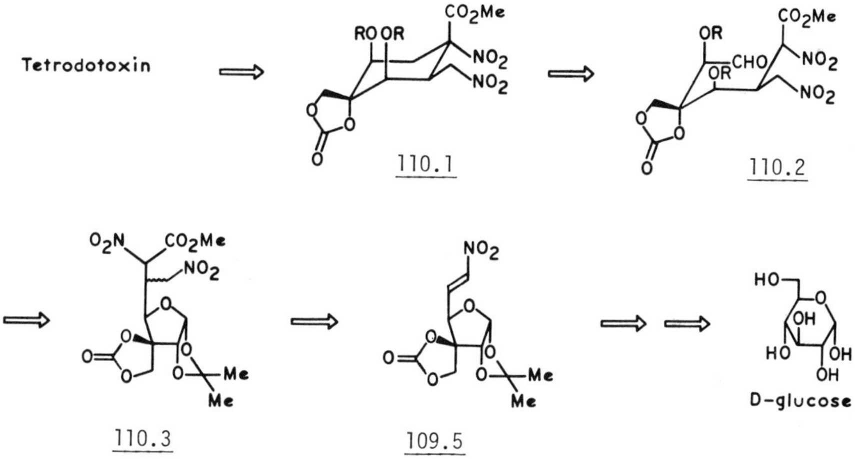

109.5 was homologated by Michael addition with methyl nitroacetate to
give a mixture of isomers 110.3 (7:2 of 5(R) and 5(S)). Since the C-6
position is also epimeric, it was hoped that the cyclic product would in
fact correspond to 109.2 after equilibration. Indeed, further manipulation of 110.3 led to an aldehydo derivative which cyclized to give
110.1 and the epimeric C-8a derivative. However, both of these were
found to exist in the undesired conformation (equatorial carbomethoxy
group for the 110.1 isomer!).

     Yoshimura and collaborators[22] have investigated a conceptually
related route to tetrodotoxin from a carbohydrate precursor (Scheme
111). One of the key intermediates was the nitroolefin derivative 111.4
which could be obtained from D-glucose in straightforward manner.
Michael addition of lithio dithian to 111.4, separation of epimers, followed by acid hydrolysis gave 111.3(and its epimer), which were cyclized
under mildly basic conditions. The desired isomer 111.2 was thus iso-

Scheme 111

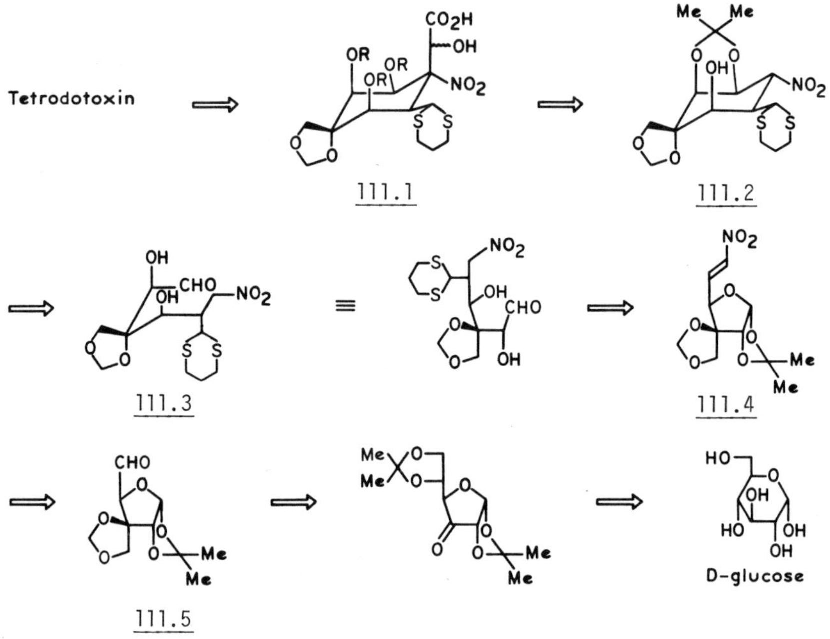

Tetrodotoxin $\Longrightarrow$

111.1   111.2

111.3 $\equiv$ 111.4

111.5   D-glucose

lated in fair yield from 111.3 (D-gluco configuration). Note that the
absolute stereochemistry in this derivative is secured based on its mode
of formation, since only two new asymmetric centers are generated and
the unwanted isomers can be separated. Condensation of 111.2 with
methyl glyoxylate gave a product(30%) which was assigned structure
111.1, but as with the Woodward approach, the conformation may also turn
out to be a stumbling block. The total synthesis of (±)-tetrodotoxin by
another approach has been reported by Kishi.[23]

g. (+)-Chrysanthemic acid - The enantiomeric chrysanthemic acids
have been synthesized from a single carbohydrate progenitor, namely D-
glucose.[24] The synthesis of the biologically important dextroro-
tatory isomer[25] will be discussed here (Scheme 112). The hidden carbohydrate-type symmetry can be unveiled in the retrosynthetic analysis although it should be appreciated that D-glucose is merely used as a
source for the partial carbon framework of the target. Note that all

Scheme 112

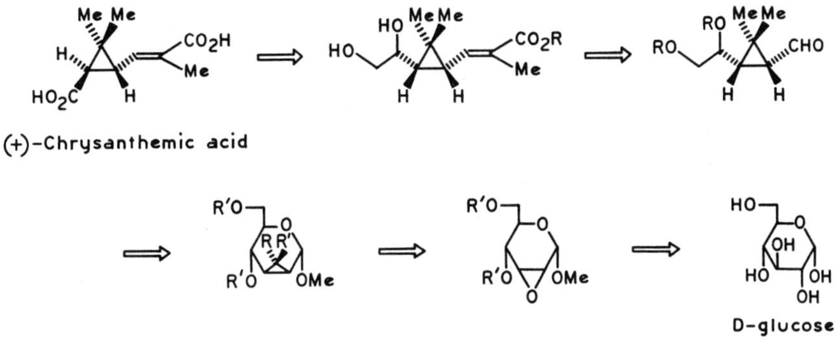

(+)-Chrysanthemic acid

D-glucose

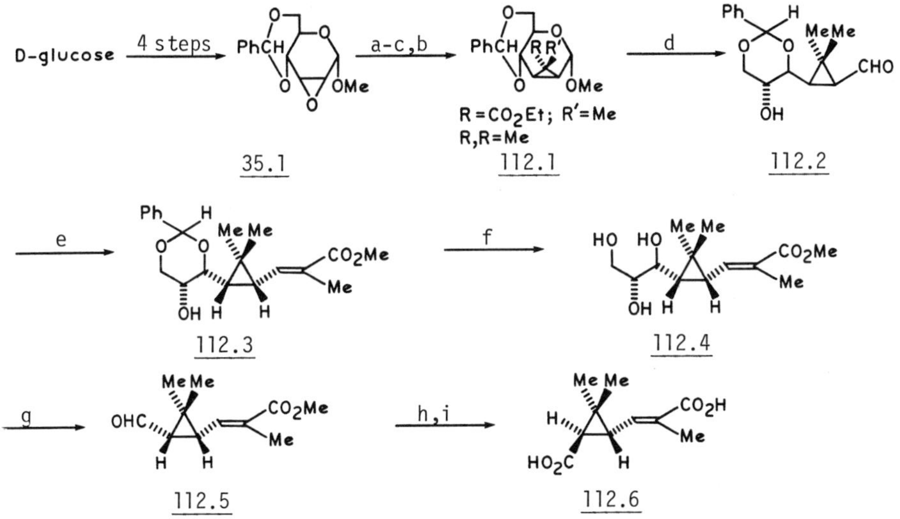

D-glucose $\xrightarrow{\text{4 steps}}$ **35.1** $\xrightarrow{\text{a-c,b}}$ **112.1** $\xrightarrow{\text{d}}$ **112.2**

R = CO₂Et; R'= Me
R,R = Me

$\xrightarrow{e}$ **112.3** $\xrightarrow{f}$ **112.4**

$\xrightarrow{g}$ **112.5** $\xrightarrow{h,i}$ **112.6**

a. (EtO)₂P(O)CHMeCO₂Et, BuLi; b. LAH; c. MsCl, DMF; d. aq. dioxane, reflux
e. Ph₃P=C(Me)CO₂Me, CH₂Cl₂; f. TsOH MeOH; g. NaIO₄; h. NaOMe, MeOH; i. Ag₂O,
aq. NaOH

the hydroxyl groups are stripped away and only two asymmetric centers are utilized in the generation of the immediate precursors. In essence a chiral cyclopropane unit is erected using the D-glucose framework. A key transformation consists in the "cyclopropanation" of the known epoxide derivative by adaptations of reactions previously studied by Meyer zu Reckendorf[26] and Fraser-Reid and coworkers.[27] Thus, treatment of 35.1 with diethyl ethylphosphonate anion gave the cyclopropane derivative 112.1 in which the methyl group had an endo orientation. The gem-dimethyl group was introduced by reductive techniques and further manipulation gave the key aldehydo derivative 112.2. Note that glycoside 112.1 could be easily hydrolyzed under neutral conditions (aq. dioxane), based on previous findings of Fraser-Reid and coworkers concerning the intermediacy of "cyclopropylcarbinyl-oxocarbonium" ions[28]. Compound 112.2 can in fact be epimerized to give the C-2 epimeric aldehyde corresponding to a *trans*-orientation of substitutents and eventually leading to the enantiomeric levorotatory chrysanthemic acid. Wittig reaction of 112.2 followed by deblocking and oxidation generated the aldehyde 112.5 which was epimerized and oxidized to (+)-chrysanthemic acid. Note how the diol unit corresponding to C-4/C-5 in the original D-glucose was eventually used to generate an aldehyde group. Thus, only four of the original six carbon atoms of D-glucose were maintained in the target, and the conformational bias present in the epoxide 35.1, superbly exploited in the construction of the cyclopropane unit. It should also be evident that intermediate 112.3 offers stereochemical duality since the two extremities can be manipulated independent of one another. All possible isomers of chrysanthemic acid can therefore be made from the same progenitor and the synthesis can be adapted to isotopically labeled analogs as well. Other syntheses of the chrysanthemic acids are also known.[29]

   h.   (-)-Quinic and (-)-Shikimic acids – These important bio-synthetic intermediates[30] have been prepared in optically pure form from D-arabinose[31] (Scheme 113). Examination of the triol structure common to the two targets reveals a D-arabino configuration (or D-lyxo!). Thus, if an extra carbon atom can be inserted between the two extremities of a pentitol derivative, there would result the trihydroxy-cyclohexane skeleton of the target compounds. Thus, D-arabinose was converted to the tribenzylated alditol by standard procedures, then, using an intriguing reaction, it was converted into the "Wittig reagent"

Scheme 113

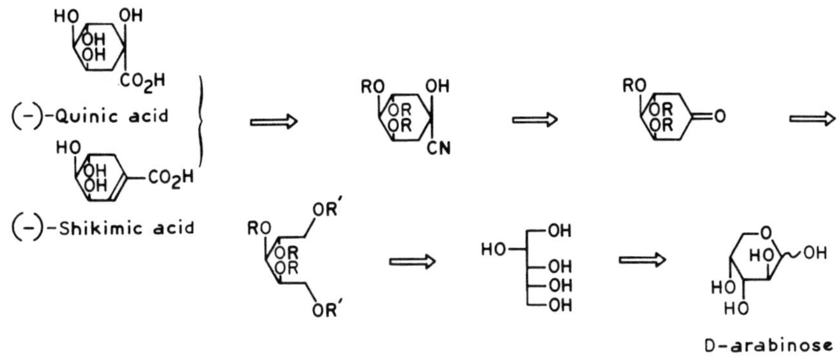

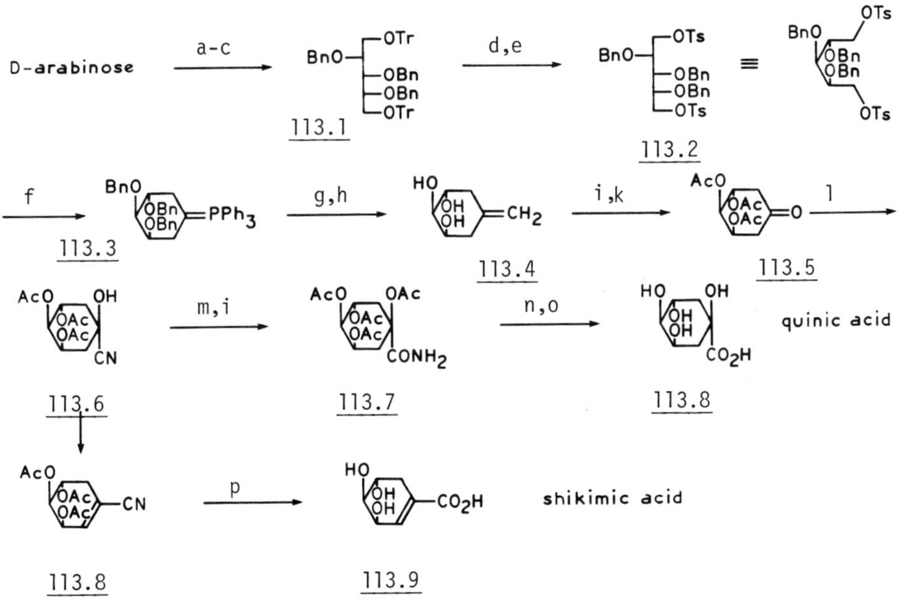

a. Ra-Ni, H_2; b. TrCl, pyr.; c. BnCl, KOH; d. aq. AcOH; e. TsCl, pyr.;
f. $Ph_3P=CH_2$; g. CH_2O; h. Na, liq. NH_3; i. Ac_2O, pyr.; j. NaOMe, MeOH;
k. $NaIO_4$, OsO_4; l. HCN; m. HBr, AcOH; n. N_2O_3; o. aq. AcOH; p. $POCl_3$,
pyr.

113.3 based on previous studies in Bestmann's laboratory.[32] This
reaction provides an unusual approach to the synthesis of cyclohexane
derivatives from alditols. Treatment of 113.3 with formaldehyde gave
the exocyclic methylene derivative 113.4, which was debenzylated (note
method) and oxidized to the known ketone 113.5 previously prepared by
Grewe.[33] The steps leading to (-)-quinic acid and (-)-shikimic acid
followed procedures reported in the literature. Note however that the
POCl₃ procedure leads stereospecifically to shikimic acid. Syntheses of
racemic quinic acid[34] and optically active shikimic acid[35] are available
by other routes. Chemical and biochemical conversions of carbohydrate
derivatives into cyclohexane derivatives related to shikimic acid have
been described.[36]

    i. Access to carbocyclic derivatives from carbohydrate precur-
sors. - In addition to the methods described in the previous syntheses
of carbocyclic compounds in this section, there exist others that should
be potentially applicable to other targets as well. In the opinion of
the writer, there is a need to further develop this aspect of the utilization of carbohydrates in synthesis, since for example a hexose
contains the same number of carbon atoms as a cyclohexane and could be a
useful precursor. A potential problem of course may reside in the pro-
pensity for β-elimination in aldehydo forms. However, this could be
turned to advantage in some instances, provided that the remaining functionality is conveniently located.

    The classical route to cyclohexane derivatives such as the inosi-
tols for example, from carbohydrates has involved the well-known cyclization of 6-nitro-6-deoxyhexoses[37] or the cyclization of nitro alkanes
with appropriate dialdehydes.[38] Some recent methods for the synthesis
of polyhydroxycyclohexanes such as the carbene insertion reaction with a
1,7-diazoketone derivative of a sugar,[39] acid-catalyzed cyclization of
analogous 1,7-dibromoketone derivatives,[40] and the photochemical cyclization of diethyl dithioacetal derivatives of unsaturated sugars,[41] and
mercuric chloride-induced rearrangement of 5,6-enopyranosides are worthy
of mention here.[42] Another source of cyclohexane derivatives which
relies on a different methodology involves the Lewis acid catalyzed
Diels-Alder reaction of butadiene with carbohydrate enones such as
114.1.[43] which gives access to bicyclic intermediate 114.2 which have
obvious utility in the synthesis of natural products of the iridoid
family and related ones (Scheme 114).

Scheme 114

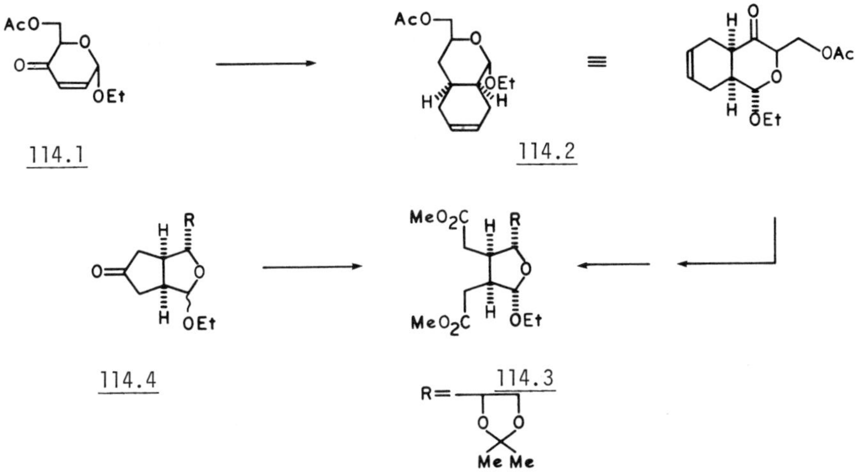

114.1   114.2

114.4   114.3

R=

Cyclopentanones constructed "around" a sugar molecule as in 114.4 are
also possible[43] by manipulation of the cyclohexane portion of 114.2 and
Dieckmann condensation via 114.3. Ferrier and coworkers[10,44] have also
reported on the formation of a cyclopentanone derivative via intramolecular cyclization of a sugar epoxide. A versatile approach to
functionalized chiral cyclopentane derivatives is based on a 1,3-dipolar
nitrone-olefin intramolecular cycloaddition.[45] Fraser-Reid and
coworkers[46] have outlined routes for the synthesis of chiral cyclo-
butanes fused to a carbohydrate framework (Scheme 115). Thus intermediates 115.2 and 115.3 produced by photoannulation of vinyl acetate to
the enone 115.1 and further manipulation, could be considered as useful
precursors to (+)- and (-)-grandisol.

     Finally, (+)-truxinic acid as well as racemic isomers have been
prepared by irradiation of D-mannitol hexacinnamate or derivatives of
L-erythritol 1,4-dicinnammate.[47] In such "photochemical asymmetric
syntheses", the 2+2 cycloaddition of cinnamic acid residues are
controlled by a "template" effect of a kind different that the one
discussed in this book.

Scheme 115

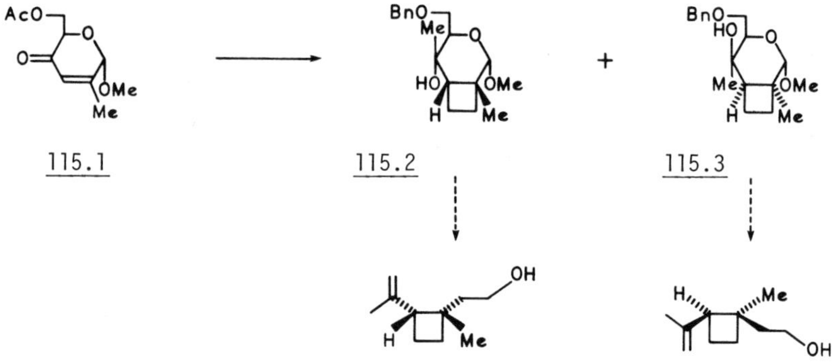

115.1   115.2   115.3

(1S:2R)-(+)-grandisol   (1R:2S)-(-)-grandisol

115.4   115.5

REFERENCES

1. For some recent reviews on synthesis, see J. S. Bindra and R.
    Bindra, Prostaglandin Synthesis, Academic Press, New York, N.Y.,
    (1977); A. Mitra, The Synthesis of Prostaglandins, John Wiley, New
    York, N.Y., (1977).

2. G. Stork and S. Raucher, J. Am. Chem. Soc., 98, 1583 (1976).

3. J.N. Baxter and A.S. Perlin, Can. J. Chem., 38, 2217 (1960);
    B.R. Baker and K. Hewson, J. Org. Chem., 22, 966 (1957).

4. See for example, E.J. Corey and G. Moinet, J. Am. Chem., Soc., 95,
    6831 (1973).

5. G. Stork and T. Takahashi, J. Am. Chem. Soc., 99, 1275 (1977).

6. J.G. Miller, W. Kurz, K.G. Untch and G. Stork, J. Am. Chem. Soc.,
    96, 6774 (1974).

7. See for example, A.F. Kluge, K.G. Untch and J.H. Fried, J. Am.
    Chem. Soc., 94, 9256 (1972).

8. G. Stork, T. Takahashi, I. Kawamoto and T. Suzuki, J. Am. Chem.
    Soc., 100, 8272 (1978).

9. F.W. Eastwood, K.J. Harrington, J.S. Josan and J.L. Pura,
    Tetrahedron Lett., 5223 (1970), see also S. Hanessian,

A. Bargiotti and M. LaRue, ibid., 737 (1978) and references cited therein.

10. R.J. Ferrier and P. Prasit, J.C.S. Chem. Comm., 983 (1981).

11. H.P. Weber, D. Hauser and H. P. Sigg, Helv. Chim. Acta, 54, 2763 (1971); H.P. Sigg , ibid., 47, 1401 (164); Y. Suzuki, H. Tanaka, H. Aoki and T. Tamura, Agr. Biol. Chem., 34, 395 (1970).

12. H. Ohrui and H. Kuzuhara, Agric. Biol. Chem., 44, 907 (1980).

13. A. Honda, K. Hirata, H. Sueoka, T. Katsuki and Y. Yamaguchi, 22nd Symposium on the Chemistry of Natural Products, Fukuoka, 1979, pp. 370.

14. E.J. Corey, R.H. Wollenberg and D.R. Williams, Tetrahedron Lett., 2243 (1977); P.A. Bartlett and F.R. Green III., J. Am. Chem. Soc., 100, 4858 (1978); T. Kitahara, K. Mori and M. Matsui, Tetrahedron Lett., 3021 (1979); A.E. Greene, C. Le Drian and P. Crabbé, J. Am. Chem. Soc., 102, 7583 (1980); M. Honda, K. Hirata, H. Sueoka, T. Katsuki and M. Yamaguchi, Tetrahedron Lett., 22, 2679 (1981).

15. K. Umino, T. Furumai, M. Matsuzawa, Y. Awataguchi, Y. Ito, and T. Okuda, J. Antibiotics, 26, 506 (1973).

16. K. Umino, N. Takeda, Y. Ito and T. Okuda, Chem. Pharm. Bull. 22, 1233 (1974).

17. T. Date, K. Aoe, K. Kotera and K. Umino, Chem. Pharm. Bull., 22, 1963 (1974).

18. J.P.H. Verheyden, A.C. Richardson, R.S. Bhatt, B.D. Grant, W.L. Fitsch and J.G. Moffatt, Pure Appl. Chem.,56, 1363 (1978); for a recent synthesis of (±)-pentenomycin and its epimers, see A.B. Smith, III, S.J. Branca, N.N. Pilla and M.A. Guaciaro, J. Org. Chem., 47, 1855 (1982).

19. See for example, T. Goto, Y. Kishi, S. Takahashi and Y. Hirata, Tetrahedron, 21, 2059 (1965); R. B. Woodward, Pure Appl. Chem., 9, 49 (1964); and references cited therein.

20. Ph.D. Thesis, R.D. Sitrin, Harvard University, 1972; E. Vieira, Jr., 1969.

21. Ph.D. Thesis, J. Upeslacis, Harvard University, 1975.

22. M. Funabashi, J. Wakai, K. Sato and J. Yoshimura, J. Chem. Soc., Perkin I, 13 (1980); see also, J. Yoshimura, K. Kobayashi, K. Sato and M. Funabashi, Bull. Chem. Soc. Japan, 45, 1806 (1972).

23. Y. Kishi, T. Fukuyama, M. Aratani, F. Nakatsubo, T.Goto, S. Inoue, H. Tanimo, S. Sugiura and H. Kakoi, J. Am. Chem. Soc., 94, 9219 (1972).

24. B.J. Fitzsimmons and B. Fraser-Reid, J. Am. Chem. Soc., $\underline{104}$, 6123 (1979).

25. L. Crombie and M. Elliott, Fortschr. Chem. Org. Naturst., $\underline{19}$, 120 (1961).

26. W. Meyer zu Reckendorf and U. Kamprath-Scholtz, Chem. Ber., $\underline{105}$, 673 (1972).

27. B. Fraser-Reid and B.J. Carthy, Can. J. Chem., $\underline{50}$, 2928 (1972).

28. B.K. Radatus and B. Fraser-Reid, J. Chem. Soc., Perkin I., 1872 (1975) and references cited therein.

29. See for example, R.W. Mills, R.D.H. Murray and R.A. Raphael, J. Chem. Soc., Perkin I., 133 (1973); E.J. Corey, and M. Jautelat, J. Am. Chem. Soc., $\underline{89}$, 3912 (1967) and references cited therein.

30. See for example F. Lingens, Angew. Chem. Int. Ed. Engl. $\underline{7}$, 350 (1968); H. Plieninger, Ibid., $\underline{1}$, 367 (1962).

31. H.J. Bestmann and H.A. Heid, Angew. Chem. Int. Ed. Engl. $\underline{10}$, 336 (1971).

32. H.J. Bestmann and E. Kranz, Chem. Ber., $\underline{102}$, 1802 (1969).

33. R. Grewe and E. Vaugermain, Chem. Ber., $\underline{98}$, 104 (1965).

34. R. Grewe, W. Lorenzen and L. Vinig, Chem. Ber., $\underline{87}$, 793 (1954); E.E. Smissman and M.A. Oxman, J. Am. Chem. Soc., $\underline{85}$, 2184 (1963); R. Wolinsky, R. Novak and R. Vasileff, J. Org. Chem $\underline{29}$, 3596 (1964).

35. E.E. Smissman, J.T. Suh, M. Oxman and R. Daniels, J. Am. Chem. Soc., $\underline{81}$, 2909 (1959); $\underline{84}$, 1040 (1962); R. McGruidle, K. Overton and R. Raphael, J. Chem. Soc., 1560 (1960); R. Grewe and I. Hinricho, Chem. Ber., $\underline{97}$, 443, (1964).

36. B. Ganem, Tetrahedron, $\underline{34}$, 3353 (1978).

37. J.M. Grosheintz and H.O.L. Fischer, J. Am. Chem. Soc., $\underline{70}$, 1479 (1948); see also T. Iida, M. Funabashi, and J. Yoshimura, Bull. Chem. Soc. Japan, $\underline{46}$, 3203, 3207 (1973); J. Kovar and H.H. Baer, Carbohydr. Res. $\underline{45}$, 101 (1975).

38. For a review, see F. Lichtenthaler, Angew. Chem., Int. Ed. Engl., $\underline{3}$, 211 (1964); H.H. Baer, Advan. Carbohydr. Chem. Biochem., $\underline{24}$, 67 (1969).

39. D.E. Kiely and C.E. Cantrell, Carbohydr. Res., $\underline{23}$, 155 (1972). C.F. Cantrell, D.E. Kiely, G.E. Abruscato and J. M. Riordan, J. Org. Chem., $\underline{42}$, 3562 (1972).

40. D.E. Kiely and J. Riordan, ACS Symposium Series, No. 125, Aminocyclitol Antibiotics, 96 (1980); see also C.E. Cantrell,

D.E. Kiely, R.A. Hearn and C.E. Bugg, Tetrahedron Lett., 4379 (1973).

41. A.A. Othman, N.A. Al-Masudi and U.S. Al-Timari, J. Antibiotics, 31, 1007 (1978).

42. R.J.Ferrier, J.Chem. Soc., Perkin J. 1455, 1979; R.J. Ferrier and P. Prasit, Carbohydr. Res., 82, 263 (1980).

43. See ref. 46. p. 53; see also D. Horton and T. Machinami, J.C.S. Chem. Comm., 88 (1981).

44. R.J. Ferrier and V.K. Srivastava, Carbohydr. Res., 59, 333 (1977).

45. B. Bernet and A. Vasella, Helv. Chim. Acta, 62, 1990, 2400, 2411 (1979).

46. See for example, B. Fraser-Reid and R.C. Anderson, Fotschr. Chem. Org. Naturst., 39, 1 (1980).

47. B.S. Green, Y. Rabinsohn and M. Rejto, J.C.S. Chem. Comm., 313 (1975); B.S. Green, A.T. Hagler, Y. Rabinsohn and M. Rejto, Israel J. Chem., 15, 124 (1976).

CHAPTER FOURTEEN

HETEROCYCLIC MOLECULES

a. (-)-Frontalin - The structure of the bark beetle pheromone
has been shown to be 1,5-dimethyl-6,8-dioxabicyclo[3,2.1]octane by a
synthesis of the racemic compound[1] (Scheme 116). At first glance,
frontalin may well appear to belong to the class of compounds which
possess apparent carbohydrate-type symmetry, by virtue of the presence
of a tetrahydropyran (or tetrahydrofuran) ring. However the approach
used in two independent syntheses of this compound from carbohydrate
precursors adopted a strategy that was not based on the elaboration of
substituents on a preformed oxygenated heterocycle. Thus, in the
retrosynthetic analysis shown in Scheme 116, it can be seen that only
the left hand portion of the molecule (C-4 — C-7) is derived from a
carbohydrate. The strategy calls for the creation of the tertiary
alcohol corresponding to C-1 in the target by using a carbohydrate
"template". Elaboration of the remaining carbon atoms and relying on a
stereoselective acetal formation leads to the intended target. In the
Fraser—Reid synthesis[2] (Scheme 116) the tertiary center is introduced
via an oxymercuration-demercuration reaction on the 2-C-methylene
derivative obtained from the ketone 116.2, to produce 116.3 as the
exclusive isomer. Further manipulation produces the aldehyde 116.5
which is chain-extended to 116.6 and eventually transformed into
(-)-frontalin. This strategy is also useful in the synthesis of the
enantiomeric compound as well, since a Grignard reaction on 116.2
produces predominantly the (R)-isomer. The Ohrui—Emoto synthesis[3] of

214

Scheme 116

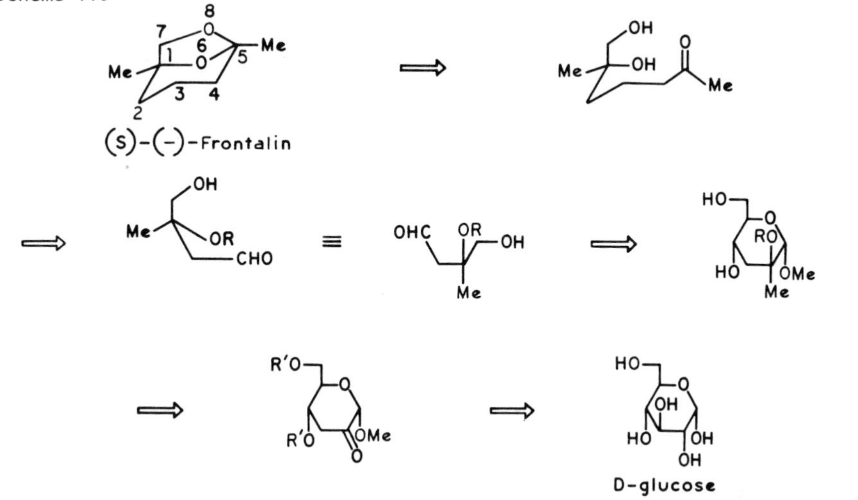

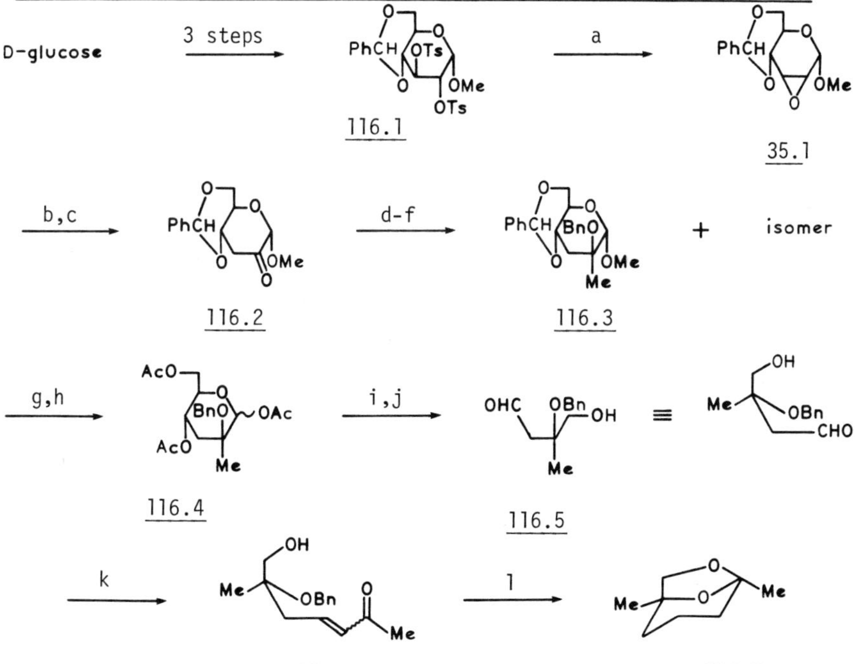

D-glucose $\xrightarrow{\text{3 steps}}$ **116.1** $\xrightarrow{a}$ **35.1**

$\xrightarrow{b,c}$ **116.2** $\xrightarrow{d-f}$ **116.3** + isomer

$\xrightarrow{g,h}$ **116.4** $\xrightarrow{i,j}$ **116.5**

$\xrightarrow{k}$ **116.6** $\xrightarrow{l}$ **116.7**

a. NaOMe, $CHCl_3$, reflux; b. LAH; c. CrO_3, pyr.; d. $Ph_3P=CH_2$, DME;
e. $Hg(OAc)_2$; f. BnBr, NaH, DMF; g. aq. H_2SO_4; h. Ac_2O, $BF_3.Et_2O$;
i. $NaBH_4$, MeOH; j. aq. $NaIO_4$; k. $Ph_3P=CHCOMe$, THF, reflux; l. Pd/C,
H_2, EtOH

(-)-frontalin (Scheme 117) generates a branched four-carbon aldehyde intermediate 117.3 via an interesting rearrangement[4] which takes place when the mesylate derivative 105.2 is treated with a methyl Grignard reagent. Note that the resulting 117.2 is produced with very high stereoselectivity, unlike the analogous Grignard reaction on a 2-keto hexopyranoside 116.2. Synthesis of racemic frontalin[1] as well as (R)-(+)- and (S)-(-)-frontalin have been reported.[5]

Scheme 117

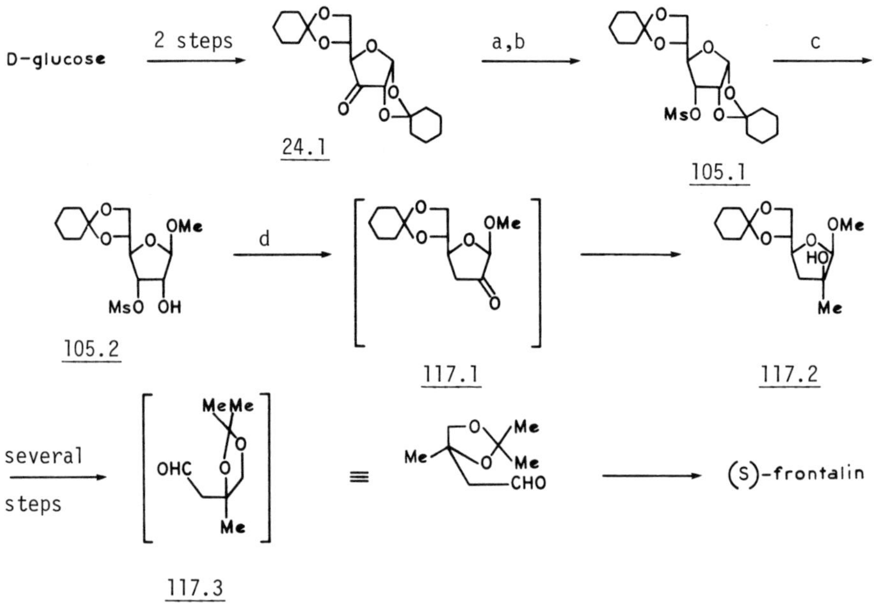

a. NaBH$_4$; b. MsCl, pyr.; c. MeOH, H$^+$; d. MeMgI;

b. (+)-Biotin – The structure of the growth promoter (+)-biotin (vitamin H) has been known since 1942 by degradative[6] and synthetic[7] studies. The diamino tetrahydrothiophene structure, with three contiguous chiral centers can be decoded in the retrosynthetic analysis shown in Scheme 118. The Ogawa synthesis[8] is based on the generation of a five-carbon aldehyde which contains the required vicinal diamino functionality by prior elaboration of a suitable carbohydrate precursor.

Scheme 118

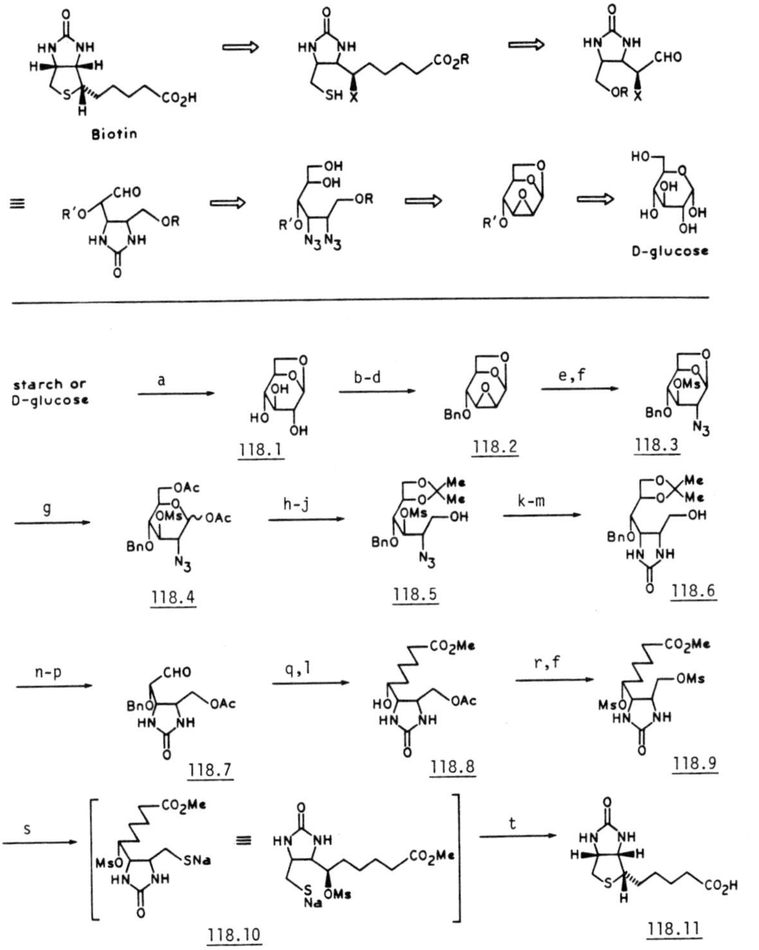

a. pyrolysis, see also Scheme 34; b. TsCl, pyr.; c. NaOMe, MeOH; d. BnBr, NaH, DMF;
e. NaN$_3$, DMF; f. MsCl, pyr.; g. Ac$_2$O, BF$_3$.Et$_2$O; h. aq. HCl, MeOH; i. NaBH$_4$;
j. Me$_2$C(OMe)$_2$, H$^+$; k. LiN$_3$, DMF; l. Pd/C, H$_2$; m. COCl$_2$, CCl$_4$, aq. NaHCO$_3$; n. Ac$_2$O,
pyr.; o. aq. AcOH; p. NaIO$_4$, aq. EtOH; q. Ph$_3$P=CH(CH$_2$)$_2$CO$_2$Me, CH$_2$Cl$_2$; r. NaOMe, MeOH;
s. Na$_2$S, MeOH, reflux; t. aq. NaOH, then H$^+$

The critical cyclization to produce the tetrahydrothiophene structure is envisaged as an intramolecular process based on an inversion of configuration by a primary thiol group, or alternatively displacement on a primary position by the thiol. Thus the chiral carbon framework of the tetrahydrothiophene ring and a one-carbon aldehyde appendage can be produced from a suitable carbohydrate precursor. The starting material in the Ogawa synthesis was 1,6-anhydro-β-D-glucopyranose which was used for the introduction of the C-3 amino group (biotin numbering) as well as the preferential substitution which allowed further elaboration. Thus, opening of the epoxide 118.2 with azide ion, followed by acetolysis gave 118.4. The latter was transformed into the alditol derivative 118.5 which was used to introduce the second amino group. Note that it was necessary to effect this operation on the acyclic mesylate 118.5 rather than the cyclic 118.4, where an S_N2 type reaction would have been exceedingly difficult. With the diamino function secured, the remainder of the synthesis consisted in generating the aldehyde 118.7, chain-extension to 118.8 and transformation into the dimesylate 118.9, which upon treatment with sodium sulfide led to (+)-biotin 118.11. As expected, displacement probably occurred at the primary site, followed by an intramolecular heterocyclization with inversion of configuration. Thus, all the hydroxyl groups of the original D-glucose were systematically removed and C-6 excised, leaving a five-carbon chain now bearing the functionality required for the elaboration of the target. Note that this synthesis is highly stereoselective and resembles a biomimetic natural process since, desthiobiotin is the biogenetic precursor of biotin.[9]

An earlier synthesis of (+)-biotin started with D-mannose[10] (Scheme 119). In this synthesis the side-chain was introduced first from the readily available 119.1, then the tetrahydrothiophene ring was formed via an intramolecular cyclization with inversion of configuration at C-4 (119.3 → 119.4). The vicinal diamino function was introduced by displacement with azide ion in a remarkably good yield since this type of reaction in a furanose derivative is normally quite difficult. As in the previous synthesis the C-1 — C-5 segment of D-mannose was used to build the functionalized heterocyclic ring in the target.

In a more recent synthesis (Scheme 120), the same group[11] recognized that C-2 in 2-amino-2-deoxy-D-glucose has the same sense of chirality as that of C-4 in biotin. By replacing the C-3 hydroxyl group by an amino group with inversion of configuration, one would have the

Scheme 119

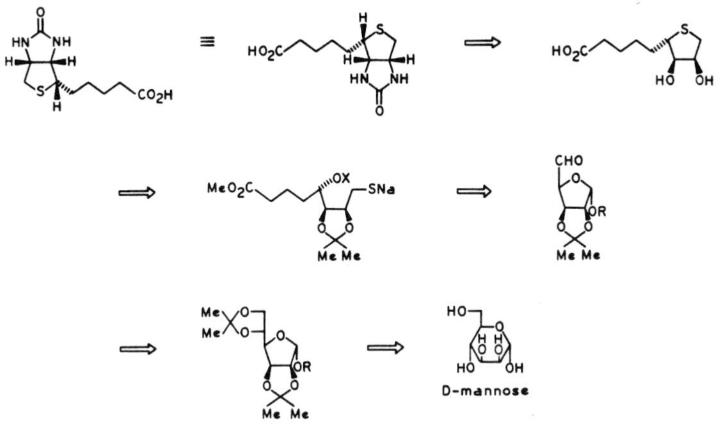

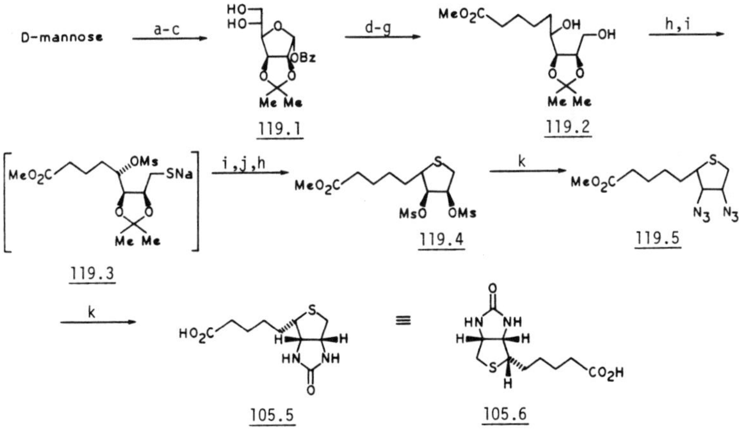

a. acetone, H$^+$; b. BzCl, pyr.; c. aq. AcOH; d. NaIO$_4$; e. Ph$_3$P=CH(CH$_2$)CO$_2$Me;
f. Pd/C, H$_2$, MeOH; g. NaBH$_4$, MeOH; h. MsCl, pyr.; i. Na$_2$S, HMPA; j. aq. HCO$_2$H
k. NaN$_3$,HMPA;l. Pd/C, H$_2$, MeOH, Ac$_2$O; m. aq. Ba(OH)$_2$, reflux; n. COCl$_2$

Scheme 120

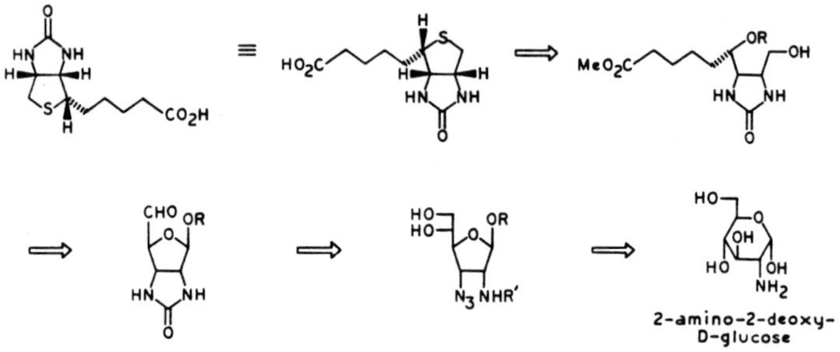

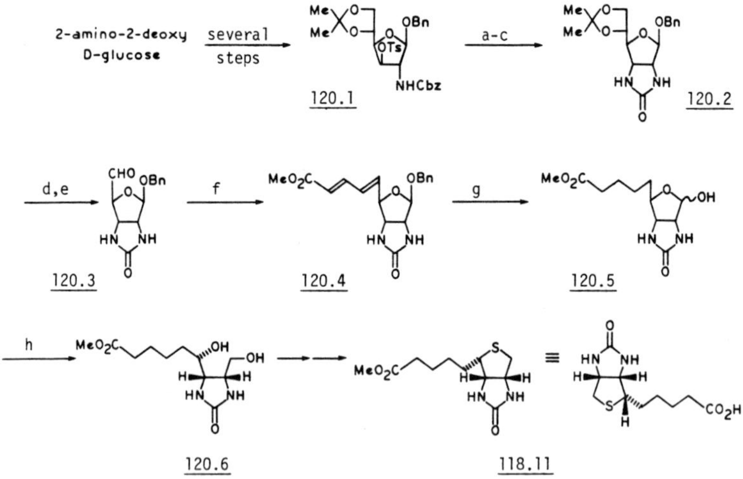

2-amino-2-deoxy-
D-glucose

120.1

120.2

120.3

120.4

120.5

120.6

118.11

a. NaN$_3$, DMF, reflux; b. Ra-Ni, H$_2$; c. COCl$_2$, NaH, DMF; d. aq. AcOH; e. NaIO$_4$,
aq.acetone; f. Ph$_3$P=CH-CH=CHCO$_2$Me; g. Pd/C, H$_2$, EtOH; h. NaBH$_4$, MeOH

diamino unit of biotin on a six-carbon framework (such as in 120.2). In
order to attach the acid side-chain it was necessary to prepare a
derivative of 2-amino-2-deoxy-D-glucose in the furanose form such as
120.1, since cleavage of the potential diol would generate the required
aldehyde (as in 120.3). Displacement of the 3-tosylate in 120.1 with
azide ion was accomplished with surprising ease and in very good yield.
The steps leading to the target from the important aldehyde intermediate
120.3, followed a predicted course. A practical feature in this
synthesis is the relatively short number of steps in going from
2-amino-2-deoxy-D-glucose to the acyclic intermediate 120.6 (10 steps).
There are several syntheses of (+)-biotin[7] as well as of the racemic
compound.

        c. (-)-Anisomycin - The structure of anisomycin isolated from
two strains of Streptomyces was determined by a combination of chemical,
spectroscopic and finally X-ray crystallographic studies.[12] (Scheme 121,
see also Scheme 10) . Its synthesis from D-glucose by Moffatt and
coworkers[13] represents a conceptually novel approach to the construction
of pyrrolidine-type heterocycles by what could be named a "scaffold
process" (see Scheme 9). The retrosynthetic analysis in Scheme 121
shows how D-glucose was in essence used as a "scaffold" onto which the
pyrrolidine ring was erected spanning C-3 and C-6 of a D-glucofuranose
structure. Note that the vicinal diol situated in the target could
initially reflect the need for a dihydroxy chiral precursor involving at
least four carbon atoms. Moffatt and coworkers[13] considered the diol
unit in anisomycin as being situated at the C-4 — C-5 positions in a D-
glucofuranose derivative, with the proviso of an inversion of a configuration at C-5. Bridging C-3 (with inversion) and C-6 with a convenient
source of nitrogen atom, followed by cleavage at C-1 — C-2 would produce
a pyrrolidine aldehyde intermediate which could be further elaborated to
the target. Thus, intermediate 121.2, readily available from D-glucose
underwent a remarkable reaction when treated with ammonia to give
121.4. Initial cleavage of the benzoate ester was followed by epoxide
formation with inversion at C-5. Presumably epoxide opening occurred
with ammonia to give 121.3, which underwent intramolecular hetero-
cyclization at C-3 with inversion. In a single step, the pyrrolidine
ring was formed and the required inversion at C-5 was achieved. At this
juncture, the 1,2-diol unit was used to generate the aldehydo intermediate 121.5 with the required chirality and functionality. The

Scheme 121

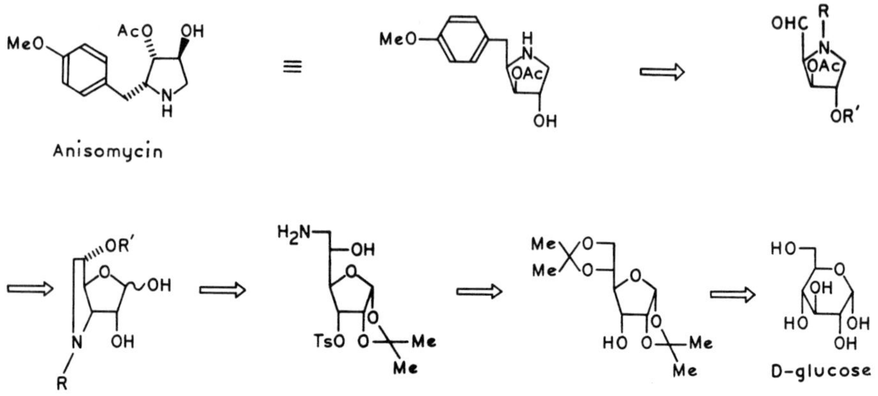

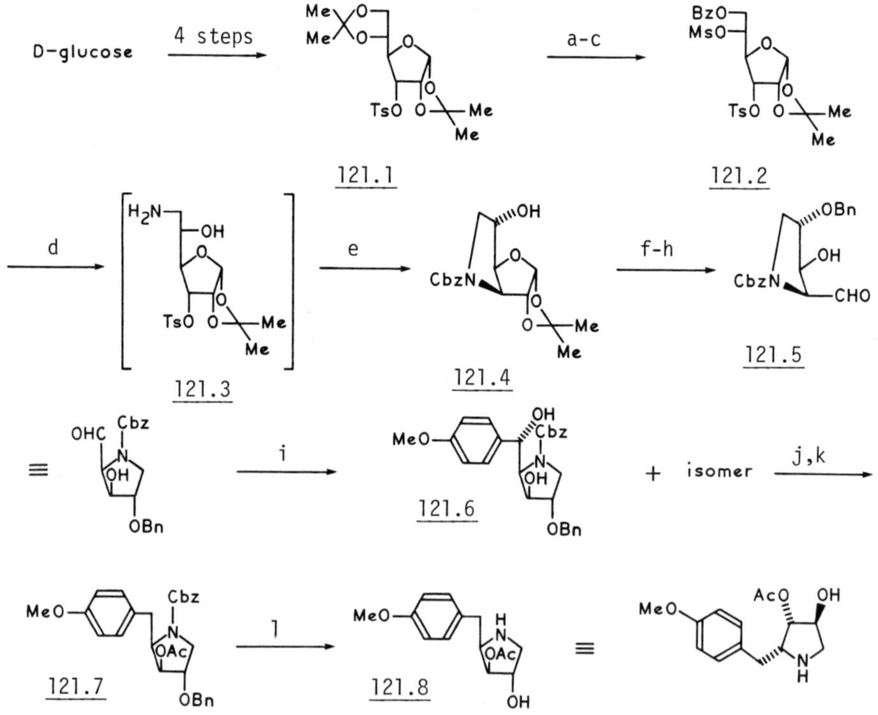

Anisomycin

D-glucose →(4 steps)→ 121.1 →(a-c)→ 121.2

→(d)→ [121.3] →(e)→ 121.4 →(f-h)→ 121.5

≡ →(i)→ 121.6 + isomer →(j,k)→

121.7 →(1)→ 121.8 ≡

a. aq. AcOH; b. BzCl, pyr.; c. MsCl, pyr.; d. NH$_3$, MeOH, 105°; e. CbzCl;
f. BnBr, NaH, DMF; g. H$_3$O$^+$; h. NaIO$_4$; i. p-MeOC$_6$H$_4$MgBr, THF; j. Et$_3$SiH,
TFA; k. Ac$_2$O, pyr; l. Pd/C, H$_2$

p-methoxybenzyl unit was then introduced by a Grignard reaction followed
by hydrogenolysis with triethylsilane and trifluoroacetic anhydride.
Note that the remaining hydroxyl group at C-3 in anisomycin was
conveniently acetylated before final debenzylation to give the target.
In the final analysis, only the original C-4 oxygen ring atom was
carried over to the target (C-3 of anisomycin). In addition, two
inversions and a C-1 excision were required, reactions which were
achieved with extremely high stereocontrol. Several syntheses of
racemic[14] and optically active[15] anisomycin are known based on different
strategies.

    d. Detoxinine - This unusual amino acid found in detoxin D_1[16] is
a selective antagonist of the antibiotic blasticidin S. Its revised
structure was determined to be (2S, 3R, 1'S)-2-(2'-carboxy 1'-hydroxy-
ethyl)-3-hydroxypyrrolidine by total synthesis of detoxinolactone
starting with D-glucose[17] (Scheme 122). The retrosynthetic analysis
shows that the target can be related to a 4-amino-3,5-dihydroxyheptanoic
acid (note position of the 5-hydroxyl group, "rule of five"). Since the
diol unit can be stereochemically related to C-3 and C-5 in a D-erythro
configuration, the main challenge resides in the elaboration of the
pyrrolidine ring. This was accomplished by the "scaffold-type process"
wherein an aminomethyl side-chain was allowed to form the heterocycle by
intramolecular attack at C-4 in an appropriate hexopyranose derivative.
Thus, the readily available 122.1 was transformed into the 6-cyano
derivative 122.3, which, upon reduction gave the bicyclic pyrrolidine
derivative 122.4. Now that the basic framework of the target was
constructed around the carbohydrate template, routine transformations
(N-acylation, hydrolysis, oxidation) gave the desired target. The total
synthesis of detoxinolactone also establishes its biochemical precursor
as L-proline. The corresponding epimeric hydroxylactone (1'R) was also
prepared from 122.1 itself.

    e. (+)-Azimic and (+)-Carpamic Acids - Azimine[18] and carpaïne[19]
are macrocyclic piperidine dilactones, consisting of the respective acid
components azimic acid and carpamic acid (Scheme 123). Their
constitutional structures have been established by degradative studies
and ultimately by synthesis of the racemic compounds. A synthesis of
the optically pure acids has been accomplished starting with D-glucose
as the chiral precursor.[20] The retrosynthetic analysis for the synthesis

Scheme 122

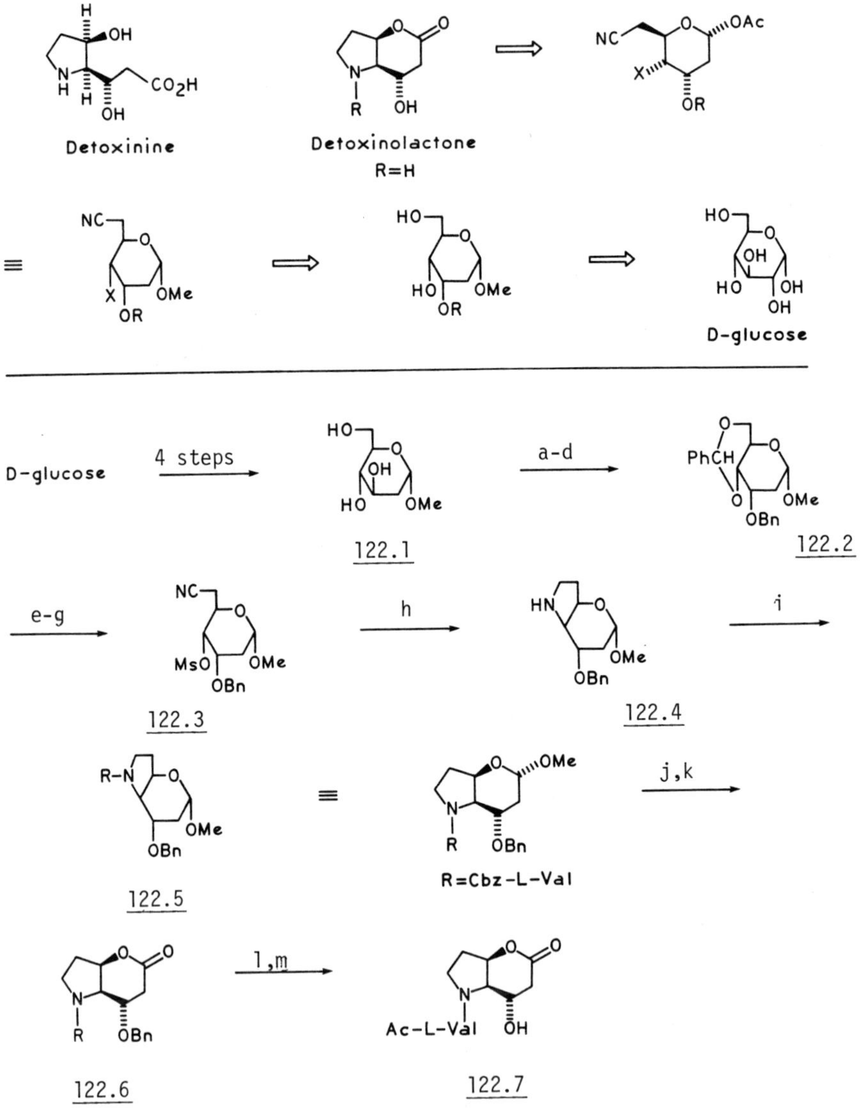

Detoxinine

Detoxinolactone
R=H

D-glucose

D-glucose $\xrightarrow{\text{4 steps}}$ 122.1 $\xrightarrow{\text{a-d}}$ 122.2

$\xrightarrow{\text{e-g}}$ 122.3 $\xrightarrow{\text{h}}$ 122.4 $\xrightarrow{\text{i}}$

122.5 $\equiv$ R=Cbz-L-Val $\xrightarrow{\text{j,k}}$

122.6 $\xrightarrow{\text{1,m}}$ 122.7

Ac-L-Val

a. PhCHO, ZnCl$_2$; b. CrO$_3$, pyr.; c. NaBH$_4$; d. BnBr, NaH, DMF; e. aq. AcOH;
f. MsCl, pyr; g. NaCN, DMSO; h. NaBH$_4$, COCl$_2$, MeOH; then KOH;
i. Cbz-L-VaL-OH; j. aq. HCl; k. PCC, CH$_2$Cl$_2$; 1. Pd/C, H$_2$; m. Ac$_2$O, MeOH

Scheme 123

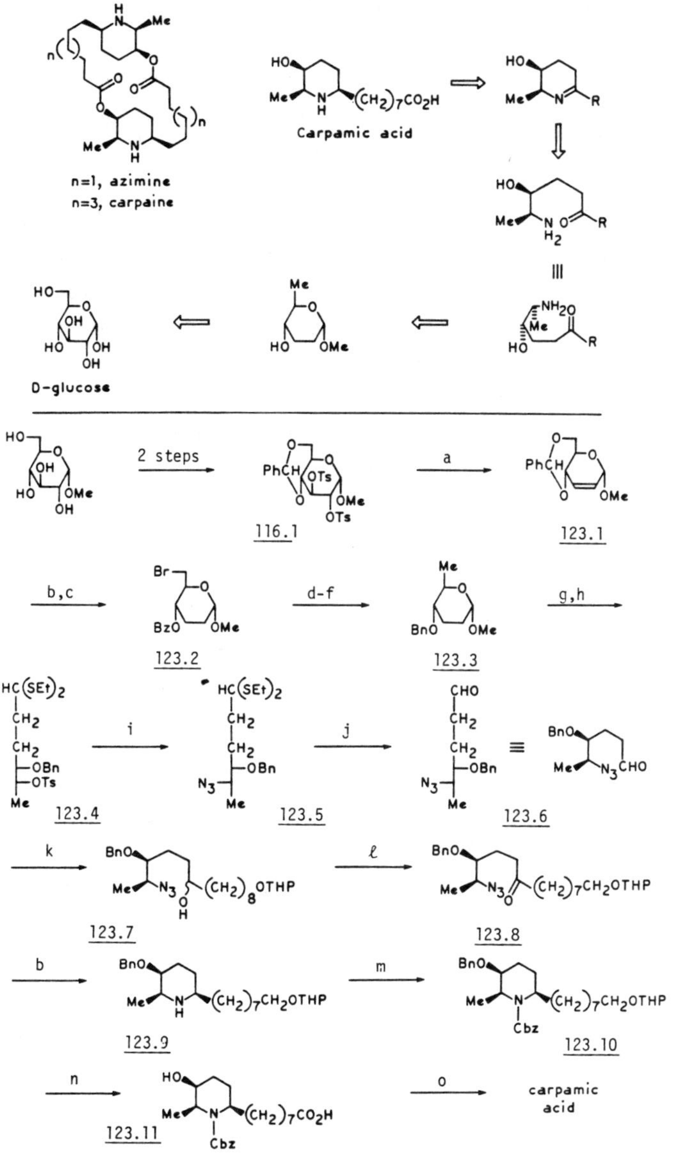

Carpamic acid

n=1, azimine
n=3, carpaine

D-glucose

116.1

a

123.1

b,c

123.2

d-f

123.3

g,h

123.4

i

123.5

j

123.6

k

123.7

ℓ

123.8

b

123.9

m

123.10

n

123.11

o

carpamic
acid

a. Zn, NaI, DMF, reflux; b. Pd/C, H_2, MeOH; c. NBS, CCl_4, reflux; d. NaOMe, MeOH; e. LAH, THF; f. BnBr, NaH, DMF; g. EtSH, H^+; h. TsCl, pyr.; i. NaN_3, DMF, 80°; j. Br_2, aq. $NaHCO_3$, ether; k. $BrMg(CH_2)_7CH_2OTHP$, THF; l. PCC, NaOAc, CH_2CH_2; m. CbzCl, aq. acetone; n. CrO_3, aq. H_2SO_4, acetone; o. 10% Pd/C, H_2.

of carpamic acid shows that the trisubstituted piperidine with an all-*cis*
stereochemistry can be derived from C-1 — C-6 of D-glucose by deoxygenation at C-2, C-3 and C-6 and replacement of the C-5 oxygen atom by
nitrogen with inversion of configuration. It is also apparent that the
C-3 hydroxyl group in carpamic acid has the same sense of chirality as
is C-4 in D-glucose. Since methyl and hydroxyl substituents at C-2 and
C-3 can be secured by manipulation of D-glucose, the main challenge
resides in the creation of the third chiral center at C-6 bearing the
acid side-chain. The strategy relied on reductive heterocyclization of
an azidoketone precursor which would be expected to provide the desired
cis orientation of substituents. Dideoxygenation of D-glucose at C-2
and C-3 can be readily achieved by application of the Tipson-Cohen
reaction[21] to the ditosylate 116.1 which gave the olefin 123.1.
Selective reduction of the double bond and routine transformations
afforded the trideoxy derivative 123.3, which was converted to the
dithioacetal 123.4. Displacement with azide ion was now possible to
give the azidoaldehyde intermediate 123.6, which was chain-extended,
oxidized and reductively cyclized to 123.9. Protection of the
piperidine nitrogen and oxidation then led to (+)-carpamic acid.
(+)-Azimic acid was similarly synthesized. It is evident that this
strategy can also be used for the synthesis of other molecules
containing substituted piperidines and pyrrolidine rings.[22] Synthesis of
racemic azimic acid has been reported[23].

    f. Streptolidine lactam - Structure elucidation studies[24] on the
streptolin-streptothricin group of antibiotics revealed the presence of
a lactam ring in the streptolidine unit of the intact antibiotic. The
lactam arises from ring closure of the corresponding amino acid,
streptolidine (Scheme 124). Examination of the structure of the lactam
indicates a parent 2,3,5-triamino-2,3,5-trideoxy aldonic acid, which can
be related to a 2,3-diaminoalditol as seen in the retrosynthetic
analysis. Indeed if the connection is made, then the synthetic[25] route
from a carbohydrate precursor is evident. The 3,4-olefin 124.2 can be
easily obtained from the readily available D-mannitol derivative 124.1
by one of several procedures.[26] Epoxidation afforded the 3,4-anhydro-
D-iditol derivative 124.3, which, after treatment with azide ion and a
second displacement gave the 3,4-diazido derivative 124.5 in which the
sense of chirality is the same as in the three contiguous centers in the
target (note the *C2* symmetry in 124.5). Conversion to the acid by

Scheme 124

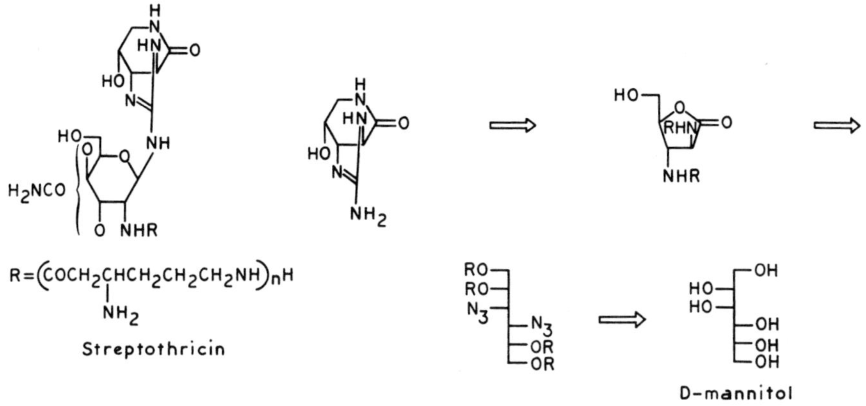

Streptothricin

D-mannitol

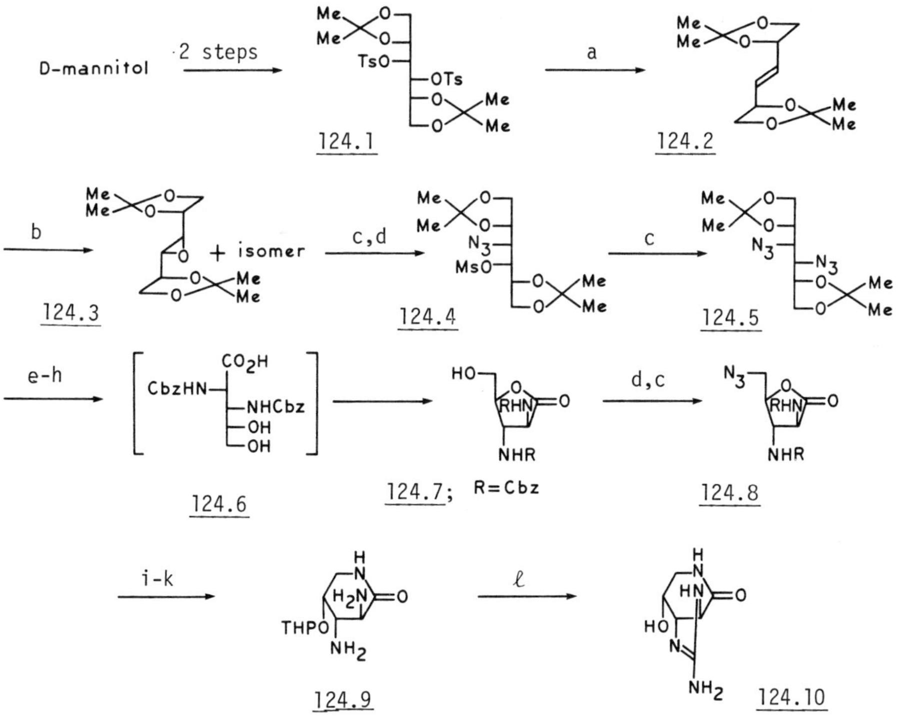

D-mannitol $\xrightarrow{\text{2 steps}}$ **124.1** $\xrightarrow{\text{a}}$ **124.2**

b $\rightarrow$ **124.3** + isomer $\xrightarrow{\text{c,d}}$ **124.4** $\xrightarrow{\text{c}}$ **124.5**

e-h $\rightarrow$ [CbzHN-**124.6**] $\rightarrow$ **124.7**; R=Cbz $\xrightarrow{\text{d,c}}$ **124.8**

i-k $\rightarrow$ **124.9** $\xrightarrow{\ell}$ **124.10**

a. Zn, NaI, DMF, reflux; b. m-CPBA; c. NaN$_3$, DMF, NH$_4$Cl, 120°; d. MsCl, pyr.;
e. Pd/C, H$_2$, MeOH; then CbzCl; f. aq. AcOH; g. NaIO$_4$, aq. acetone; h. Br$_2$,
aq. dioxane; i. Ra-Ni, H$_2$, MeOH; J. dihydropyran, H$^+$; k. Pd/C, H$_2$; l. CNBr

oxidative cleavage at one end only, followed by elaboration of the terminal position gave streptolidine lactam. Streptolidine has been also synthesized from D-ribose and D-xylose but by somewhat longer routes.[27,28]

g. (-)-Mesembrine - Although several syntheses of the Sceletium alkaloid mesembrine have been reported in the literature[29], no enantio-selective synthesis of the natural configuration is known. An examination of the structure in question (Scheme 125), shows that the main challenge resides in the generation of a quaternary carbon atom with the proper sense of chirality. Takano and coworkers[30] have used the (S)-benzyl 2,3-epoxypropyl ether 125.2 prepared from D-mannitol as a source of three carbon atoms in the target and a means of regioselective introduction of the aromatic moiety. Whereas the chiral center in 125.2 was not transferred per se to the target, it was useful in the elaboration of an intermediate lactone structure 125.4, in which the relative disposition of groups allowed a stereoselective preferential alkylation at the benzylic carbon from the least hindered side to give 125.5. Hence the original epoxide was utilized as a "template" upon which to build the carbon skeleton and capitalize on steric effects in the introduction of a new chiral center. Once this process was successfully accomplished the original chiral center was destroyed. The remainder of the sequence was accomplished by well-known procedures to give (-)-mesembrine 125.8 and an immediate precursor 125.7 which could be transformed to the natural product by a one-step ring closure.

h. Miscellaneous heterocyclic molecules - From some of the preceding syntheses it is clear that if the C-4 or C-5 oxygen in a sugar derivative can be replaced by a hetero atom such as nitrogen or sulfur, then in principle, it should be possible to construct corresponding heterocycles with varying degrees of substitution on the carbon framework. Thus, sugars containing nitrogen, sulfur and phosphorus in the ring in place of oxygen are known,[31] and can be synthesized by ring closure of the corresponding heterosubstituted aldehydes. Nojirimycin[32] which is an analog of D-glucose in which the ring oxygen is replaced by nitrogen with such a unique structure. It has been synthesized from D-glucose by a series of reactions which led to the preparation of a 5-amino-5-deoxy-D-glucose bisulfite adduct. Nojirimycin is unstable and loses water to give 3-hydroxyl-6-hydroxymethylpyridine. It can also be

Scheme 125

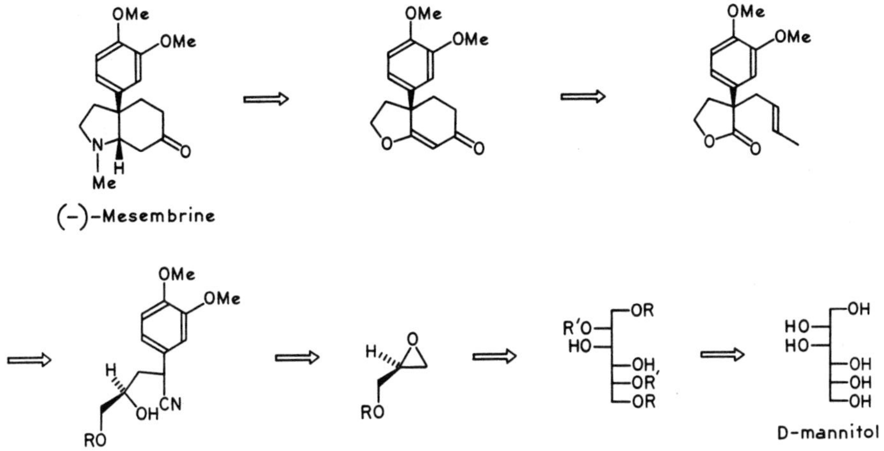
(−)-Mesembrine
D-mannitol

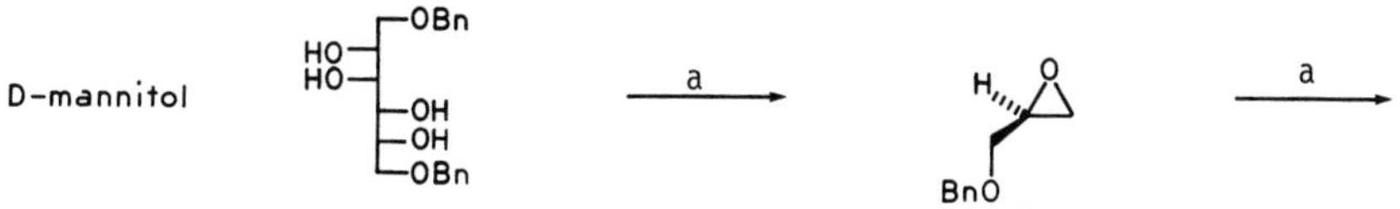
D-mannitol
a
a

125.1 125.2

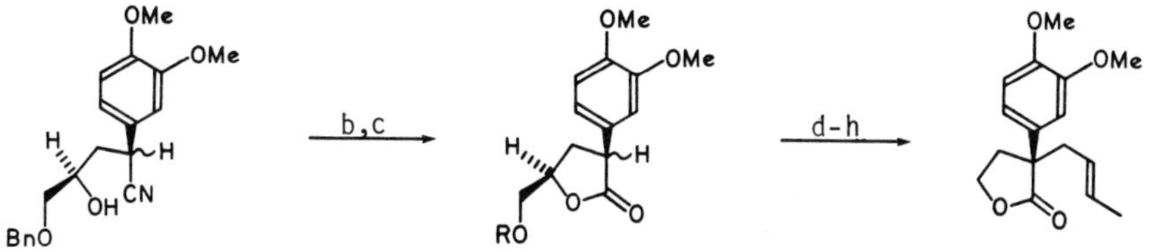
b,c
d-h

125.3 125.4 125.5

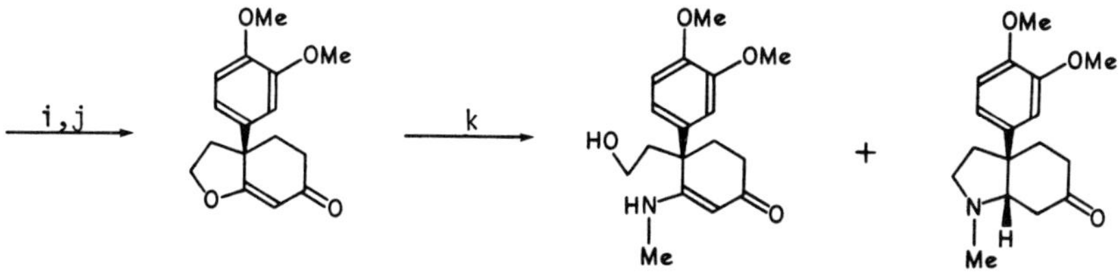
i,j
k
+

125.6 125.7 125.8

a. several steps; b. 3,4-dimethoxybenzyl cyanide, LDA, THF; c. aq. KOH, EtOH, reflux; d. HCl-EtOH; e. LDA, crotyl bromide, THF; f. HCl -EtOH, reflux; g. aq. KOH, MeOH; CO_2 gas, then $NaIO_4$; h. $NaBH_4$; i. $PdCl_2$-CuCl, wet DMF, O_2; j. t-BuOK, THF, reflux; k. aq. $MeNH_2$, sealed tube, 180°; 125.7 → 125.8 with Ph_3P, $EtO_2CN{=}NCO_2Et$.

catalytically reduced to the corresponding piperidine.

　　　　Chiral, hydroxylated piperidines can be prepared from the
corresponding azidoaldehydes by reductive cyclization.[33,34]   Chiral
hydroxylactams can also be readily prepared from appropriate azido
lactones by reductive processes.   Thus, five, six and seven-membered
polyhydroxy lactams with predisposed orientations of hydroxyl groups can
be obtained.[33-35] 11-Thia-[36] and related hetero-prostaglandin
intermediates can be envisaged from carbohydrate precursors and the idea
can be extended to related heterocyclic derivatives in the prostaglandin
and prostacyclin class of compounds.   Miscellaneous heterocyclic
compounds including aromatic types can be prepared from carbohydrates
either by incorporating a segment of the carbon chain or by cyclization
with aldehydo forms.[31]   Often such derivatives contain carbon appendages
with one or more hydroxyl groups and can be utilized as useful
'chirons'which contain a specific heterocyclic moiety.

　　　　i.   Approaches to the synthesis of β-lactam antibiotics -
Synthetic studies in the field of β-lactam antibiotics have presented
the organic chemist with much challenge.[37] Although the penicillins and
related compounds do not possess numerous chiral centers, their
structures are complex due to the presence of hetero atoms in addition
to the delicate balance of reactive sites such as the azetidinone ring.
In order to secure at least one chiral center, some previous syntheses
have utilized optically active amino acids as building blocks,[37] but to
the best of our knowledge no carbohydrate approach has been reported.
Inspection of the structure of penicillin G (Scheme 126) reveals that
cleavage of the azetidinone ring in a retrosynthetic analysis, leads to
a 5(R), 6(R) benzyl penicilloate which in fact was a critical intermediate in the classical Sheehan synthesis of penicillin.[38]   Since the
sense of chirality at C-6 in penicillin G is the same as that of C-2 in
2-amino-2-deoxy-D-glucose, a possible synthetic approach can be seen via
a stereocontrolled thiazolidine ring formation between the amino sugar
and D-penicillamine.   In such an approach, and assuming that the correct
chirality is obtained at C-5, the carbon chain of the sugar portion
could be sacrificed producing the penicilloate skeleton.   In fact,
extensive studies by Bognar and coworkers[39] have demonstrated the
formation of thiazolidines from carbohydrates, including derivatives of
D-penicillamine. Scheme 126 depicts the approach[40] which utilizes the 2-
phenylacetamido derivative 126.1, readily prepared from 2-amino-2-deoxy-

D-glucose, as the source of the three carbon unit needed in the target. Condensation with D-penicillamine gave isomeric thiazolidine derivatives which could be separated and converted to the polyol 126.2 (and its isomer) individually. Oxidative cleavage leads to the corresponding 5(R), 6(R) and 5(S). 6(R) benzyl penicilloates which are available from the antibiotic.[41]

Scheme 126

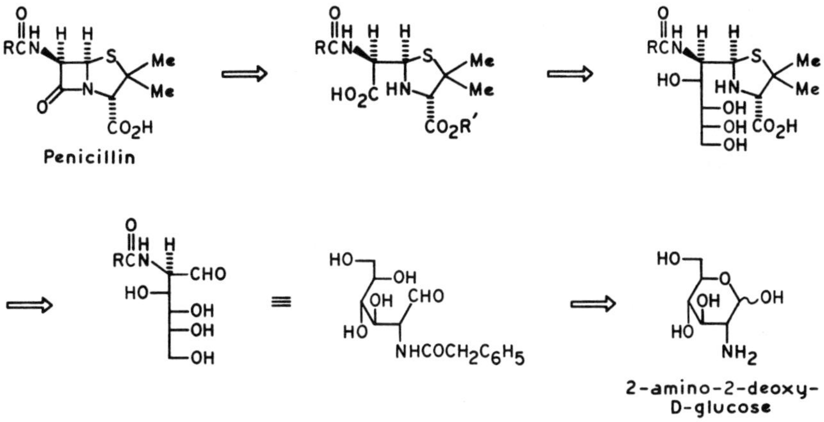

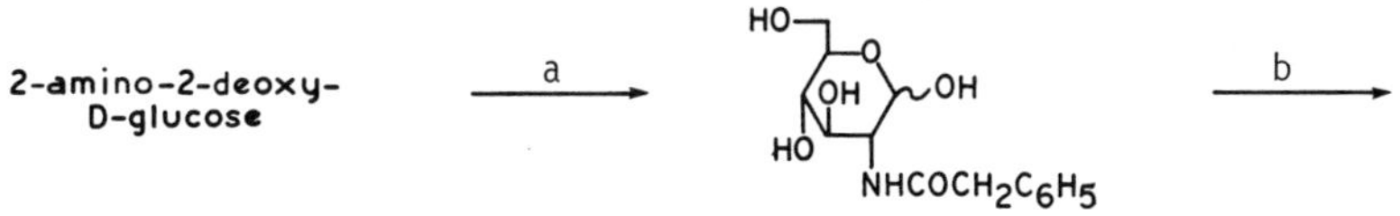

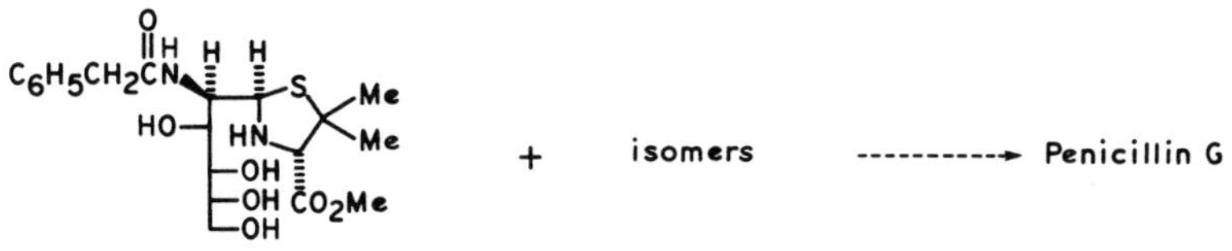

a. $C_6H_5CH_2COCl$, base; b. D-penicillamine, pyr. or EtOH, 80^0; c. CH_2N_2

The thienamycin structure presents a different synthetic challenge
and elegant syntheses as well as approaches to the antibiotic and its
derivatives are available.[42] Our retrosynthetic analysis of thienamycin
locates hidden carbohydrate-type symmetry which could greatly simplify
the construction of its carbon backbone incorporating three chiral
centers (Scheme 127).[43] Thus, retrosynthetic cleavage of the azetidinone
ring, excision of the C-3 carbon and the aminothiol side-chain leaves a
six-carbon backbone (C-1 — C-9) which can be considered at different
oxidation levels at C-2 (see Scheme 11). The perspectives shown in
Scheme 127 reveal the convergence with a trideoxyamino sugar derivative. In fact the sense of chirality at C-8 in the target coincides
with C-5 of a D-hexose, hence the problem resolves itself in the generation of a C-4 carboxyl branched-chain sugar. Note that the "rule of
five" can be nicely applied here and in principle the substitution
pattern as well as the absolute configuration can be elaborated using an
alkyl D-hexopyranoside of choice. Scheme 127 illustrates how D-glucose
was transformed into the critical 4-keto intermediate 127.4 by routine
reactions. Except for the carboxyl group, all other functional groups
are incorporated in latent or apparent form. Several routes can be envisaged to incorporate the carboxyl group at C-4 in this intermediate
via the use for example, of ketene dithioacetals in cyclic or acyclic
precursors. In the latter case there is a greater possibility for the
formation of products isomeric at C-6 hence, a possible route to the
olivanic acids. Intermediate 127.4 was converted into the ketenedithio-
acetal derivative 127.5 which upon reduction with lithium aluminum
hydride gave the α-orientated side-chain at C-4, presumably by virtue of
a highly stereoselective attack of hydride ion from the β-face. This
was anticipated to occur by complexation with the newly generated amino
group. Standard transformations led eventually to the chiral lactone
127.8 which had been previously converted into thienamycin (racemic
form).[42] Two other carbohydrate-based approaches have been recently reported.[44,45] It is evident that such a strategy could be applicable to
the synthesis of carbapenams, carbapenems, carbacephems and related
β-lactam antibiotics.[37] The synthesis of the L(3R,5R) analog of
Ro22-5417, a new clavam antibiotic from D-xylose has been reported by
Weigele and coworkers.[46] The complete structure of the antibiotic was
established as L(3S,5S)-3-(7-oxo-1-aza-4-oxabicyclo[3.2.0]hept-3-yl-
alanine.

Scheme 127

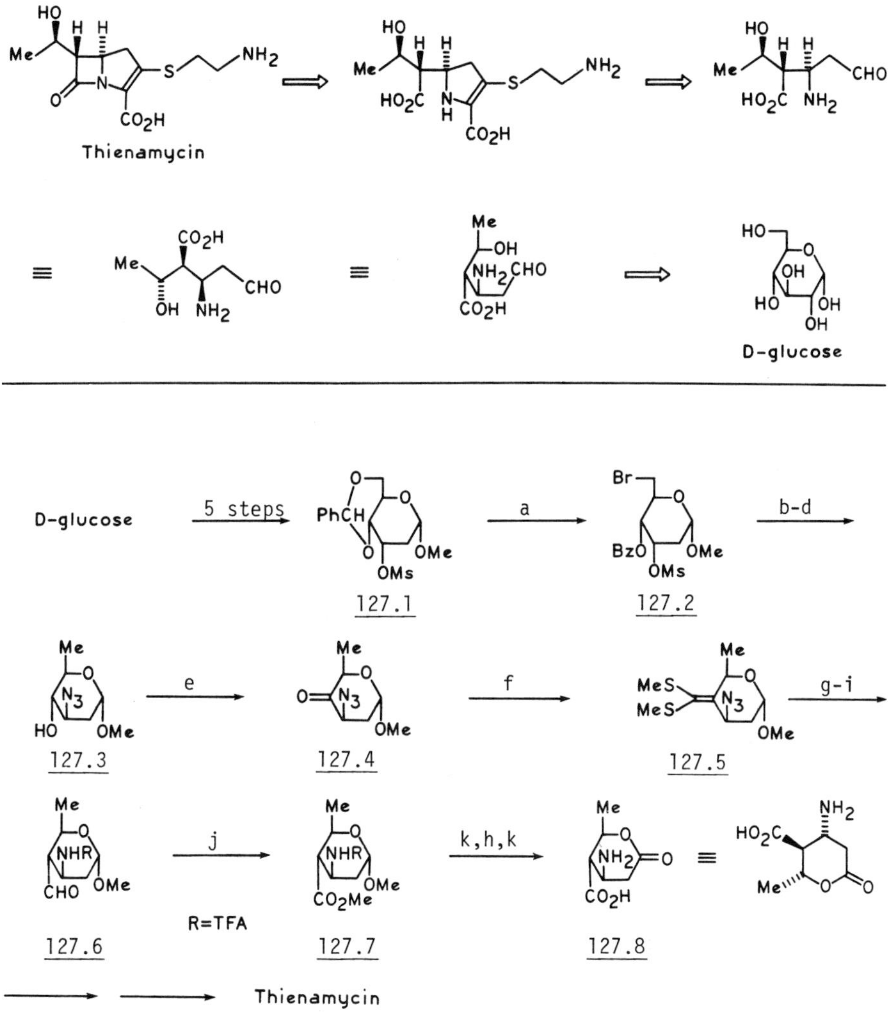

Thienamycin
D-glucose
5 steps
a
b-d
e
f
g-i
j
k,h,k
R=TFA
127.1
127.2
127.3
127.4
127.5
127.6
127.7
127.8
Thienamycin

a. NBS, CCl_4, reflux; b. Pd/C, H_2; c. Bu_4NN_3, toluene, reflux; d. NaOMe, MeOH; e. PCC, CH_2Cl_2, mol. sieves; f. $(MeO)_2P(O)CH(SMe)_2$, BuLi, THF; g. LAH, THF; h. TFAA, pyr; i. $HgCl_2$, HgO, aq. acetone; j. CrO_5, acetone, then CH_2N_2; k. aq. HCl, THF; l. aq. Br_2, MeCN

   j. A precursor to the acyclic side-chain in Vitamin E - The
acyclic side-chains in vitamin E containing two asymmetric centers has
been prepared from D-glucose by a combination of operations in which the
carbohydrate framework was used as a template in two ways. First to
generate a five-carbon segment of the target, and second for C-C bond
formation based on chirality transfer.[47] Relating the acyclic segment
of vitamin E to a carbohydrate precursor is somewhat remote since
extensive deoxygenation is involved. However, Trost and coworkers[47]
have capitalized on the predictable stereoselectivity of π-allyl
palladium reactions, particularly with a chiral appendage such as a
cyclic carbohydrate derivative (Scheme 128). Thus, D-glucose was
transformed into the deoxygenated derivative 128.1, which was in turn
transformed to the olefin 128.2. Manipulation of the oxidation state
and protection gave 128.3 which was subjected to a critical alkylation
with sodio malonate via a π-allyl palladium intermediate to give 128.4,
with remarkable stereocontrol. Elaboration of the second chiral center
was affected by chain-extension to 128.6 and a stereospecific cuprate
reaction on a secondary tosylate to give 128.3 together with some
elimination. Routine decarboxylation gave the known, optically active
acid 128.8. Note that all the original hydroxyl groups of the
carbohydrate were sacrificed in this operation with the resultant
incorporation of two C-methyl groups on an acyclic hydrocarbon chain in
which five carbon atoms can claim descendance from D-glucose. Other
syntheses and synthetic approaches to vitamin E are also known.[48]

REFERENCES

1. B.P. Mundy, R.D. Otzenberger and A.R. De Bernardis, J. Org. Chem.,
    36, 2390 (1971).

2. S. Jarosz, D.R. Hicks and B. Fraser-Reid, J. Org. Chem., 47, 935
    (1982); D.R. Hicks and B. Fraser-Reid, J.C.S. Chem. Comm., 869
    (1976); for a synthesis of (+)-exo-brevicomin, a pheromone related
    in structure to frontalin, see A.E. Streck and B. Fraser-Reid, J.
    Org. Chem., 46, 932 (1982).

3. S. Ohrui and S. Emoto, Agr. Biol.Chem., 40, 2267 (1976).

4. M. Kawana and S. Emoto, Tetrahedron Lett., 39, 3395 (1975).

5. K. Mori, Tetrahedron 31, 1381 (1975).

6. V.Du Vigneaud, D.B. Melville, K. Folkers, D.E. Wolf, R. Mozuigo,
    J.C. Keresztesy and S.A. Harris, J. Biol. Chem., 146, 475 (1942);

Scheme 128

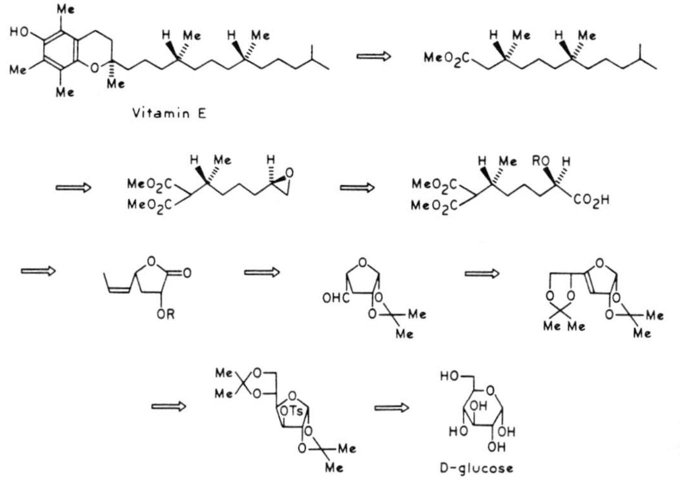
Vitamin E
D-glucose

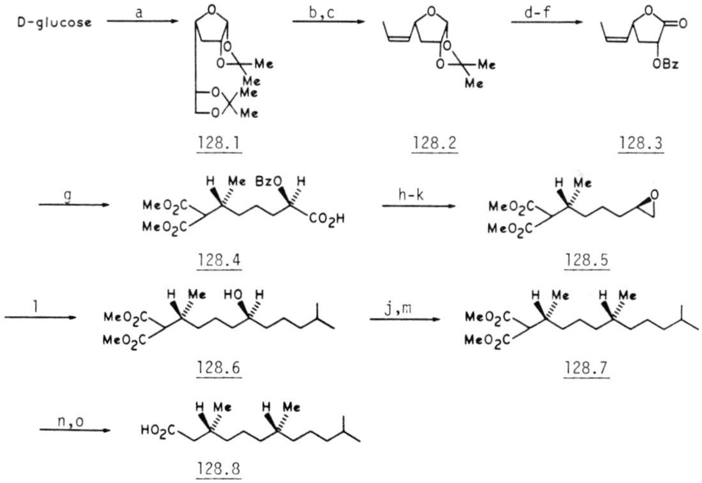
D-glucose
a
b,c
d-f
128.1
128.2
128.3
g
128.4
h-k
128.5
l
128.6
j,m
128.7
n,o
128.8

a. See schemes 64, 89; b. aq. $NaIO_4$; c. $Ph_3P{=}CHCH_3$, THF; d. H_3O^+; e. Ag_2CO_3, celite, benzene; f. BzCl, pyr; g. $CH_2(CO_2Et)_2$, $Pd(PPh_3)_4$, THF; h. PtO_2, H_2, AcOH; i. BH_3-SMe_2, ether; j. TsCl, pyr; k. NaOMe, MeOH; l. isopentenyl copper cyanide, ether; m. MeCuLi, ether; n. KOAc, DMSO, 140°; o. aq. KOH, MeOH see also C. Bonnemere, J.A. Hamilton, L.K. Steinrauf and J. Knappe, Biochem., 4, 240 (1965); J. Trotter and J.A. Hamilton, Biochem., 5, 713 (1966).

7. For a summary of syntheses, see P. Rossy, F.G.M. Vogel, W. Hoffmann, J. Paust and A. Nürrenbach, Tetrahedron Lett., 22, 3493 (1981); P.N. Confalone, G. Pizzolato, E.G. Baggiolini, D. Lollar and M.R. Uskoković, J. Am. Chem. Soc., 99, 7020 (1977); see also, S. Lavielle, S. Bory, B. Moreau, M.J. Luche and A. Marquet, J. Am. Chem. Soc., 100, 1558 (1978).

8. T. Ogawa, T. Kawana and M. Matsui, Carbohydr. Res., 57, C31 (1977).

9. R.J. Parry and M.G. Kunitani, J. Am. Chem. Soc., 98, 4024 (1976) and references cited therein.

10. H. Ohrui and S. Emoto, Tetrahedron Lett., 2765 (1975).

11. H. Ohrui, N. Sueda and S. Emoto, Agr. Biol. Chem., 42, 865 (1978).

12. B.A. Sobin and F. W. Tanner, J. Am. Chem. Soc., 76, 4053 (1954); J.P. Schaefer and P.J. Wheatley, J. Org. Chem., 33, 166 (1968) and references cited therein.

13. J.P.H. Verheyden, A.C. Richardson, R.S. Bhatt, B.D. Grant, W.L. Fitch and J.G. Moffatt, Pure Appl. Chem., 51, 1363 (1978).

14. A.B. Smith, III, S.J. Branca, N.N. Pilla and M.A. Guaciaro, J. Org. Chem., 47, 1855 (1982); see also S. Oida and E. Ohki, Chem. Pharm. Bull., 17, 1405 (1969); C.M. Wong, J. Buccini, I. Chang, J. Te Raa and R. Schwenk, Can. J.Chem., 47, 2421 (1969).

15. C.M. Wong, J. Buccini and J. Te Raa, Can. J. Chem., 46, 3091 (1968); I. Felner and K. Scheuker, Helv. Chim. Acta, 53, 754 (1970).

16. H. Yonehara, H. Seto, A. Shimazu, S. Aigawa, T. Hidaka, K. Kakinuma and N. Otake, Agric. Biol. Chem., 37, 2771 (1973); K. Kakinuma, N. Otake and H. Yonehara, Tetrahedron Lett., 2509 (1972).

17. K. Kakinuma, Tetrahedron Lett., 21, 167 (1980).

18. T.M. Smalberger, G.J.H. Rall and H.L. de Waal, Tetrahedron 24, 6417 (1968) and references cited therein.

19. J.L. Coke and W.Y. Rice, J.Org. Chem., 30, 3420 (1965) and references cited therein.

20. S. Hanessian and R. Frenette, Tetrahedron Lett., 3391 (1979).

21. R.S. Tipson and A. Cohen, Carbohydr. Res., 1, 338 (1965-66).

22. See for example G. Fodor, J.-P. Fumeaux and V. Sankanau,

Synthesis, 464 (1972); D. Gross, Fortschr. Chem. Org. Naturst., 29, 1 (1971).; R.K. Hill in, The Alkaloids, S.W. Pelletier, Ed., Van Nostrand Reinhold, New York, N.Y., 1970, p.395.

23. E. Brown and D. Dhal, Tetrahedron Lett., 1029 (1974).

24. E.E.Van Tamelen, J.R. Dyer, H.A.Whaley, H.E.Carter and G.B. Whitfield, Jr., J. Am. Chem. Soc., 83, 4295 (1961); H.E. Carter, R.K. Clark, Jr., J.W. Rothrock, W.R. Taylor, C.A.West, G.B. Whitfield and W.G. Jackson, J. Am. Chem. Soc., 76, 566 (1954); D.B. Borders, K.J. Sax, J.E. Lancaster, W.K. Hausmann, L.A. Mitscher, E.R. Wetzel and E.L. Patterson, Tetrahedron, 26, 3123 (1970).

25. M. Kinoshita and Y. Suzuki, Bull. Chem. Soc. Japan, 50, 2375 (1977).

26. See for example,S. Hanessian, A. Bargiotti and M. LaRue, Tetrahedron Lett., 787(1978).

27. S. Kusumoto, S. Tsuji and T. Shiba, Tetrahedron Lett., 1417 (1974); Bull. Chem. Soc. Japan, 47, 2690 (1974); S. Kutsumoto, S. Tsuji, K. Shima and T. Shiba, ibid., 49, 3611 (1976).

28. T. Goto and T. Ohgi. Tetrahedron Lett. 1413 (1974).

29. For a review, and pertinent syntheses, see, R.V. Stevens, in The Total Synthesis of Natural Products, J.ApSimon, ed., Wiley-Interscience, New York, N.Y., vol. 3, 1977, p.443; S.F. Martin, T.A. Puckette and J.A. Colapret, J. Org. Chem., 44, 3391 (1979) and references cited therein.

30. S. Takano, Y. Inamura and K. Ogasawara, Tetrahedron Lett., 22, 4479 (1981).

31. For a review, see H. Paulsen, Angew. Chem. Int. Ed. Engl., 5, 495 (1966).

32. S. Inouye, T. Tsuruoka, T. Ito and T. Niida, Tetrahedron 24, 2125 (1968).

33. C.C. Deane and T.D. Inch, Chem. Comm., 813 (1969), see also T.D. Inch, Advan. Carbohydr. Chem. Biochem., 27, 191 (1972).

34. S. Hanessian, Chem. Ind., 2126 (1966).

35. S. Hanessian, J. Org. Chem., 34, 675 (1969); S. Hanessian and T.H. Haskell, J. Heterocyclic Chem., 1, 55, 57 (1964).

36. I.K. Boessenkool and G.J. Lourens, S. African J.Chem., 33, 113 (1980)..

37. For a recent review, see Chemistry and Biology of β-lactam Antibiotics, R.B. Morin and M. Gorman, eds., Academic Press, New York, N.Y., vol. 1-3, 1982.

38. J.C. Sheehan and K.R. Henery-Logan, J. Am. Chem. Soc., 79,
    1262 (1957); 81, 3089 (1959) and references cited therein.
39. R. Bognár, Z. Györgydeák and L. Szilágyi, Ann. 701 (1979); 1536
    (1977) and previous papers.
40. S. Hanessian, S. Sahoo, H. Wyss, W. Streicher, E.Cullen and G.
    Patil, unpublished results.
41. R. Busson, P.J. Claes and H. Vanderhaeghe, J. Org. Chem., 41, 2556
    (1976).
42. S.M. Schmitt, D.B.R. Johnson and B.B. Christensen, J. Org. Chem.,
    45, 1142 (1980) and previous paper; D.G. Melillo, I. Shinkai, T.
    Lies, K. Ryan and M. Sletzinger, Tetrahedron Lett., 21, 2783
    (1980); see also D.G. Melillo, T. Liu, K. Ryan, M.Sletzinger and I.
    Shinkai, Tetrahedron Lett., 22, 913 (1981); T. Kametani, S.P. Huang,
    S. Yokohama, Y. Suzuki and M. Ihara, J. Am. Chem. Soc., 102, 2060
    (1980); J.J. Tufariello, G.E. Lee, P.A. Senaratne and M. Al-Nuri,
    Tetrahedron Lett. 4359 (1979). M. Shiozaki, N. Ishida, T. Hiraoka
    and H. Yanagisawa, Tetrahedron Lett., 22, 5205 (1981). T.N.
    Salzmann, R.W. Ratcliffe, B.G. Christensen and F.A. Bouffard,
    J. Am. Soc., 102, 6161 (1980).
43. S. Hanessian, G. Rancourt, D. Desilets and R. Fortin, Can. J. Chem.,
    60, 2292 (1982); S. Hanessian and R. Fortin, Abstract 181st National
    Am. Soc. Meeting, Atlanta, March 29-April 3, 1981, CARB 27.
44. P. Durette, Carbohydr. Res., 100, C27 (1982).
45. N. Ikota, D. Yoshino and K. Koga, Chem. Pharm. Bull., 30, 1929
    (1982).
46. J.-C. Mueller, V. Toome, D.L. Pruess, J.F. Blount and M. Weigele,
    J. Antibiotics, 36, 217 (1983).
47. B.M. Trost and T.P. Klun, J. Am. Chem. Soc., 103, 1864 (1981).
48. N. Cohen, R.J. Lopresti, C. Neukom and G. Saucy, J. Org. Chem.,
    45, 582 (1980), and earlier references in this series; H. Mayer,
    P. Schudel, R. Rüegg and O. Isler, Helv. Chim. Acta, 46, 650
    (1963); C. Fuganti, P. Guselli, J. Chem. Soc., Chem. Comm., 995
    (1979); M. Schmid and R. Barner, Helv. Chim. Acta, 62, 464 (1979);
    R. Zell, Helv. Chim. Acta, 62, 474 (1979); H. Heitzer, Synthesis,
    888 (1979).

CHAPTER FIFTEEN

MACROLIDES AND ANSA COMPOUNDS

The constitutional structures, important biological properties and intriguing conformational features of the so-called macrolide antibiotics, have been the subject of elegant studies over the years.[1] We should ponder over R.B. Woodward's words written in 1956[2] concerning the prospects of achieving a synthesis of a macrolide: "Erythromycin, with all its advantages, looks at present quite hopelessly complex, particularly in view of its plethora of asymmetric centers..." It is therefore a great victory for the organic chemist to have conquered this structure by synthesis some 25 years later particularly that one of the two groups involved was that of the late R.B.W.'s.[3,4] These landmark achievements in synthesis have been paralleled by important studies on the biological front, in quest of antibiotics with broader antibacterial spectra. It is evident that work aimed at the assembly of the multifunctional carbon backbone of these molecules and segments thereof, constitutes a great challenge to the synthetic organic chemist. Indeed, elegant methodology has been developed as a result of this impetus and applied to the total synthesis of macrolides.[1] Concurrent with these studies, we have seen the emergence of ingenious methods for the formation of macrocyclic lactones and lactams.[1]

a. Erythronolide A - Assembly of the carbon skeleton - The aglycone portion of the commercially important antibiotic erythromycin A, has been the object of intense studies over the years and on many fronts. It is among a large number of natural products which are

Scheme 129

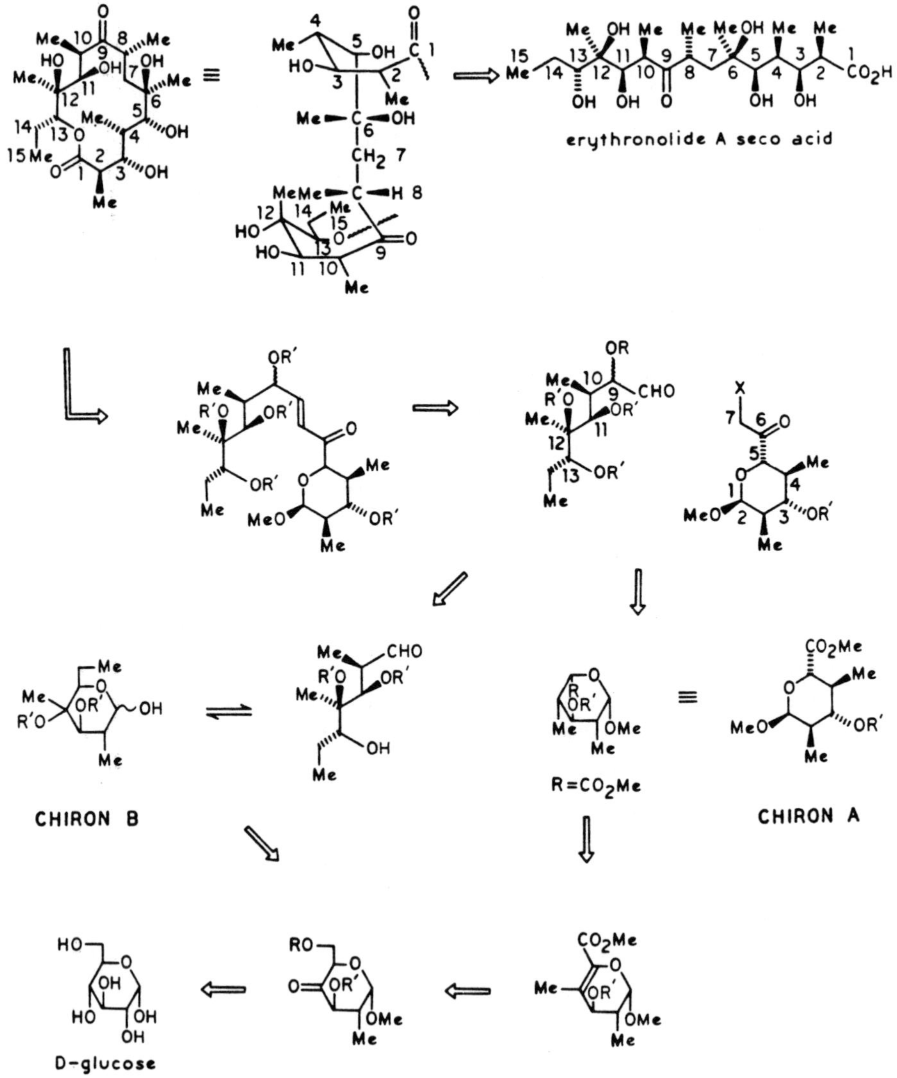
erythronolide A seco acid
CHIRON B
R=CO2Me
CHIRON A
D-glucose

biosynthetically derived from the so-called propionate pathway.[1] Examination of its structure (Scheme 129) reveals the presence of alternating
C-methyl and hydroxyl groups on a 15-carbon backbone. There are ten
asymmetric centers, two tertiary sites, and a ketone function in a
14-membered lactone. Since the synthetic approach to be described here
can be considered as a prototype study applicable to other members of
the macrolide family, a cursory look at the conceptual part of the
strategy may be in order. Erythronolide A possesses hidden
carbohydrate-type symmetry in its most intriguing yet simple form.[5]
When the two-dimensional perspective structure is "unfolded" to
accommodate other features,there emerge two "carbohydrate-like" segments
(Scheme 129, see also Scheme 7) encompassing C-1 — C-6 and C-9 — C-15.
These can be related to two possible 'chirons' A and B which can be
derived from a suitable carbohydrate by systematic introduction of
functional groups.[6] The acyclic equivalents of these 'chirons' are also
shown and could be correlated with segments of erythronolide A seco
acid. Viewed in a vertical perspective (not shown) the seco acid is
nothing but a large Fischer projection which has been useful in
establishing configurational correlations and biogenetic interrelation-
ships among the macrolide antibiotics.[1] It is here that the visual
dialogue alluded to in Chapter three manifests itself in terms of the
obvious progeny with carbohydrate-type precursors when considering
various synthetic approaches. There are several features in the
structures of 'chirons' A and B that are worthy of comment; i. the
absolute configuration at C-2/C-3 in both are the same, hence the
possibility of utilizing common synthetic intermediates; ii. the
anomeric carbon atom in each corresponds to C-1 and C-9 respectively of
the target (note sp^2 centers); iii. eight out of ten asymmetric
carbon atoms of the target are encompassed in the structures of the
'chirons'; iv. the ring oxygens corresponds to C-5 and C-13 hydroxyl
groups in the target respectively and they are conveniently "protected"
until such time as they are needed. Note how the "rule of five" can be
superbly applied to the C-1—C-6 and C-9—C-15 segments of the
target. With this type of stereochemical decoding done, there remains
to plan a synthetic strategy for the generation of these 'chirons', for
their homologation to acyclic intermediates and for joining them to give
the acyclic carbon framework of erythronolide A.

The retrosynthetic analysis in Scheme 129 shows the type of bond-
breaking and bond forming operations envisaged.[6] Thus, an enone encom-

passing the entire carbon framework of the target and containing the
eight asymmetric centers can be envisaged to arise from the union of a
nucleophilic component (a β-ketophosphonate) obtained by homologation of
'chiron' A (C-1 — C-6), and an electrophilic component (an aldehyde)
produced from 'chiron' B (C-9 — C-15) by a one-carbon extension.

The C-methyl group at C-8 can be introduced by conjugate addition
on the enone, while the tertiary center at C-6 can be introduced by
organometallic methodology. Note that the C-5 and C-13 hydroxyl groups
remain differentiated all along since the former is still "protected" as
the ring oxygen in the enone. 'Chirons' A and B arise from a common 4-
keto intermediate which is easily obtained from D-glucose. Note however
that 'chiron' A formally belongs to the L-series while B is in the D-
series. For the latter, it is simply necessary to introduce a stereocontrolled branching at C-4 and a chain extension at C-5, operations
which are well in hand by existing methodology. Access to 'chiron' A
was envisaged via an endocyclic unsaturated ester (Scheme 129) which
would be expected to undergo stereoselective catalytic hydrogenation
because of the α-orientation of the anomeric substituent (anomeric
stereoselection)[6].

Scheme 130 shows how D-glucose was transformed into a valuable 4-
keto intermediate 130.3 which could be used for the synthesis of both
'chirons'.[6] The plan for a stereocontrolled introduction of the C-2/C-3
substituents was based on notions of conformational bias. Thus,
oxidation of 35.2 and epimerization led to 130.1, generating the first
asymmetric center in the intended target (C-2). The hydroxyl group at
C-3 was generated by epimerization of the intermediate ketone 130.2.
For purposes of operational convenience the hydroxyl group was protected
as the methyl ether in this and subsequent intermediates, which
unexpectedly conferred a high degree of crystallinity. Thus, starting
with D-glucose it was possible to operate on large scale without
recourse to chromatography for many steps. Treatment of 130.3 with
methyl lithium[7] led to a preponderance of 130.4 (after methylation and
separation by crystallization) which now had three asymmetric centers
convergent with C-10 — C-13 of the target. Standard homologation then
gave 'chiron' B as the dimethyl ether (Scheme 131). The mother liquors
from the methylation, containing the C-4 epimeric alcohol as well as
130.4 were used to generate 'chiron' A, rendering the entire sequence
highly efficient. Thus, treatment of the aldehyde 131.1 with calcium
hydroxide (other bases gave mixtures!) followed by oxidation and

Scheme 130

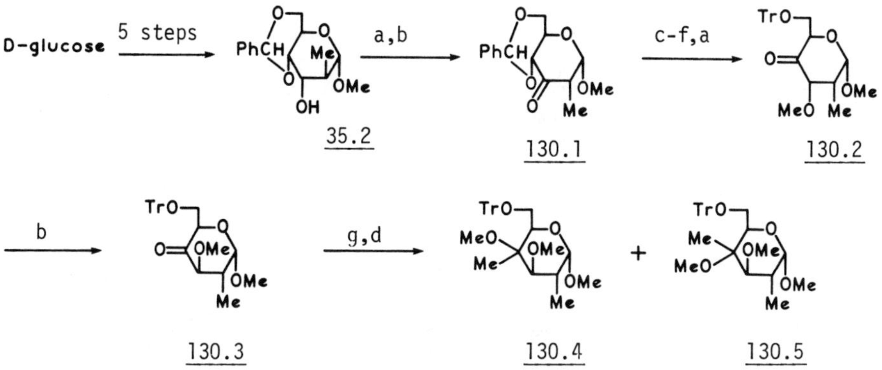

a. DMSO, Ac_2O; b. NaOMe, MeOH; c. $NaBH_4$, MeOH; d. MeI, NaH, DMF;
e. $Pd(OH)_2/C$, H_2; f. TrCl, pyr.; g. MeLi

esterification gave 131.2, which as expected was reduced catalytically
to afford 'chiron' A. Homologation with dimethyl- methylphosphonate
anion gave the phosphonate 131.4 in good yield. Homologation of
'chiron' B was done by treatment of the lactol with vinylmagnesium
bromide, benzylation and ozonolysis to give the aldehyde 131.7. The
stage was now set for the critical union of intermediates 131.4 and
131.7, which was successfully accomplished to give the enone 132.1[8] as a
1:1 mixture of epimeric alcohols at C-9. Even though they could be
separated by chromatography, the sequence was continued on the mixture,
since C-9 is the site of a carbonyl group in the final target.
Treatment of 132.1 with lithium dimethylcuprate followed by methyl
lithium and methylation gave 132.2. Hydrolysis, oxidation and further
hydroxyl protection gave 132.3 which is shown in its acyclic as well as
macrolide perspectives (Scheme 132).

Thus, the entire carbon backbone of 9-dihydroerythronolide A seco
ester was assembled from D-glucose. All functional groups were introduced with virtually complete regiospecificity and the sense of chirality of at least eight centers was assured. Only the stereochemistry at
carbon atoms 6 and 8 are questionable, although the latter could be
subject to equilibration in the final product. The sense of chirality
at C-6 is as yet unknown and cannot be changed. However, this center is
created late in the sequence and could be studied further. Note that a
Grignard reaction was not used because it was anticipated that by virtue

Scheme 131

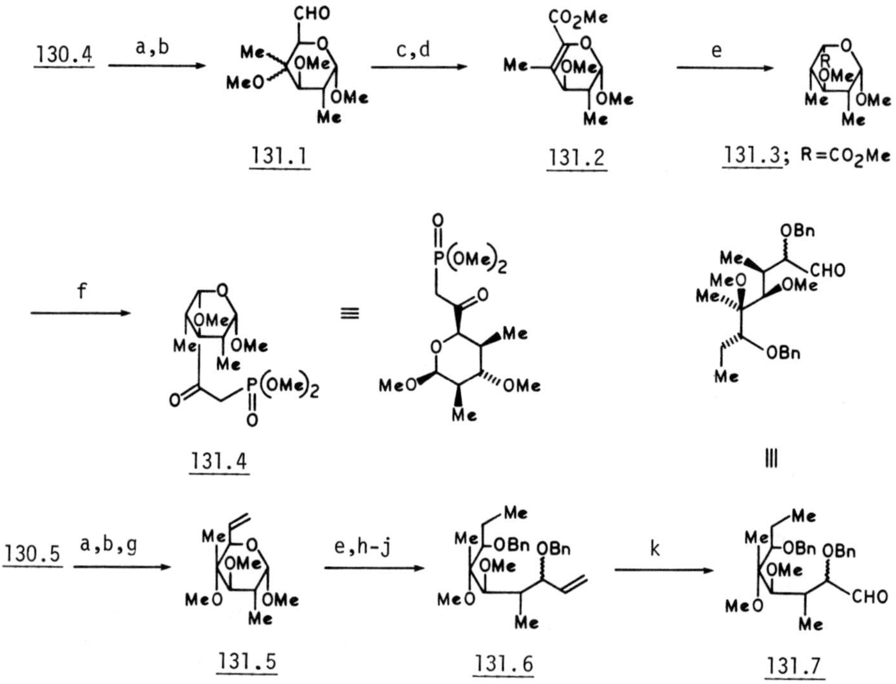

a. $Pd(OH)_2/C$, H_2; b. Collins; c. $Ca(OH)_2$; d. NaCN, MnO_2, MeOH; e. Pd/C, H_2;
f. $(MeO)_2P(O)CH_2Li$, THF; g. $Ph_3P=CH_2$; h. aq. AcOH; i. $CH_2=CHMgBr$, ether;
J. BnBr, NaH, DMF; k. O_3, CH_2Cl_2

of coordination with the ring oxygen in 132.1 (after cuprate addition),
the stereochemical outcome would be such that the desired isomer would
be the minor one. Previous work[9] and model studies[6] have borne this
out. For example, it was shown that a Grignard reaction on 133.3,
obtained from 130.4 by a sequence similar to the one that gave 131.3,
led to a 6:1 mixture of allylic alcohols in favor of the correct isomer
133.4 (Scheme 133). Presumably a coordinated species such as 133.5
could be involved. This could also constitute another strategy for the
synthesis of erythronolide and its precursors, in which the stereochemistry at C-6 is secured early in the sequence. It should also be
pointed out that 'chirons' A and B could be obtained as the free hydroxy

Scheme 132

$131.4 + 131.7$ $\xrightarrow{a,b}$ 132.1 $\xrightarrow{c,d}$ 132.2

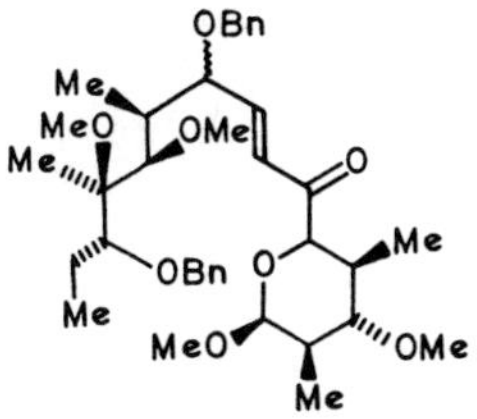

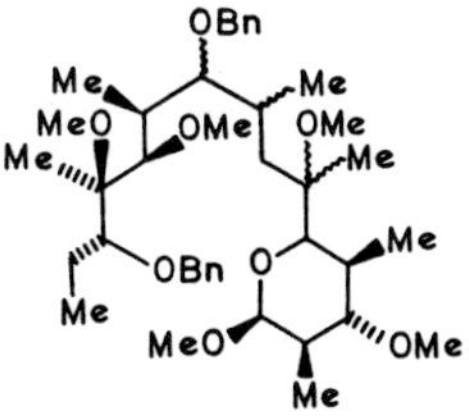

$\xrightarrow{e-g,\ d}$ 132.3 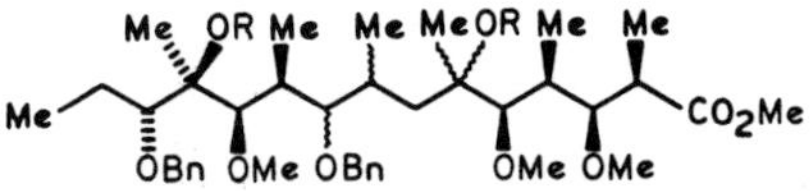 $\equiv$ $R = Me$

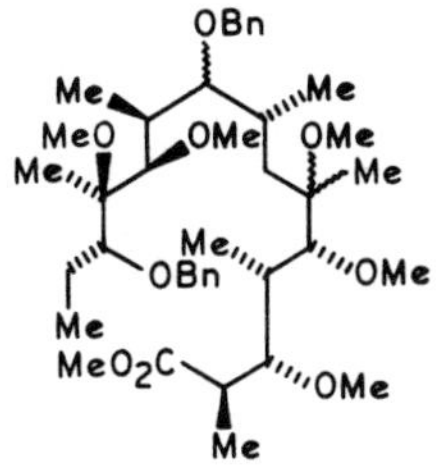

a. NaH, THF; b. Me$_2$CuLi; c. MeLi, ether; d. MeI, NaH, DMF; e. aq. AcOH;
f. PCC, CH$_2$Cl$_2$; g. aq. KOH; h. Dowex-50(H$^+$), then CH$_2$N$_2$

derivatives[8] since intermediate 130.4 could be demethylated with lithium
in ethylamine.[10] From the preceeding studies it is evident that the
carbohydrate route to macrolide aglycones is a viable and practical one
and it should be applicable to the construction of other structures in
this series, as well as molecules derived from propionate precursors in
general. Other chiral precursors[11,13] useful in the construction of
six-carbon segments of such molecules can be prepared by regio- and
stereocontrolled introduction of C-methyl and hydroxyl groups based on
the concepts discussed above. Impressive syntheses of erythronolide
A,[3,4] and a stereoselective synthesis of the chiral sequence of erythronolide A[14] have been recently reported based on different strategies.

   b. Carbonolide B - Synthesis of the C-1 — C-6 segment - Examination of the right-hand portion of carbonolide B, the aglycone portion of
carbomycin B (magnamycin B,[15] Scheme 134) shows a segment C-1—C-6

Scheme 133

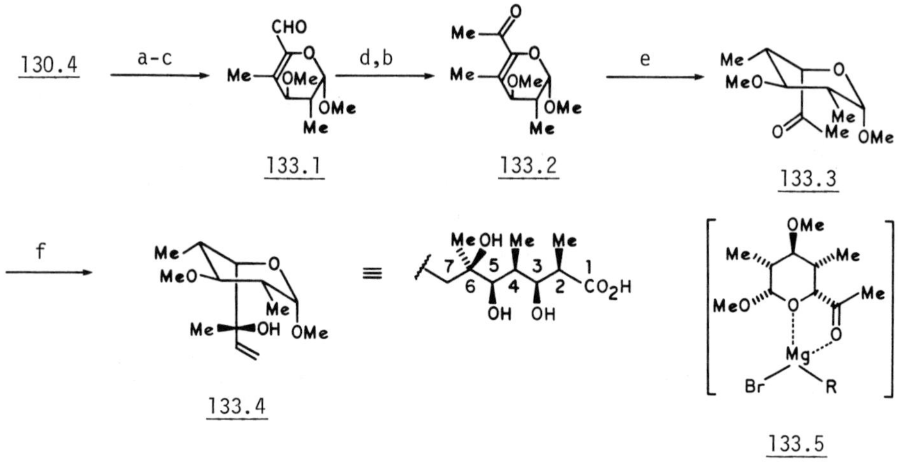

a. $Pd(OH)_2/C$, H_2; b. Collins; c. $Ca(OH)_2$; d. MeLi; e. Pd/C, H_2; f. $CH_2=CHMgBr$

which can be formally related to a synthetic carbohydrate precursor
based on the concepts developed in section a. There are basically two
modes of bond disconnection to uncover a carbohydrate portion depending
on the location of latent sp^2 center and its relationship to a C-5
hydroxyl group ("rule of five"). Thus, if one starts with C-1 of the
target and traces a path along the carbon framework, what emerges is a
2-deoxy-3-0-methyl-L-xylo-hexose precursor for C-1 — C-6 portion
(Expression A). One can also consider C-6 of the target as a latent
aldehyde in which case C-1 — C-6 of a carbohydrate precursor would
constitute the reverse order of carbons atoms. Thus, C-1 in the target
becomes C-6 of the sugar and the precursor would have to be a 5-deoxy-3-
0-methyl-D-xylo-hexose (Expression B). Note that the C-3 — C-5 portion
in the target converges directly with the C-2 — C-5 triol unit in D-
glucose, hence the possibility of an attractive synthetic approach. By
considering head-to-tail transpositions one can arrive at expressions A'
and B' which are superposable with B and A respectively, except for the
oxidation states at the extremities. Because of the symmetry relationships in this segment, the concept of stereochemical duality becomes
particularly attractive in choosing carbohydrate templates. Finally,
another segment can be discussed which still encompasses the three
chiral centers at C-3 — C-5, if the six-carbon carbohydrate chain is

Scheme 134

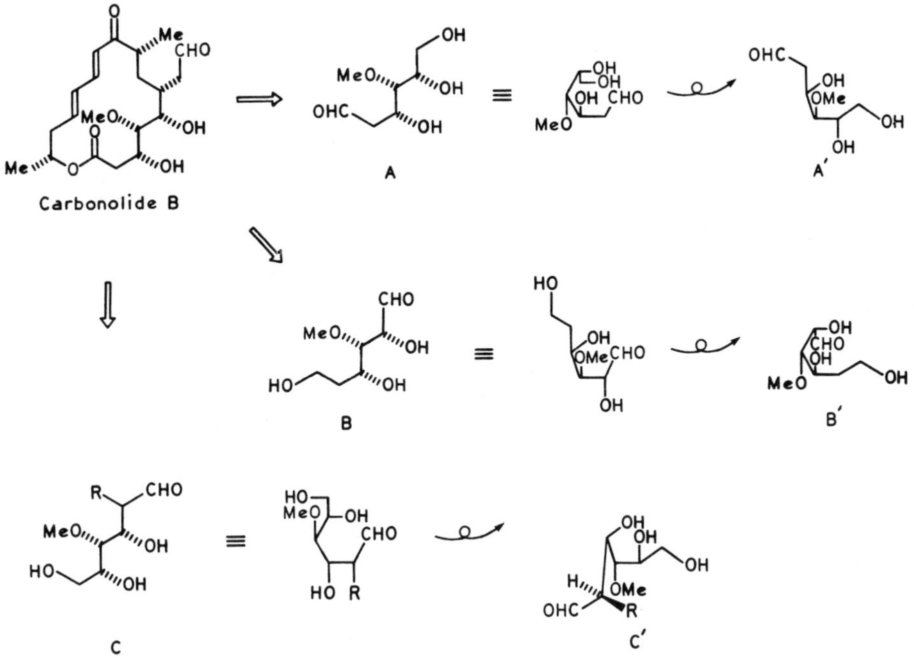

Carbonolide B

taken from the aldehyde carbon to C-2, with the option of a branching carbon chain at C-6. (Expressions C and C'). The approach chosen by Ziegler and coworkers[16] proceeds through an intermediate related to B (Scheme 135). Thus, the readily available 5,6-unsaturated derivative 135.1[17] was hydroborated and the resulting alcohol allylated to give 135.2. Conversion to the alditol 135.3 and acetal formation gave a mixture rich in the desired isomer 135.4 which was oxidized to the aldehyde 135.5. The latter compound corresponds to C-1 — C-6 of the target and harbors three of its six asymmetric centers. Note that C-2 — C-4 of the starting D-glucose were "transferred" to the target without loss of stereochemistry.

c.   Carbonolide B - The total synthesis of carbonolide B and a formal synthesis of the parent antibiotic have been recently achieved based on an approach that utilizes carbohydrate precursors representing

Scheme 135

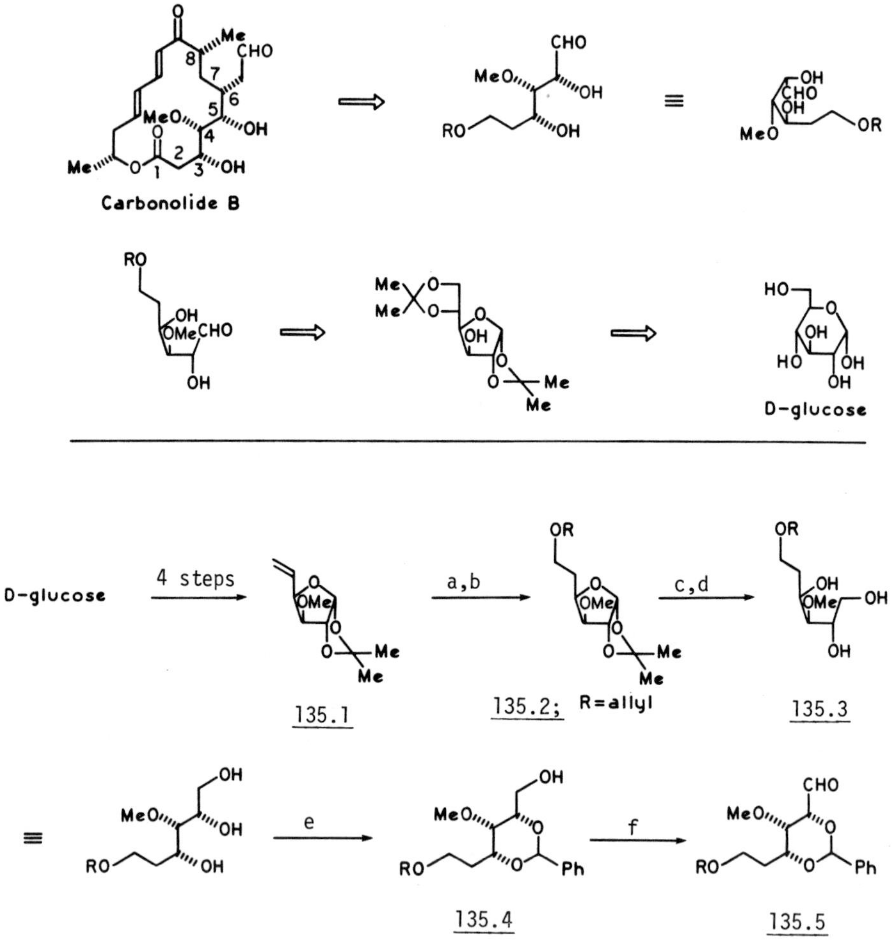
Carbonolide B
D-glucose
4 steps
a,b
c,d
e
f
135.1
135.2; R=allyl
135.3
135.4
135.5

a. B_2H_6, then H_2O_2, NaOH; b. allyl bromide, NaOH, DMF; c. aq. H_2SO_4; d. $NaBH_4$; e. PhCHO, H_2SO_4; f. Collins

Scheme 136

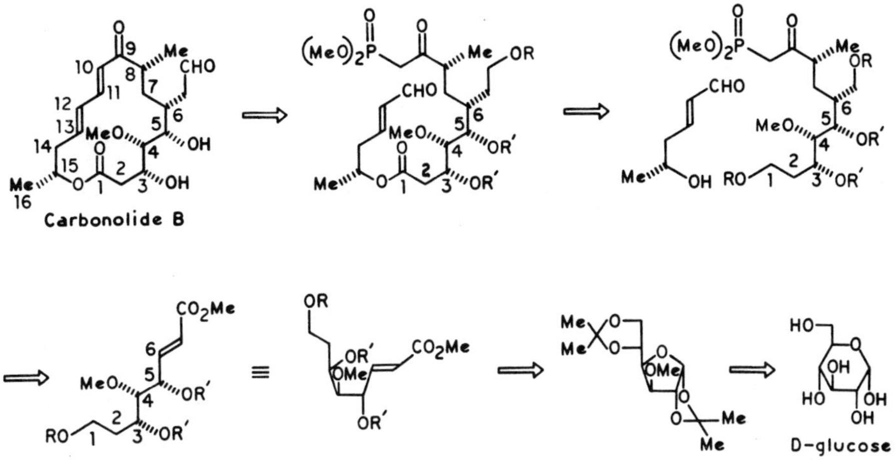

Carbonolide B

the C-1 — C-6 segment of the target.[18] A key cyclic intermediate was
constructed, which, in previous studies had been converted to the parent
antibiotic.[19] It should be pointed out that carbomycin B and magnamycin
B are the same and that the 9-dihydro derivative is the antibiotic
leucomycin A₃ (also known as josamycin). Scheme 136 illustrates the
retrosynthetic analysis and the important bond disconnections. Thus,
the C-10 — C-11 bond can be envisaged to arise from an intramolecular
β-keto phosphonate condensation of a preformed ester. As previously
explained, the C-1 — C-6 segment can be easily derived from D-glucose
(see preceeding section). A key intermediate is an α,β-unsaturated
ester 137.3 which can be subjected to conjugate addition of lithium
methallyl cuprate to provide 137.4 which contains two latent functionalities at C-6 of the target (Scheme 137). Interestingly, the cuprate
reaction led to a preponderance of the desired isomer 137.4, the three-
dimensional structure of which was confirmed by X-ray analysis of a
derivative. Elaboration of the olefinic linkage to an aldehyde, and
condensation with dimethyl methylphosphonate anion followed by oxidation
gave the important β-ketophosphonate intermediate 137.7. The stage was
now set for esterification with a chiral fragment representing C-11 —
C-16 which was obtained from (R)-β-hydroxybutyric acid. Scheme 138
illustrates the completion of the synthesis. The ester 138.2 was

Scheme 137

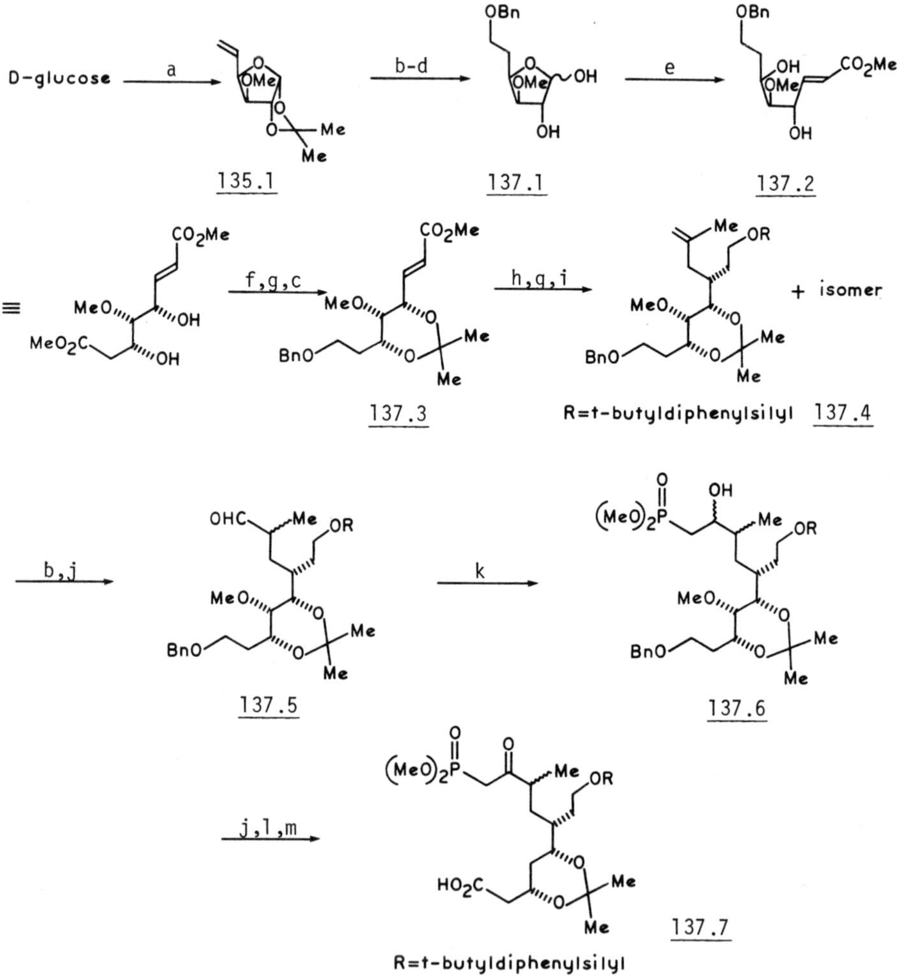

a. 4 steps; b. disiamylborane, THF, NaOH, H_2O_2; c. BnBr, NaH, DME, 60^0;
d. IR-120, H^+; e. $Ph_3P=CHCO_2Me$, toluene; f. $Me_2C(OMe)_2$, TsOH; g. LAH; h.
lithium methallylcuprate, THF, -40^0; i. t-butyldiphenylsilyl chloride,
DMF, imidazole; j. PCC; k. $(MeO)_2P(O)CH_2Li$; l. Pd/C, H_2, EtOAc; m. Jones
oxid.

Scheme 138

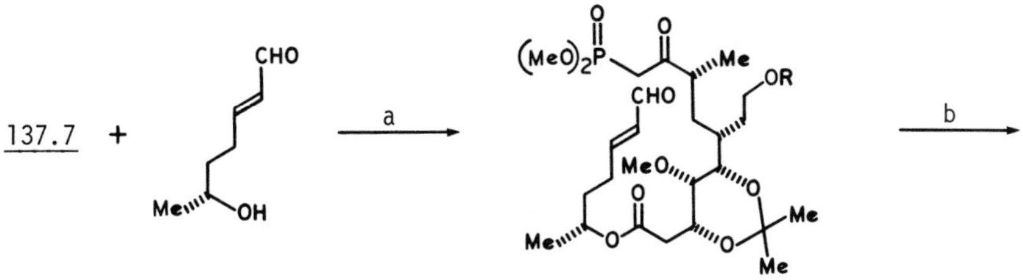

137.7 +   138.1 $\xrightarrow{a}$ 138.2 $\xrightarrow{b}$

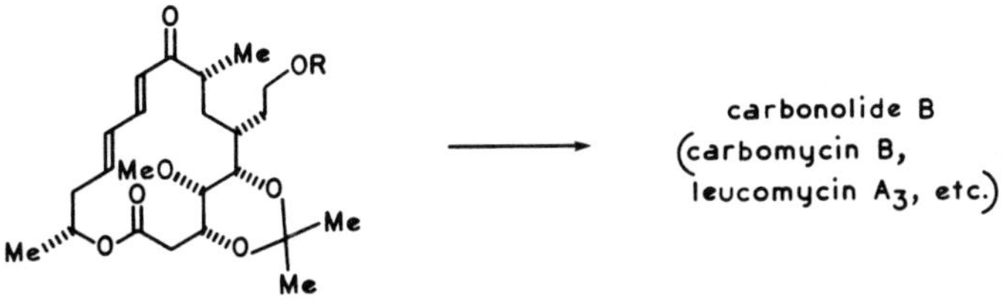

138.3 $\longrightarrow$ carbonolide B
(carbomycin B,
leucomycin A_3, etc.)

a. DCC, ether, DMAP; b. Na, toluene, high dilution

subjected to a critical intramolecular condensation using high dilution
techniques to give 20% of the desired cyclic lactone 138.3. Note that
the stereochemistry of each center, including C-8 is correct, based on
the strategy used. Intermediate 138.3 was also obtained from the parent
antibiotic[20] and was used for confirming the structure of the synthetic
product. Furthermore its ethylene acetal derivative had been
previously transformed into the parent antibiotic so that the obtention
of 138.3 by synthesis constitutes in effect a total synthesis of
carbomycin B (or magnamycin B). Carbomycin B has also been transformed
into leucomycin A_3 (josamycin).[19]

     d.   Carbomycin B (Magnamycin B), Leucomycin A_3 (Josamycin) - The
total synthesis of the title macrolide antibiotics, including the
attachment of the sugar units has been accomplished by Japanese
chemists.[19,21]   The approach is based on the generation of two segments,
C-1 — C-6 and C-10 — C-16 of the target from carbohydrate precursors.
The retrosynthetic analysis (Scheme 139) illustrates the bond

Scheme 139

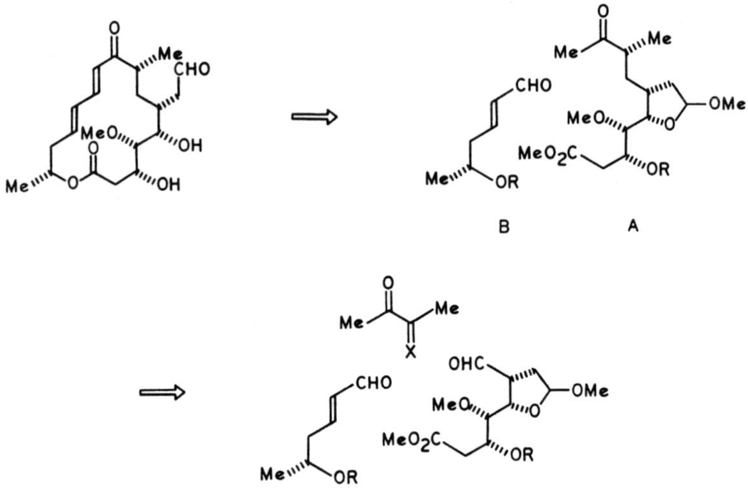

Scheme 140

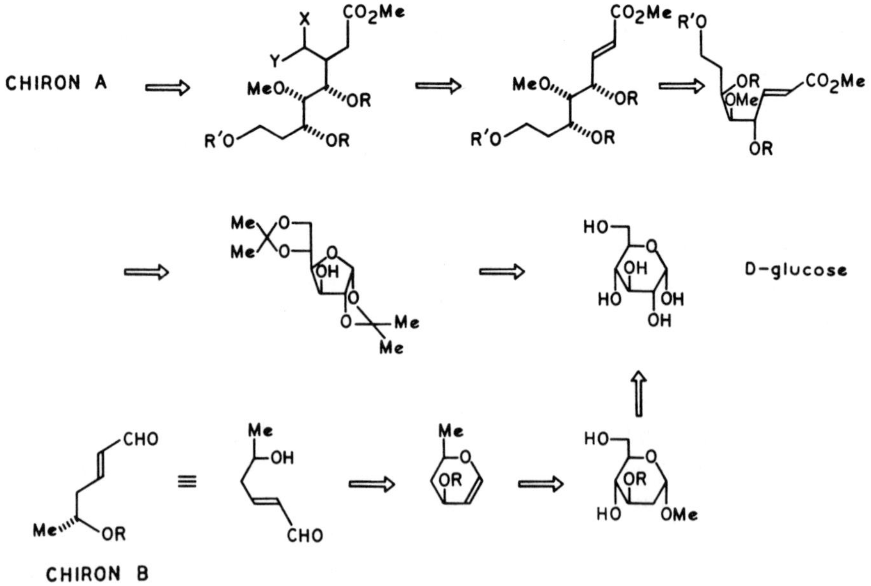

disconnections and bond forming strategies. Thus disconnection at
C-10 — C-11 provides two 'chirons' A and B which can be further
simplified to three fragments. The crucial bond-forming reactions rely
on a Michael addition at C-6, and β-ketophosphonate (C-7 — C-8)
condensation, and a kinetically controlled aldol condensations (C-10 —
C-11). Scheme 140 illustrates the approaches to 'chirons' A and B both
derived from D-glucose by independent routes. Advantage is taken of the
fact that the sense of chirality of at C-3 — C-5 in the right hand
segment of the target can be related to the triol unit in D-glucose (see
Scheme 134 for an analysis). With the template constituted of an
α,β-unsaturated ester containing the triol unit, a key reaction consists
in a Michael addition which introduces the branch points at C-6. Chiron
B, a 6-(R)-hydroxy-2-E-hexenal, can be related to a 6-deoxy-D-hexose
("rule of five") which has been extensively deoxygenated and stereospecifically dehydrated. Scheme 141 shows the route to 'chiron' A.
Note how 141.2 was chain-extended to 141.3 which now corresponds to C-1
- C-6 of the target, the two-carbon appendage being used for the
acetaldehyde side-chain. Note also, as in the previous strategies
(sections b and c), that C-6 of the sugar precursor has become C-1 of
the target and in performing this operation the triol unit can be
situated with the desired sense of orientation. Intermediate 141.3
provided the necessary functionality for a 1,4-conjugate addition which
was effected with the anion of methyl methylthiomethyl sulfoxide.
Remarkably, this Michael addition was highly stereoselective, generating
141.4. Further manipulation of functional groups led to the lactone
141.5 which harbors C-1 — C-7 of the target with four chiral centers.
The next three carbon atoms (C-8 — C-10) were incorporated by a Wittig
reaction followed by hydrogenation and separation of isomers at C-8 and
oxidation at C-1 to generate 141.8 ('Chiron' A). Scheme 142 shows the
synthesis of 'chiron' B. Note the transformation of the glycal
derivative 142.3 into the desired hexenal 142.4 in the presence of
mercuric acetate.[22] The critical union of the two 'chirons' via a
kinetically controlled aldol condensation is seen in Scheme 143. Note
that A and B were reacted without prior hydroxyl protection to give
143.1 which was found to be identical with a sample obtained by
degradation of the antibiotic. Lactonization and manipulation of the
remaining functional groups led to the dimethyl acetal of carbonolide B
which was identical to material obtained from the aglycone. As
previously stated, the aglycone had been converted into carbomycin B and

Scheme 141

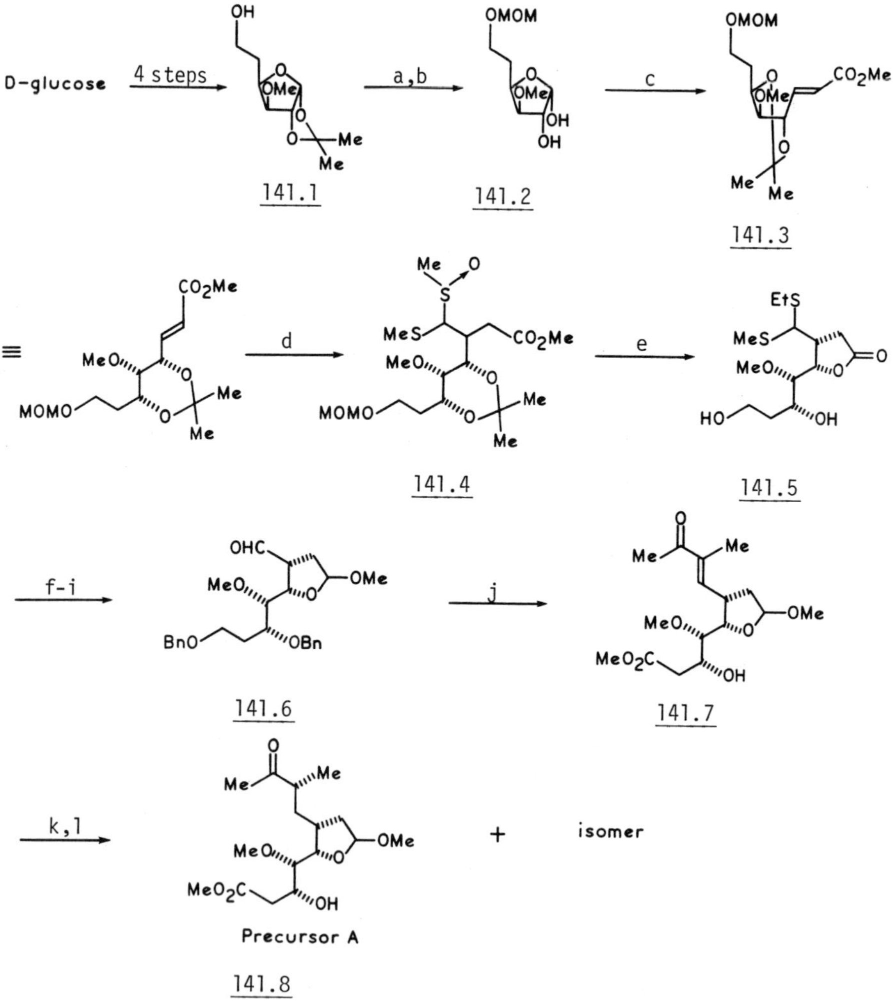

a. ClCH$_2$OMe, base; b. TsOH, DMF; c. Ph$_3$P=CHCO$_2$Me, then acetone, H$^+$;
d. MeS(O)CH$_2$SMe, BuLi, THF; e. EtSH, BF$_3$.Et$_2$O; f. DIBAL; g. Amberlyst
15(H$^+$), MeOH; h. BnBr, NaH, DMF; i. HgCl$_2$, CdCO$_3$, aq. acetone;
j. Wittig; k. Pd black, H$_2$; l. Pd-black, O$_2$, then CH$_2$N$_2$

Scheme 142

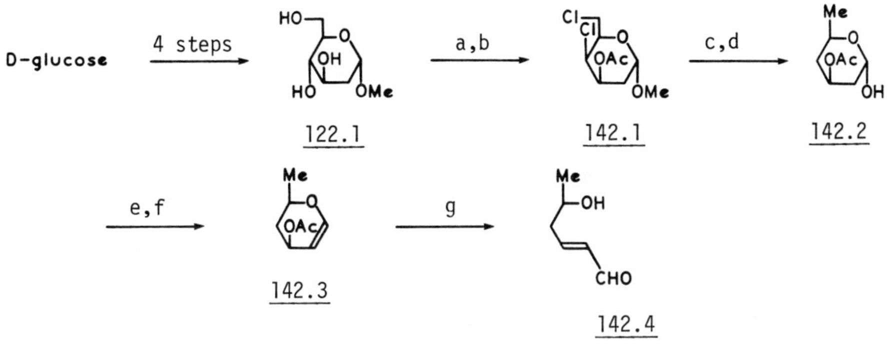

a. SO_2Cl_2, pyr.; b. Ac_2O, pyr.; c. Bu_3SnH, AiBN; d. aq. acid; e. TsCl, pyr.;
f. Et_3N; g. aq. $Hg(OAc)_2$, acetone

leucomycin A_3, so that the present synthesis constitutes a total
synthesis of these antibiotics. The general strategy has been extended
to the total synthesis of tylonolide, the aglycone of tylosin[23] as well
as the total synthesis of tylosin[24].

    e. Antibiotic A26771B – A stereospecific total synthesis of the
title 16-membered macrocyclic lactone antibiotic[25] has been recently
accomplished by Tatsuta and coworkers.[26] Examination of the structure in
question reveals little if any carbohydrate-type symmetry (Scheme 144).
There are however two chiral centers bearing a hydroxyl group each, and
one could consider a carbohydrate route to these segments particularly
if efficient methods can be found for deoxygenation and related
structural modification of appropriate precursors. If one considers
convergence of C-1 — C-6 of the target with a hexose precursor, the
sense of chirality at C-5 would correspond to that of an L-sugar. On
the other hand, the C-15 center can be related to C-5 of a D-hexose (or
C-4 of a D-pentose, etc). Tatsuta and coworkers considered
5-deoxy-D-xylo-hexose as being convergent with the C-1 — C-6 segment of
the target, in such a way that C-1 of the sugar corresponds to C-6 of
the target. They envisaged the construction of three segments
encompassing C-1 — C-6, C-7 — C-12 and C-13 — C-16 and their union by
Wittig methodology. Scheme 145 shows the synthesis of the C-1 — C-6

Scheme 143

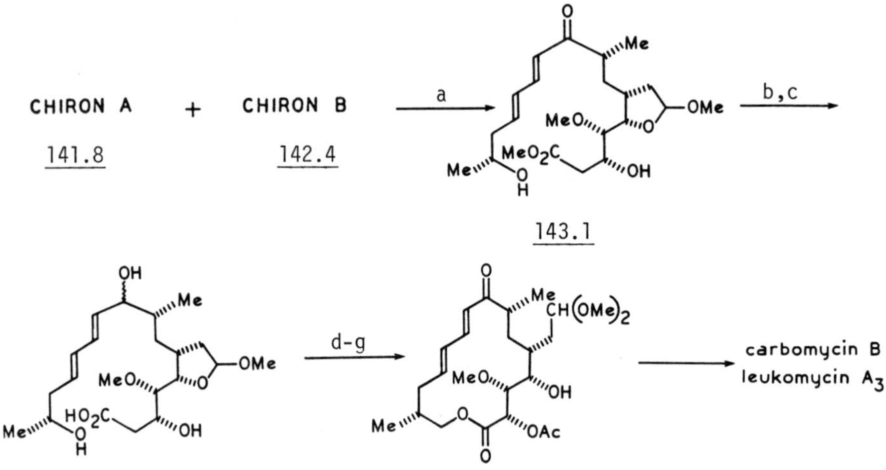

CHIRON A   +   CHIRON B

141.8   142.4

143.1

143.2   143.3

a. LDA; b. NaBH$_4$; c. aq. KOH; d. diethylphosphochloridate, then PhSTl;
then CF$_3$CO$_2$Ag, Masamune lactonization; e. CrO$_3$, HMPA; f. Ac$_2$O, pyr.;
g. Amberlyst 15(H$^+$), MeOH

segment from D-glucose. Thus, the aldehydo derivative 145.3 could be
easily obtained from D-glucose[27] by established procedures. Note that
the C-5 carbon corresponds to C-2 of the target. Wittig methodology led
to 145.4, and then to 145.5 using another carbohydrate-derived three
carbon 'chiron' containing the C-15 hydroxyl group. With the carboxyl
group in place at C-1, it was now possible to attempt a controlled
β-elimination with a bulky base to introduce the intended E-olefininc
linkage at C-2 — C-3. Activation of the carbonyl group and
lactonization gave the 16-membered lactone, which was selectively
succinoylated at C-5 (and not the allylic C-4 hydroxyl), then oxidized
to give the intended target. Although only the C-2 hydroxyl group of
the original D-glucose was preserved, the six-carbon framework of the
sugar template was efficiently used, particularly in planning to
introduce E-olefinic linkage by β-elimination.

    f. N-Methylmaysenine - The macrocyclic lactam maytansine[28] has

Scheme 144

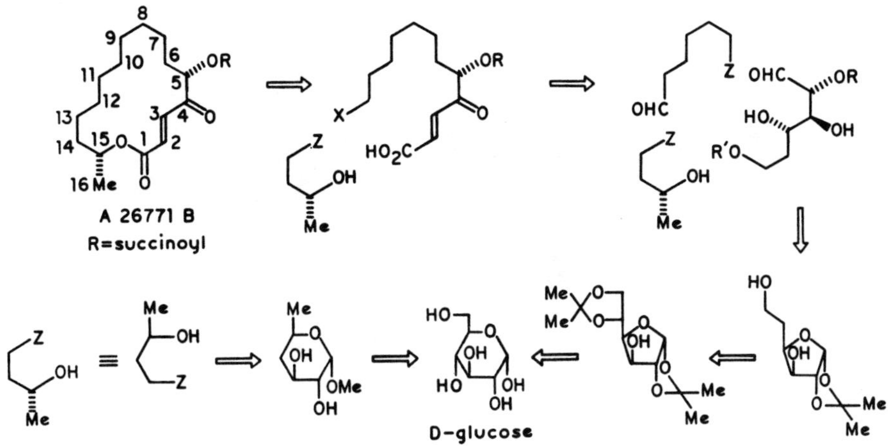

A 26771 B
R=succinoyl

D-glucose

been the subject of intense studies on chemical and biological fronts.
N-methylmaysenine (Scheme 146) can be considered as a pivotal intermediate for the synthesis of the parent antitumor agent maytansine which
contains a 4,5-epoxide and a 3-ester. In Scheme 146 we can see the
synthetic route devised by Corey and coworkers[29] which generates the C-5
— C-9 segment of N-methylmaysenine from D-glucose. The strategy is
based on the recognition of a 2,4-dideoxy-4-C-methyl dialdehyde unit
corresponding to the mentioned segment and its derivation by systematic
functionalization of D-glucose. In essence, none of the hypothetical
possibilities outlined on page 31 which take advantage of the "rule of
five" situating sp^2 centers in relation to chiral centers bearing
oxygen, were used in the Corey strategy. Rather, D-glucose was used as
a source for a five-carbon chain in which two chiral centers were introduced as required for C-6 — C-7 of the target. Thus, the readily
available 2-deoxy-D-arabino-hexose derivative 142.1 was transformed into
the epoxide 146.1, which, when treated with lithium dimethylcuprate, led
almost exclusively, to the desired C-methyl derivative 146.2. Note that
the stereochemistry at C-3/C-4 is convergent with C-6/C-7 of the target
and that the merits of the "chiral template" concept are evident.
Elaboration of required functionality afforded the important acyclic
intermediate 146.4 which was used in the construction of the macrocyclic

Scheme 145

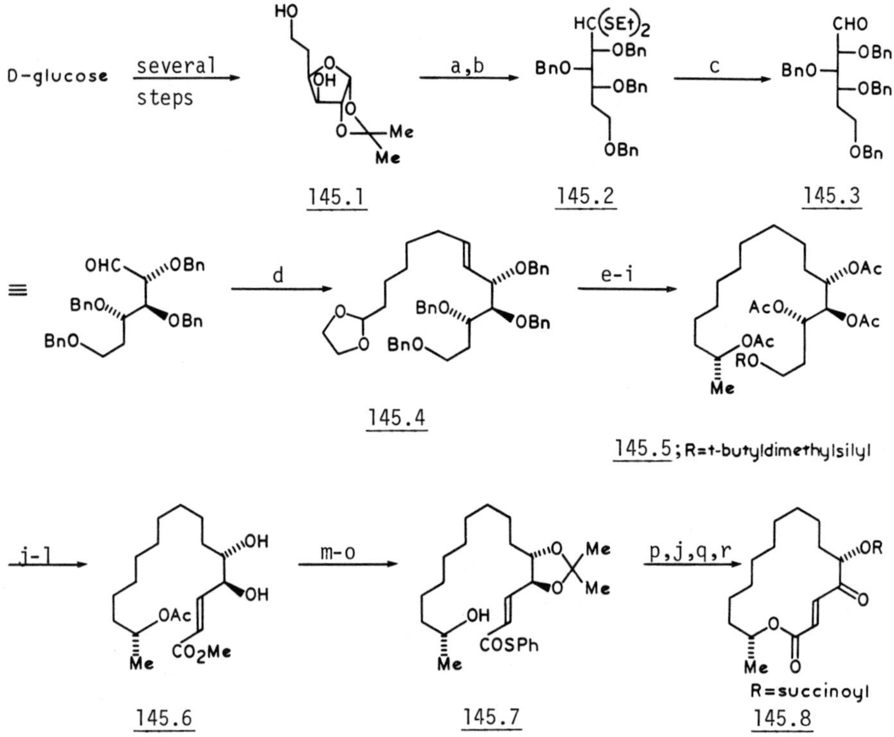

145.1   145.2   145.3

145.4

145.5; R=t-butyldimethylsilyl

145.6   145.7   145.8
                                         R=succinoyl

a. EtSH, BF$_3$, Et$_2$O; b. BnBr, NaH, DMF; c. HgCl$_2$, CaCO$_3$, aq. acetone;
d. Ph$_3$P=CH(CH$_2$)$_4$CH (with dioxolane), THF, e. aq. TFA; f. Me—CH(OH)—=PPh$_3$, THF;
g. Pd black, MeOH; h. Me$_2$t-BuSiCl, imidazole, DMF; i. Ac$_2$O, pyr.;
j. CF$_2$HCO$_2$H, aq. MeCN; k. Pt-C, O$_2$, then CH$_2$N$_2$; l. t-BuOK, aq.
t-BuOH; m. Me$_2$C(OMe)$_2$, H$^+$; n. LiOH, aq. THF; o. diethylphosphoro-
chloridate, Et$_3$N, DMF, then PhSeTl; p. CF$_3$CO$_2$Ag, benzene; q. succinic
anhydride, base, CCl$_4$

Scheme 146

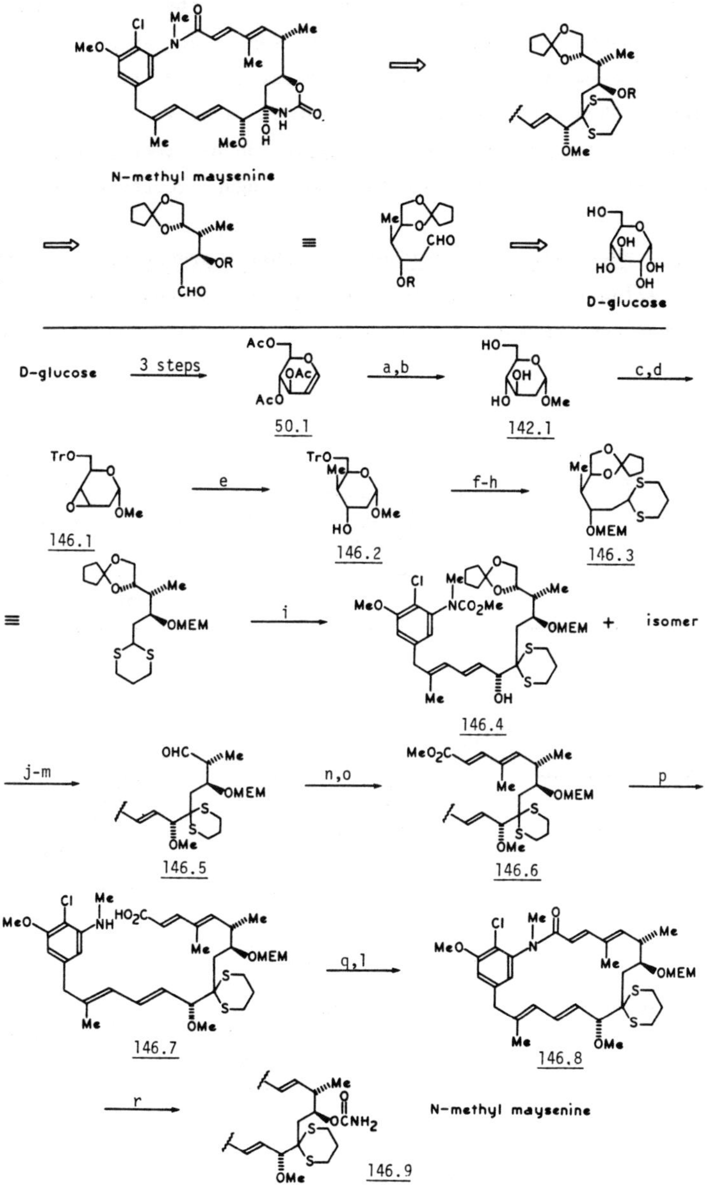

N-methyl maysenine

D-glucose

a. NaOMe, MeOH; b. Hg(OAc)$_2$, MeOH, then NaCl, then NaBH$_4$; c. TrCl, pyr; d. NaH, HMPA, then triisopropylbenzenesulfonylimidazole; e. MeLi, CuI, toluene, -78°; f. HS(CH$_2$)$_3$SH, HCl; g. 1-ethoxycyclopentene, BF$_3$.Et$_2$O; h. MEMCl, base; i. BuLi, TMEDA, THF, -78°; then RCHO; j. MeI, NaH, THF; k. MeSLi, DMF; l. aq. HClO$_4$; m. Pb(OAc)$_4$, MeCN, -25°; n. α-trimethyl-silylpropionaldehyde N-tert-butylimine, -110°; o. (MeO)$_2$P(O)CH$_2$CO$_2$Me, BuLi, THF; p. Bu$_4$NOH, aq. THF; q. mesitylenesulfonyl chloride, base, benzene; r. p-nitrophenylchloroformate, pyr.

structure. Note once again the versatility of the carbohydrate route to natural product synthesis, in providing regio- and stereocontrolled substitution patterns using cyclic derivatives, and the flexibility in converting them into useful acyclic compounds at strategic points in the synthetic plan. The synthesis of maytansine has been recently announced as a sequel to this work[30] and by other elegant routes as well.[31,32]

g. **Maytansinoids – Synthesis of chiral precursors** – Schemes 147 and 148 illustrate two different routes for the construction of functionalized segments of maytansine. Scheme 147 shows an attempt to

Scheme 147

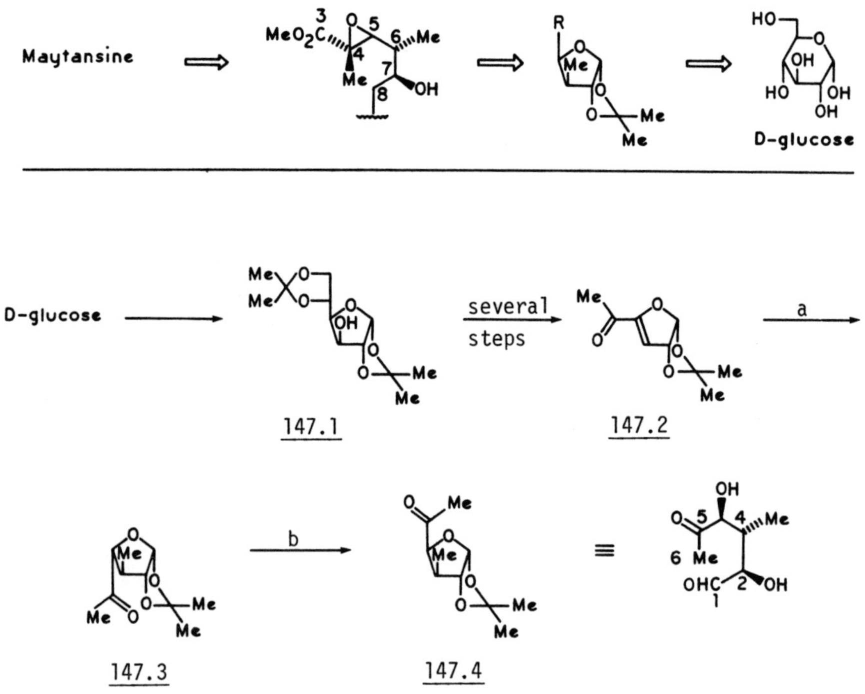

a. Me$_2$CuLi, ether; b. NaOMe, MeOH

Scheme 148

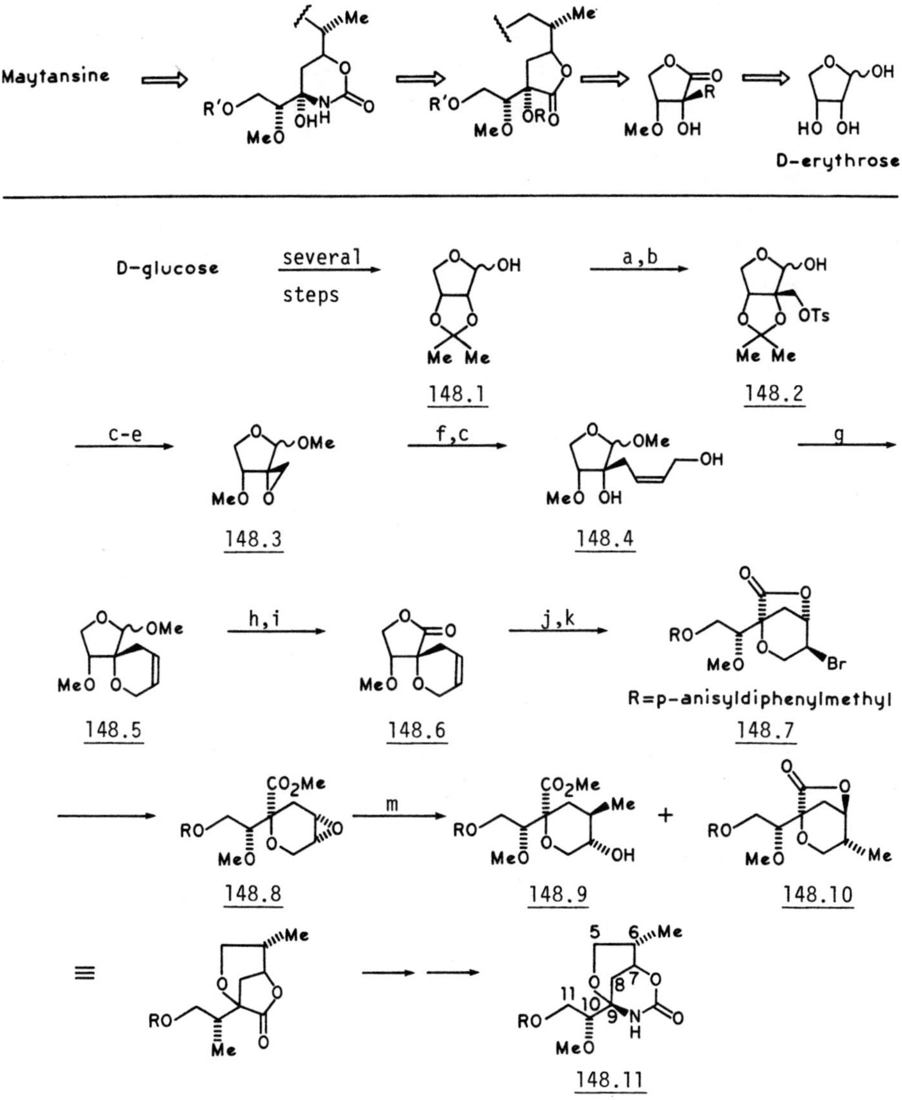

Maytansine

D-erythrose

D-glucose $\xrightarrow[\text{steps}]{\text{several}}$ **148.1** $\xrightarrow{a,b}$ **148.2**

$\xrightarrow{c-e}$ **148.3** $\xrightarrow{f,c}$ **148.4** $\xrightarrow{g}$

148.5 $\xrightarrow{h,i}$ **148.6** $\xrightarrow{j,k}$ **148.7**

R=p-anisyldiphenylmethyl

$\longrightarrow$ **148.8** $\xrightarrow{m}$ **148.9** + **148.10**

$\equiv$ $\longrightarrow$ $\longrightarrow$ **148.11**

a. 37% CH_2O, K_2CO_3, MeOH; b. TsCl, pyr.; c. TsOH, aq. MeOH, reflux; d. KOH, MeOH; e. MeI, NaH, THF; f. $Me_2AlCH_2C\equiv CCH_2OTHP$, NaH, toluene; g. Ph_3P, DEAD, LiBr, DMF; h. aq. TFA; i. DMSO, Ac_2O; j. aq. KOH, then p-anisyldiphenylmethyl chloride, pyr.; k. Br_2, KBr, aq. KOH; l. LiOH, MeOH; m. Me_2CuLi, ether

create a C-3 — C-8 segment by transforming D-glucose into the C-3 methyl derivative 147.4[33]   The latter has the correct absolute stereochemistry relative to C-5 — C-7 of the target. A key reaction was the 1,4-conjugate addition of a cuprate to the α,β-unsaturated ketone 147.2 to give 147.3. Attainment of the desired stereochemistry necessitated a further epimerization to 147.4. Scheme 148 shows how the C-5 — C-11 segment of maytansine was constructed from D-glucose.[34] A critical reaction involves the incorporation of the tertiary hydroxyl group at C-9 starting with the formaldehyde adduct 148.3, which already contains the C-10 methoxyl group (maytansine numbering). Intermediate 148.6 sets the stage for the introduction of two additional asymmetric centers via epoxide 148.8 to give 148.10 and the unwanted regioisomer 148.9. The former, shown in two perspectives, now has the correct absolute stereochemistry at C-6 — C-11 and can be regarded as a valuable synthetic intermediate in the quest for another maytansine synthesis. A five-carbon acetylenic acetal derivative corresponding to C-1–C-5 of maytansine, incorporating the C-3 hydroxyl group has been recently synthesized from S-malic acid and from D-ribonolactone.[35] There are several other reports on the synthesis of segments of this target.[36]

h. Rifamycin S - Synthetic approaches to the C-19 — C-29 aliphatic segment - Rifamycin S is a commercially important antitubercular agent whose structure was elucidated by Prelog and coworkers[37]. It has been assigned to the so-called "ansa" group of antibiotics and it has been the subject of extensive chemical modification in order to expand its biologial activity. The aliphatic C-19 — C-29 portion of rifamycin S (Scheme 149) is biosynthetically derived in major part via the propionate route as is evident from the presence of alternating C-methyl and hydroxyl groups. Viewed in acyclic perspective, one can see a plane of symmetry bisecting the molecule through C-23 and this has been the conceptual basis of a published synthetic strategy[38] as well as Kishi's elegant synthesis of the racemic[39] and (+)-rifamycin S.[40] Masamune and coworkers have utilized their elegant aldol methodology to construct the ansa chain of rifamycin S,[40] while Still and Barrish[41] have described stereoselective syntheses of 1,3-diol derivatives and demonstrated their application to the ansa bridge of the same antibiotic. 'Chirons' derived from carbohydrates had been previously proposed as segments of polyketide-type natural products,[6] and their extension to ansa chains was therefore evident. Thus, cleavage at

Scheme 149

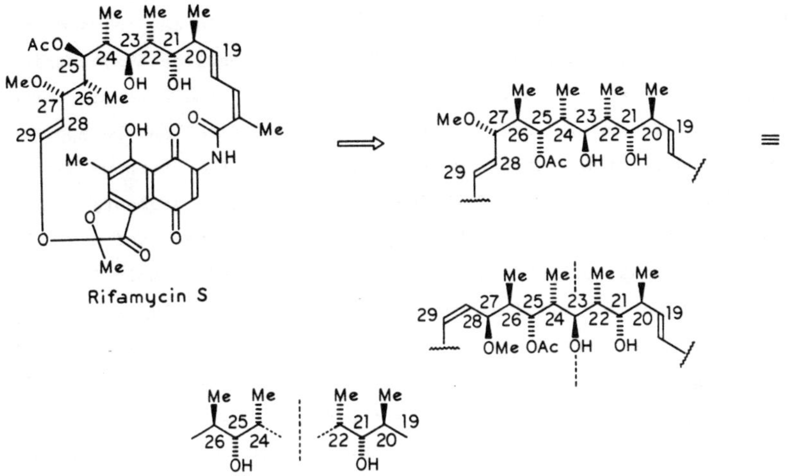

Rifamycin S

the C-23 - C-24 bond is the basis of another synthesis of the aliphatic
portion of rifamycins starting with D-glucose by Kinoshita and
coworkers[43] (Scheme 150). The sense of chirality at C-20/C-21 and

Scheme 150

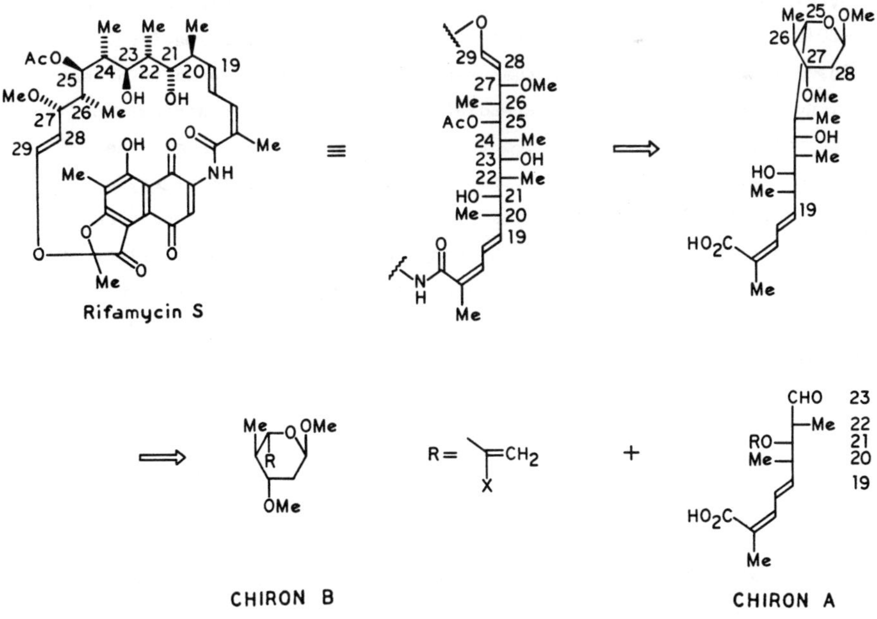

Rifamycin S

CHIRON B   CHIRON A

C-26/C-25 respectively is the same and these centers can be secured in
early intermediates, common to two segments representing C-19 — C-23 and
C-24 — C-29 of the target. The retrosynthetic analysis in Scheme 150
illustrates how the C-24 — C-29 portion of rifamycin can be used in a
masked glycoside form, reminiscent of the erythronolide A synthesis from
D-glucose (see section a). In this case however, the anomeric carbon of
D-glucose corresponds to C-29, and the "rule of five" can be applied in
reverse order, with 0-25 being the ring oxygen. The "bottom" part of
the chain comprises C-19 — C-23 (excluding the diene segment), which can
be secured from D-glucose with loss of two carbon atoms and addition of
one (CHO). Thus, only the substituents at C-20/C-21 remain intact from
a "chiral template" originating with D-glucose. Coupling was envisaged
to take place between 'chiron' A and the vinyl lithium species derived
from 'chiron' B. The coupled product, previously isolated by
degradation of rifamycin S was considered by Kinoshita and coworkers to
be a useful intermediate in the synthesis of the antibiotic, although no
routes were suggested.

Scheme 151 shows the synthesis of 'chiron' A[43] in the form of a
masked diol derivative, 151.7. The retrosynthetic analysis reveals that
only the C-19 — C-22 portion of the target comes from a D-glucose framework and corresponds to C-2 — C-5. Note that the C-methyl group at C-22
corresponds to the original terminal carbon atom of D-glucose. The
C-20/C-21 centers are introduced by manipulating a conformationally
biased intermediate. The actual synthetic sequence[42] shown in Scheme
151 generates a furanose intermediate 151.2 which was chain-extended and
the mixtures of epimers carried through to the end. It should be
pointed out that the diol unit is introduced as a masked aldehyde
precursor (C-19), hence the presence of isomers is of no consequence.
The predisposed orientation of hydroxyl groups and the possibility of
differential functionalization, produced the epoxide 151.5 which was
reacted with dithian anion to give the desired isomer 151.6 and its
regioisomer.

Access to 'chiron' B is shown in Scheme 152[44]. The retrosynthetic
analysis shows how the three contiguous asymmetric carbon atoms in the
target can be related to early intermediates obtained from D-glucose.
The outcome of the operation is such that the "rule of five" still
applies, but C-1 of D-glucose corresponds to C-29 of the target. Intermediate 43.7 which was used for the synthesis of 'chiron' A was

Scheme 151

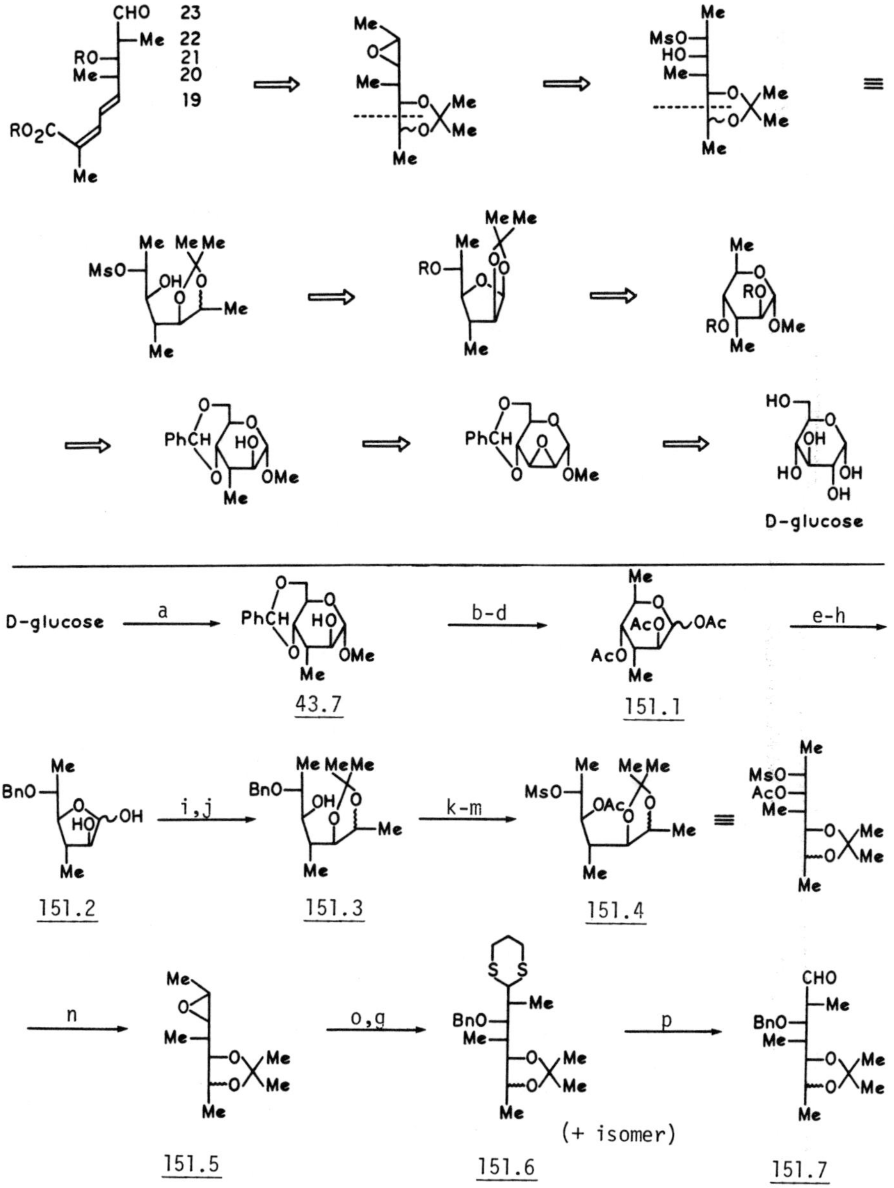

a. See Scheme 43; b. NBS, CCl_4; c. LAH; d. Ac_2O, H_2SO_4; e. aq. NaOH, MeOH; f. acetone, $FeCl_3$; g. BnBr, DMF, NaH; h. aq. AcOH; i. MeMgI, ether; j. acetone, H^+; k. Ac_2O, DMAP, EtOAc; l. Pd-black, H_2; m. MsCl, pyr.; n. NaOMe, MeOH; o. dithiane, BuLi, THF; p. $HgCl_2$, HgO, aq. acetone

Scheme 152

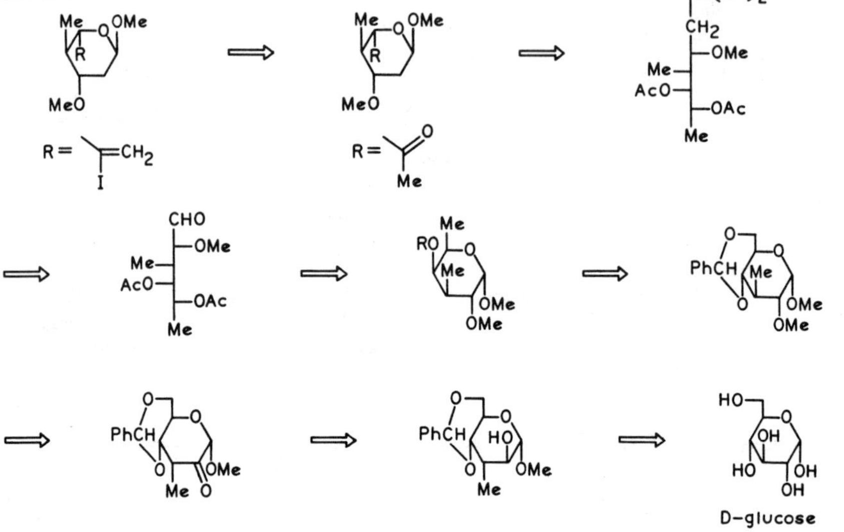

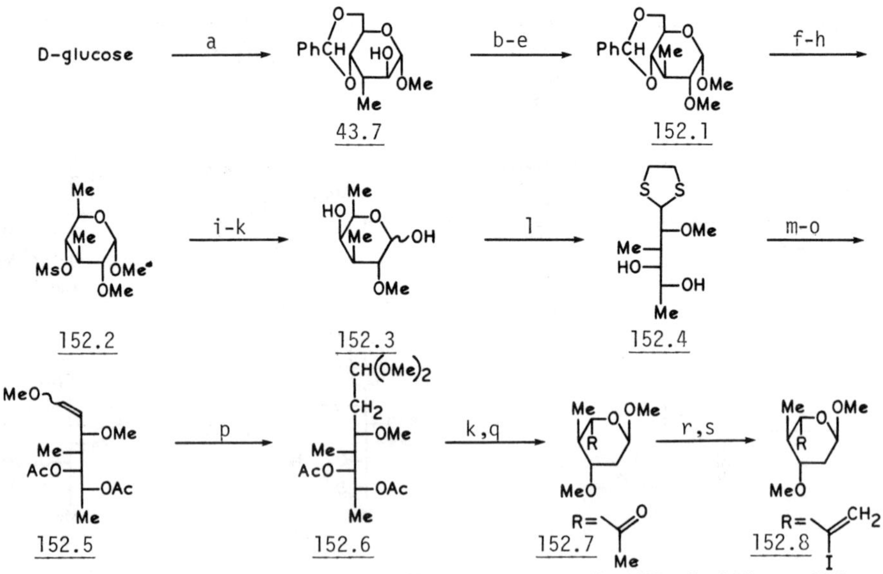

a. See Scheme 43; b. DCC, DMSO, TFAA, pyr.; c. Et$_3$N, DMF; d. LAH; e. MeI,
DMF, NaH; f. NBS, CCl$_4$; g. LAH; h. MsCl, pyr.; i. NaOBz, DMSO, 120°;
j. Ac$_2$O, H$_2$SO$_4$; k. aq. NaOH, dioxane; l. ethylenedithiol, HCl; m. Ac$_2$O,
pyr., DMAP; n. HgCl$_2$, HgO, aq. acetone; o. Ph$_3$P=CHOMe, DMSO, ether;
p. NBS, MeOH, NaHCO$_3$, MeOH, HCl; q. CrO$_3$, acetone, aq. H$_2$SO$_4$; r. H$_2$NNH$_2$
hydrate, EtOH, Et$_3$N; s. I$_2$, THF, Et$_3$N

efficiently manipulated based on the notions of conformational bias
discussed earlier to produce 152.2. Configurational inversion at C-4,
transformation to the acyclic derivative 152.4, and chain-extension gave
152.6. At this point, it was necessary to reform cyclic intermediates
which was easily achieved by methanolysis. Note that the asymmetric
carbons corresponding to C-25 — C-27 are secured and that O-25 is
temporarily protected as the ring oxygen. The original C-5 hydroxyl
group of D-glucose was transformed to a carbonyl function and the
terminal unit converted into the vinyl iodide 152.8. Model reactions
between the lithio derivative of 152.8 and benzaldehyde demonstrated the
feasibility of chain-extension at that site.

Coupling 'chirons' A and B was effected via the above described
lithio derivative to give a mixture of isomers 153.1.[43] (Scheme 153)
Following chromatographic separations and different hydrogenation
procedures, it was possible to isolate 153.8. Selective cleavage of the
isopropylidene group, periodate oxidation and Wittig reaction gave 153.4
which proved to be identical to a sample obtained from the degradation of
rifamycin S.

Another element of symmetry in rifamycin S can be uncovered by bond
disconnection at the C-24 — C-25 juncture to produce two six-carbon
segments, C-19 — C-24 and C-25 — C-29 which can be related to two
hexoses, ('chirons' A and B)[44] (Scheme 154). In both cases the "rule of
five" is applicable and the possiblity of using common intermediates
starting from D-glucose should be obvious. Scheme 155 illustrates access
to precursor A in one acyclic modification 155.10. The strategy is the
same as in the construction of chiral segments of erythronolide A.
Branching at C-4 was introduced by a Wittig reaction followed by
reduction of the allylic alcohol 155.5 with palladium hydroxide on
carbon which provided the required stereoselectivity (4:1).
Intermediate 155.6 has the correct stereochemistry corresponding to
C-20 — C-23 of the target. Inversion of C-5 was achieved through
transformation to an acyclic derivative 155.9 which gave the epoxide
155.10, now having C-19 —C-25 with four chiral centers convergent with
the target. This epoxide could be a useful intermediate in C-C bond
forming reaction aimed at joining the two carbohydrate-derived segments.
In another approach, it was possible to convert intermediate 155.6 into
the acyclic ketone 155.11, which is ideally suited for an aldol-type
condensation with an aldehyde derivative corresponding to 'chiron' B.
Scheme 156 shows how the C-25 — C-29 segment of rifamycin S

Scheme 153

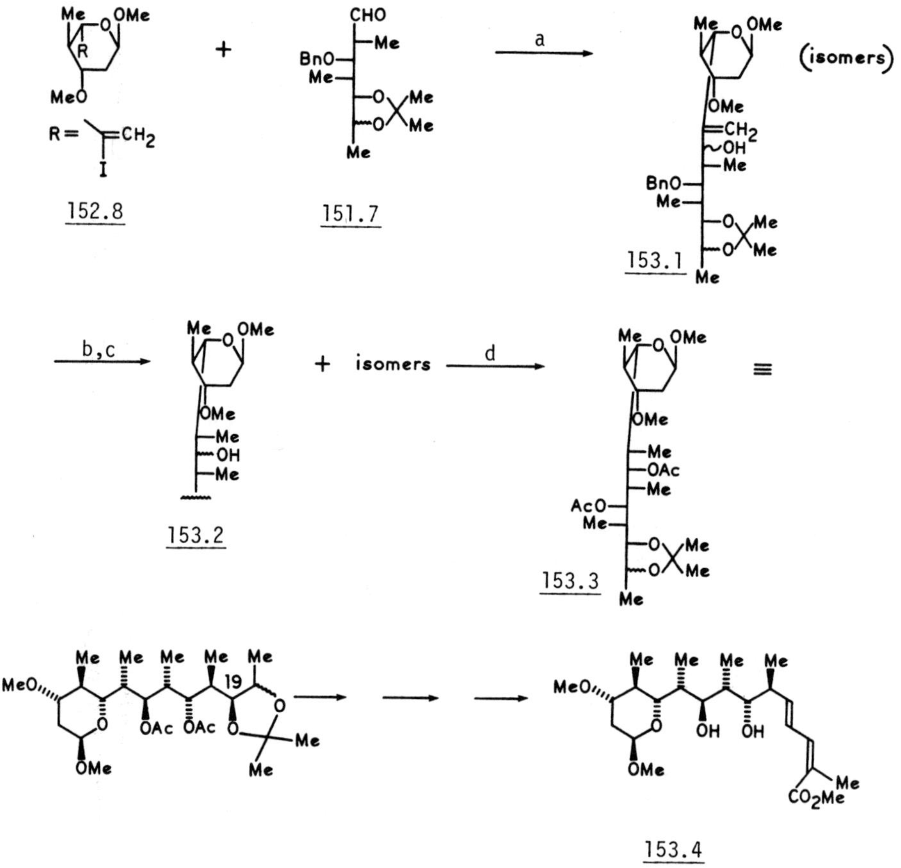

152.8   +   151.7   →ᵃ   153.1 (isomers)

153.2   + isomers →ᵈ   153.3 ≡

153.4

a. BuLi, ether; b. Pd-black, H$_2$, MeOH; c. (Ph$_3$P)$_3$RhCl, H$_2$; d. Ac$_2$O, DMAP, EtOAc

Scheme 154

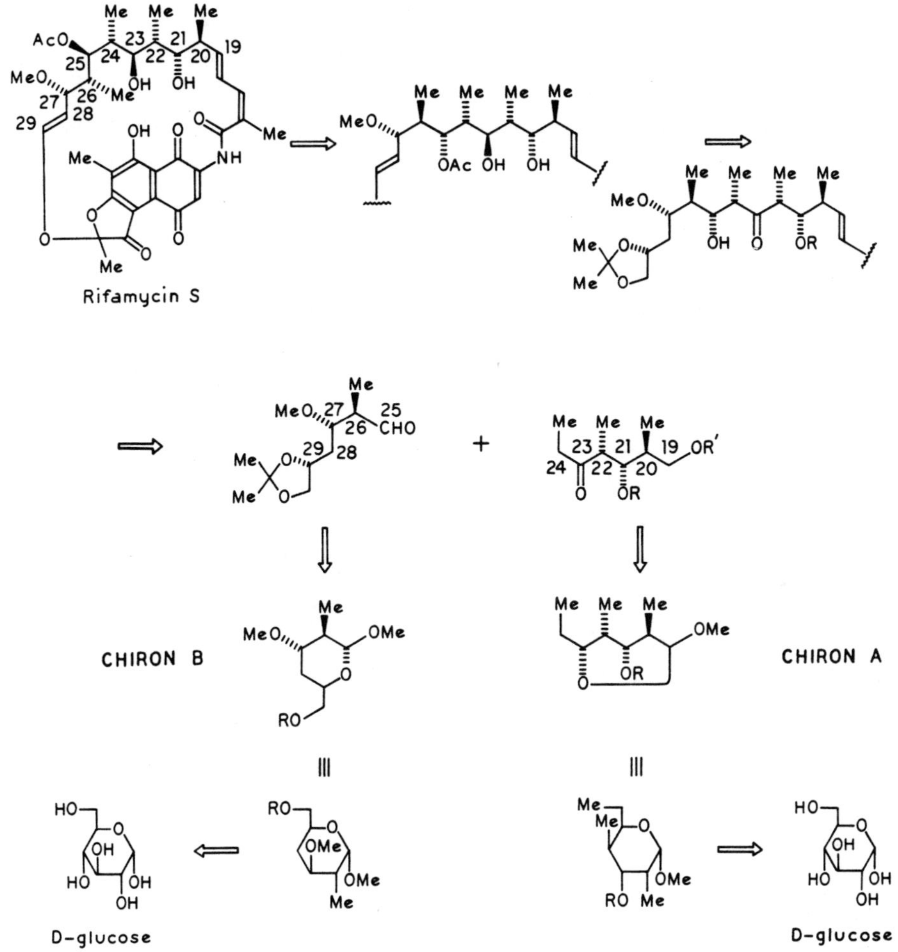

Rifamycin S

CHIRON B

CHIRON A

D-glucose

D-glucose

was synthesized. This segment can be related to a 4-deoxy-D-hexo-
pyranose derivative with the convenience that the terminal diol unit can
be converted to an aldehyde (C-29 of the target) at some stage in the
sequence. Intermediate 156.1, common to both 'chirons', was transformed
into 130.2 and the C-3 center epimerized relying on previous experience
in the erythronolide A series.[5,6,8] Deoxygenation via chlorination and
reduction gave precursor B, 156.4, which in turn was transformed into
three types of nucleophilic components representing C-25 — C-29 of the
target. The dithian derivative 156.5, the β-ketophosphonate 156.9 and
the sulfoxide 156.10 were prepared by standard procedures and their
reactivities tested in model reactions. The possibility for functional
duality is also evident by the availability of 'chiron' B in the form
of electrophilic components, such as the aldehyde 156.7 and the ester
156.8. 'Chirons' such as 155.10, 155.11 representing the right hand
segment of the aliphatic portion and 156.5 - 156.10 representing the
left hand segment combine six of the eight chiral centers in the target
and can be obtained efficiently and with high stereocontrol and complete
regiocontrol. Provided the sense of chirality at C-24/C-25 can be
correctly established in one of the several C-C bond forming reactions
that can be envisaged, the entire aliphatic segment of rifamycin S can
be constructed from carbohydrate precursors. An aldol condensation[46]
between 156.7 and the Z-enolate derived from 155.11 produced a major
isomer to which structure 157.1 was assigned based on a consideration of
possible transition states(Scheme 157). The structure of this product
was definitively established by chemical correlation with a degradation
product 158.1 of rifamycin[47] (as shown in Scheme 158). It should be
noted that reduction of 157.1 with Dibal was highly stereoselective and
produced the desired isomer 157.2 large in preponderance. Standard
transformations led to the acetal and acetate derivatives 157.4 and 157.5
which were found to be identical with samples obtained from the
degradation of the antibiotic. The C-19 —C-29 aliphatic portion of
rifamycin S was therefore assembled from D-glucose with excellent
stereocontrol and virtually complete regiocontrol. Intermediate 157.5,
now readily available from 157.3 or 157.4 can be converged with the
Kishi route[40] to (+)-rifamycin.

Scheme 155

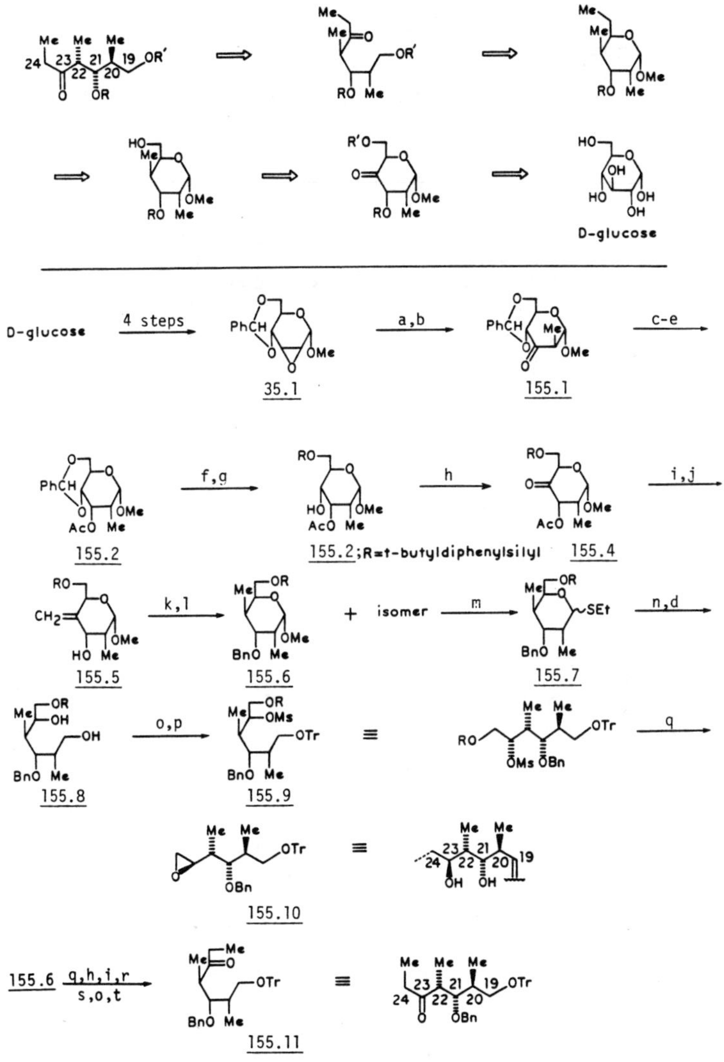

a. Me$_2$CuLi, ether; b. Ac$_2$O, DMSO; c. NaOMe, MeOH; d. NaBH$_4$; e. Ac$_2$O,
pyr; f. TsOH, MeOH; g. Ph$_2$t-BuSiCl, imidazole, DMF; h. (COCl)$_2$, DMSO;
i. Ph$_3$P=CH$_2$; j. KCN, MeOH; k. Pd(OH)$_2$/C, H$_2$; l. BnBr, KH, DMF; m. EtSH,
ZnCl$_2$; n. Br$_2$, aq. NaHCO$_3$; o. TrCl, pyr.; p. MsCl, pyr.; q. Bu$_4$NF, THF;
r. Rh/alumina, H$_2$, toluene; s. aq. AcOH; t. PCC

Scheme 156

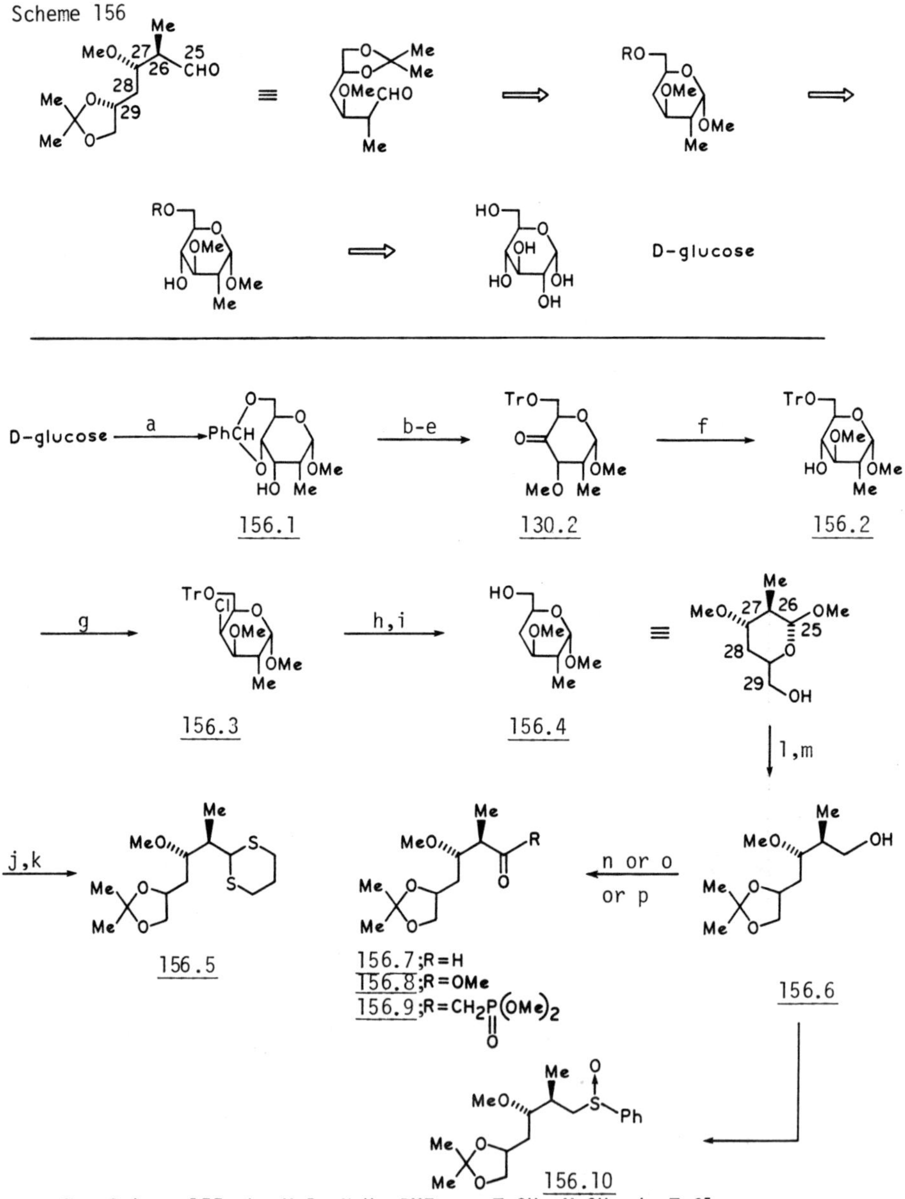

a. See Scheme 155; b. MeI, NaH, DMF; c. TsOH, MeOH; d. TrCl, pyr.;
e. $(COCl)_2$, DMSO; f. NaOMe, MeOH then $NaBH_4$; g. SO_2Cl_2, pyr.; h. Bu_3SnH,
AiBN; i. aq.AcOH; j. $HS(CH_2)_3SH$, BF_3, Et_2O; k. $Me_2C(OMe)_2$, TsOH; l. H_3O^+;
m. $NaBH_4$ then acetone, H^+; n. PCC, 4Å sieves, CH_2Cl_2 (for 151.7); o. RuO_4,
then CH_2N_2 (for 156.8); p. $(MeO)_2P(O)CH_3$, BuLi, on 156.8; q. PhSSPh, Bu_3P;
r. m-CPBA, CH_2Cl_2

Scheme 157

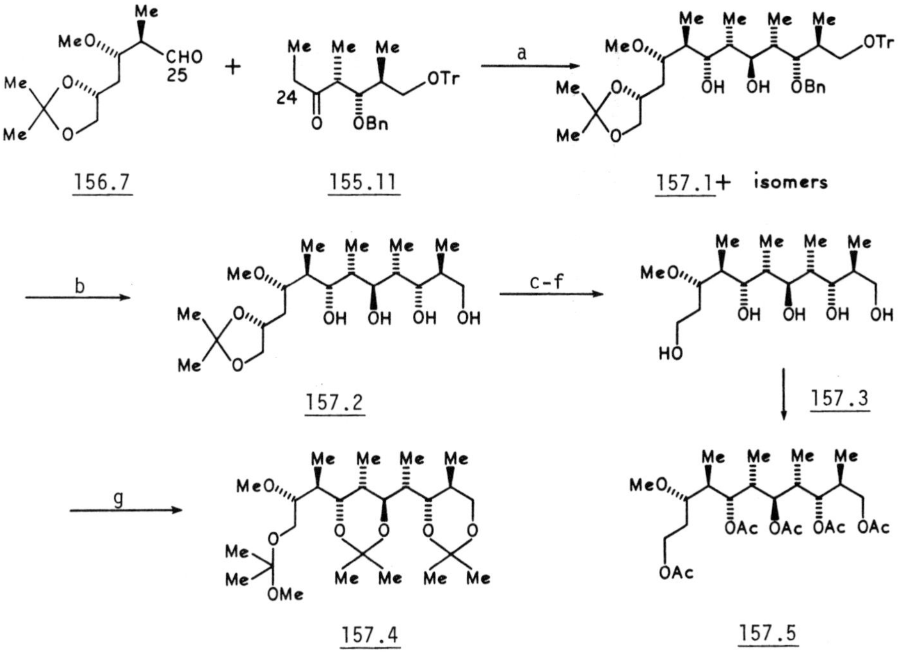
156.7
155.11
157.1 + isomers
157.2
157.3
157.4
157.5
a
b
c-f
g

a. LDA, THF, -78°; b. DIBAL, THF; c. Pd/C, H_2, MeOH; d. aq. AcOH; e. $NaIO_4$; f. $NaBH_4$, MeOH; g. $MeC(OMe)_2$, CSA

Scheme 158

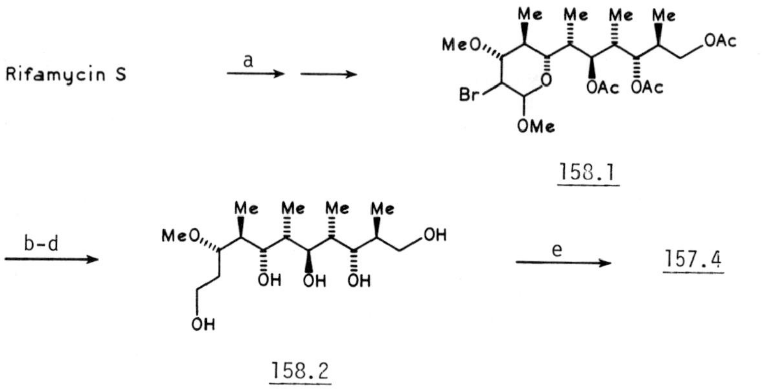

Rifamycin S →ᵃ → 158.1

→ᵇ⁻ᵈ 158.2 →ᵉ 157.4

a. see Ref. 47 ; b. Bu₃SnH, AIBN, toluene; c. aq. AcOH; d. NaBH₄, MeOH;
e. Me₂C(OMe)₂, CSA

REFERENCES

1. For recent reviews, see, S. Masamune, G. Bates and J.W. Corcoran, Angew. Chem. Int. Ed. Engl., 16, 585 (1977); K.C. Nicolaou, Tetrahedron, 33, 3041 (1977); see also, A.D. Celmer, Pure Appl. Chem., 28, 413 (1971); T.G. Back, Tetrahedron, 33, 3041 (1977).

2. R.B. Woodward, in Perspectives in Organic Chemistry, A.R. Todd, ed., Interscience, 1956, p. 155.

3. R.B. Woodward et al., J. Am. Chem. Soc., 103, 3210, 3213, 3215 (1981).

4. E.J. Corey, P.B. Hopkins, S. Kim, S. Yoo, K.P. Nambiar and J.R. Flack, J. Am. Chem. Soc., 101, 7131 (1979).

5. S. Hanessian and G. Rancourt, Pure Appl. Chem., 49, 1201 (1977).

6. S. Hanessian and G. Rancourt, Can. J. Chem., 55, 1111 (1977).

7. For a study of chain-branching reactions in the carbohydrate series in connection with a similar approach, see, M. Miljković and D. Glišin, J. Org. Chem., 40, 3357 (1975).

8. S. Hanessian, G. Rancourt and Y. Guindon, Can. J. Chem., 56, 1843 (1978).

9. See for example, M.L. Wolfrom and S. Hanessian, J. Org. Chem., 27, 1800 (1962); T.D. Inch, Carbohydr Res., 5, 45 (1967); W.C. Still and J.H. McDonald, III, Tetrahedron Lett., 21, 1031 (1980); D.J. Cram and K.R. Kopecky, J. Am. Chem. Soc., 81, 2748 (1959).

10. C. Monneret, J.C. Florent, I. Kabore and Q. Kuong-Hu, J. Carb. Nucl. Nucleotides, 1, 161 (1974).

11. S.S. Costa, A. Lagrange, A. Olesker and G. Lukacs, J.C.S. Chem. Comm., 721 (1980).

12. N.K. Kochetkov, A.F. Sviridov and M.S. Ermolenko, Tetrahedron Lett., 22, 4315 (1981).

13. M. Kinoshita, N. Ohsawa and S. Gomi, Carbohydr. Res., 109, 5 (1982); see also, H. Redlich and H.-J. Neumann, Chem. Ber., 114, 229 (1981).

14. G. Stork, I. Paterson and F.K. Lee, J. Am. Chem. Soc., 104, 4686 (1982).

15. R.B. Woodward, L.S. Weiler and P.C. Dutta, J. Am. Chem. Soc., 87, 4662 (1965); F.A. Hochstein and K. Murai, ibid., 76, 5080 (1954).

16. F. Ziegler, P.J. Gilligan and V.R. Chakraborty, Tetrahedron Lett., 3371 (1979); see also, F.E. Ziegler and P.J. Gilligan, J. Org. Chem., 46, 3874 (1981).

17. J.S. Josan and Eastwood, Carbohydr. Res., 7, 161, (1978).

18. K.C. Nicolaou, M.R. Pavia and S.P. Seitz, J. Am. Chem. Soc., 103, 1224 (1981); Tetrahedron Lett., 2327 (1979).

19. K. Tatsuta, A. Tanaka, K. Fujimoto, M. Kinoshita and S. Umezawa, J. Am. Chem. Soc., 99, 5826 (1977); L.A. Freiberg, R.S. Egan, and W.H. Washburn, J. Org. Chem., 39, 2474 (1975).

20. K.C. Nicolaou, J.P. Seitz and M.R. Pavia, J. Am. Chem. Soc., 103, 1222 (1981).

21. K. Tatsuta, Y. Amemiya, S. Maniwa and M. Kinoshita, Tetrahedron Lett., 21, 2837 (1980); see also K. Tatsuta, T. Yamauchi and M. Kinoshita, Bull. Chem. Soc. Japan, 51, 3035 (1978).

22. F. Gonzalez, S. Lesage and A.S. Perlin, Carbohydr. Res., 42, 267 (1975).

23. K. Tatsuta, Y. Amemiya, Y. Kanemura and M. Kinoshita, Tetrahedron Lett., 22, 3997 (1981); for a synthesis of tylonolide, see also, S. Masamune, L.D.-L. Lu, W.P. Jackson, T. Kaiho and T. Toyoda, J. Am. Chem. Soc., 104, 5523 (1982); see also L.D.-L. Lu, Tetrahedron Lett., 23, 1867 (1982).

24. K. Tatsuta, Y. Amemiya, Y. Kanemura, H. Takahashi and M. Kinoshita, Tetrahedron Lett., 23, 3375 (1982).

25. K.H. Michel, P.V. De Marco and R. Nagarajan, J. Antibiotics, 30, 371 (1977); for a recent synthesis of the racemic antibiotic, see M. Asaoka, N. Yanagida and H. Takei, Tetrahedron Lett., 21, 4611 (1980).

26. K. Tatsuta, A. Nakagawa, S. Maniwa and M. Kinoshita, Tetrahedron Lett., 21, 1479 (1980).

27. E.J. Hedgley, O. Meresz and W.C. Overend., J. Chem. Soc., C, 888 (1967).

28. For a recent review, see Y. Komoda and T. Kishi, in Anticancer Agents Based on Natural Product Models, J.M. Cassady and J.D. Douros, eds., Medicinal Chemistry Monographs, Academic Press, New York, N.Y., vol. 16 (1980), p.353.

29. E.J. Corey, L.V. Weigel, A.R. Chamberlain and B. Lipshutz, J. Am. Chem. Soc., 102, 1439 (1980).

30. E.J. Corey, L.V. Weigel, A.R. Chamberlain, H. Cho and D.H. Huon, J. Am. Chem. Soc., 102, 6613 (1980).

31. A.I. Meyers, P.J. Reider and A.L. Campbell, J. Am. Chem. Soc., 102, 6597 (1980); A.I. Meyers, D.L. Comins, D.M. Roland, R. Henning and K. Shimizu, J. Am. Chem. Soc., 101, 7104 (1979).

32. M. Isobe, M. Kitamura and T. Goto, J. Am. Chem. Soc., 104, 4997 (1982).

33. R. Bonjouklian and B. Ganem, Carbohydr. Res., 76, 245 (1979).

34. P.T. Ho, Can. J. Chem., 88, 858 (1980).

35. D.H.R. Barton, M. Bénéchie, F. Khuong-Huu, P. Potier and V. Reyna-Pineda, Tetrahedron Lett., 23, 651 (1982).

36. See for example, R. Bonjouklian and B. Ganem, Tetrahedron Lett., 2835 (1977); E. Götschi, F. Schneider, H. Wagner and K. Bernauer, Helv. Chim. Acta, 60, 1416 (1977); M. Samson, D. DeClercq, H. DeWilde and M. Wanderwalle, Helv. Chim. Acta, 60, 3195 (1977); D.H.R. Barton, S.D. Gero and C.D. Maycock, J. Chem. Soc., Chem. Comm., 1089 (1980).

37. For reviews, see K.L. Rinehart, Jr., Acc. Chem. Res., 5, 57 (1972); K.L. Rinehart, Jr., and L.S. Shield, Fortschr. Chem. Org. Naturst. 33, 231 (1976); W. Wehrli, Top. Curr. Chem., 72, 22 (1977); V. Prelog, Pure Appl. Chem., 7, 551 (1963); P. Sensi, Pure Appl. Chem., 41, 15 (1974).

38. E.J. Corey and T. Hase, Tetrahedron Lett., 335 (1979).

39. H. Nagoaka, W. Rutsch, G. Schmid, H. Iio, M.R. Johnson and Y. Kishi, J. Am. Chem. Soc., 102, 7965 (1980); H. Iio, H. Nagoaka and Y. Kishi, J. Am. Chem. Soc., 102, 7967 (1980).

40. H. Nagoaka and Y. Kishi, Tetrahedron, 37, 3873 (1981).

41. S. Masamune, B. Imperiali and D.S. Garvey, J. Am. Chem. Soc., 104, 5528 (1982).

42. W.C. Still and J.C. Barrish, J. Am. Chem. Soc., 105, 2487 (1983).

43. M. Nakata, H. Takao, Y. Ijuyama, T. Sakai, K. Tatsuta and M. Kinoshita, Bull. Chem. Soc. Japan, 54, 1749 (1981).

44. M. Nakata, Y. Ikeyama, H. Takao and M. Kinoshita, Bull. Chem. Soc., Japan, 53, 3252 (1982).

45. S. Hanessian, J.-R. Pougny and I.K. Boessenkool, J. Am. Chem. Soc., 104, 6164 (1982); S. Hanessian, 28th IUPAC Congress, Vancouver, Canada, Aug. 16-21 (1981) OR88.

46. For pertinent references, see D.A. Evans, J.V. Nelson and T.R. Taber, Topics in Stereochemistry, 13, 1 (1982); C.H. Heathcock, in Comprehensive Carbanion Chemistry, vol.II, T. Durst and E. Buncel, eds., Elsevier, Amsterdam, (1981); T. Mukaiyama, Org. React. 28, 203 (1982); S. Masamune, and W. Choy, Aldrichimica Acta, 15, 47 (1982); C.H. Heathcock, Science, 214, 395 (1981), etc.

47. M. Kinoshita, K. Tatsuta and M. Nakata, J. Antibiotics, <u>31</u>, 630
    (1978); see also M. Nakata, K. Tatsuta and M. Kinoshita, Bull.
    Chem. Soc. Japan, <u>54</u>, 1743 (1981).

PART SEVEN

EXECUTION

A Role for the Computer?

CHAPTER SIXTEEN

VISUAL DIALOGUE, ARTIFICIAL INTELLIGENCE OR BOTH?

As is evident, the practice of a visual dialogue with the target, combined with retrosynthetic analysis are essential for the "discovery" and location of suitable 'chirons' for the synthesis of a large variety of natural products. The question may be asked if the visual recognition process cannot be accelerated or even perfected by relying on a more unbiased technique. The computer has entered the chemist's world, and it is rapidly gaining strides on all fronts. What then would be the prospects of generating programs that could probe the depths of a target structure, decoding functional interrelations, stereochemical features, etc. and correlating them to potential, readily available chiral starting materials? In other words, could we ask the computer to look at molecules through our eyes, and to do even better? The answer to these questions is in part provided by a program which we have developed in our laboratories[1,2], and have termed Computer-Assisted Precursor Recognition (CAPR).

Impressive advances have already been made in interfacing the computer with synthetic organic chemistry.[3,4] Since the first program written at Harvard in 1967, under the direction of E.J. Corey[5] (Organic Chemical Simulation of Synthesis, OCSS), several other programs have become available.[3,4] The current synthesis program at Harvard, which is called Logic and Heuristics Applied to Synthetic Analysis (LHASA),[6] is an interactive one that allows the user to use graphical images. This program relies on retrosynthetic analysis to arrive at readily available starting materials including those from natural sources, and synthetic routes are given. Several strategies can be used based on functional group matching or disconnection of "strategic" bonds. In general, the routes do not take into account considerations of absolute configuration. Another program called SECS (for Simulation and Evaluation of Chemical Synthesis),[7] and developed by W. Todd Wipke, allows the analysis of stereochemical features of molecules in graphic form. Several other programs developed in academic institutions as well as by private concerns, are also available.[3]

CAPR is different from other existing programs in that it is concerned only with "discovering" subunits in the target framework that can be structurally, functionally and stereochemically related to a readily available chiral precursor. Thus, bond breaking in the retrosynthetic sense is done with minimum perturbation of chiral centers in the target to provide the most appropriate carbon chain that contains the maximum overlap of chiral and functional information with the precursor. Other options are also open to the user who for example, may wish to locate all possible six-carbon precursors with one or more asymmetric centers. Alternatively a three-carbon precursor with one asymmetric center may be asked. The program then operates by probing all possible disconnections that lead to the above mentioned precursors.

As a first priority, the program will not break a bond between two asymmetric carbon atoms. The carbon framework will be scanned forwards and backwards, locating "interesting" carbon atoms (potentially asymmetric or containing substituents other than hydrogen). It will then generate carbon subunits (the minimum being three carbon atoms) with one to X asymmetric centers. It will usually break a bond at the first instance where two non-asymmetric (or"non-interesting") carbon atoms are found. The subunits are then related to appropriate precursors that contain the same number of asymmetric centers, but not necessarily the same type of functionality, since this can be modified chemically, en route to the target. The length and nature of the carbon framework is an important link between the target subunits and the precursor. The nature and the number of asymmetric centers can be adjusted depending on whether one uses what is available from the particular precursor, or adds on, or eliminates. The ultimate goal is to "match" a segment of the target structure with a 'chiron' derived from an appropriate precursor.

An important feature is the fact that provisions for convergence of absolute stereochemistry between the targets and the potential precursors can be built in the program. Thus, the human challenge of a visual dialogue may be replaced by the artificial intelligence of the computer. The program is further different from the existing ones in that it does not suggest chemical routes, a task which is left to the choice of the investigator. It will however provide bibliographic information for specific transformations based on an independnet program also developed in our laboratories.[8] Thus, once the possible chiral precursors are given, one can ask for bibliographic information with specific key words, or request references in which a given precursor was used as a starting material in a synthesis.

The present protocol involves the following steps:

1. Display target structure graphically, or by an equivalent linear notation.
2. Digest functional group interrelations, connectivity and stereochemical information found in the target.
3. Probe for chains with maximum, intermediate and minimum numbers of asymmetric carbon atoms in that order.
4. Disconnect C–C bonds with minimum perturbation of chiral centers. Find three-carbon (minimum) subunits as well as longer X-carbon subunits.
5. Probe for overlapping combinations.
6. Ask for output in each category of subunits (ex. 3-carbon, 4-carbon, etc.).
7. Ask for corresponding chiral precursor and its equivalent structures.
8. Ask for literature precedents pertaining to chemistry done with each precursor of interest.

A. A new linear notation for molecules.

We have developed a simple, new linear notation that easily describes a variety of organic molecules.[1,2,9] Unlike other notations,[10] we use a menu of symbols and letters that are familiar to the chemist and are easily transcribable using a normal keyboard. For a given molecule, the atoms (except hydrogens) are numbered in a sequence of choice. Normally, carbon atoms and intervening hetero atoms in a chain are numbered in sequence and substituents follow. This mode of numbering usually follows the accepted conventions of nomenclature. A carboxylic acid may therefore be numbered 1 for example. Stereochemistry in the absolute sense ($\underline{R}$) or ($\underline{S}$) is defined next, either

by direct insertion in brackets after the asymmetric carbon in question,
or by giving up/down/left/right coordinates and allowing the computer to
assign the absolute stereochemistry by "weighing" the four neighboring
atoms.[11] Connectivity is simply shown by parentheses and a dashed
number which is that of the remote (connecting atom). Single bonds are
shown as a dash, double bonds as an equal sign and triple bonds as an
asterisk, all following the atom to which they are attached. A
vertically slanted dash indicates isolated substituent atoms. The
linear notation for thienamycin for example, is shown below, adopting
the presently accepted numbering for the bicyclic ring system. The
other atoms are numbered arbitrarily.

C1-C2=C3-N4-C5[R](-1)-C6[S]-C7(-4)/C8[R](-6)-C9/C10(-3)=O11/
O12(-10)/S13(-2)-C14-C15-N16/O17(=7)/O18(-8)

Linear notation for thienamycin

REFERENCES

1. S. Hanessian, F. Major and S. Léger, Synthetic Design of
    Enantiomerically Pure Medicinal Agents – Aspects of Precursor
    Recognition by Visual and Computer–Assisted Perception, in New
    Methods in Drug Research, A. Makryannis, ed., Prous, Barcelona,
    Spain, in press.

2. S. Hanessian, F. Major and S. Léger, Symposium on Molecular
    Substructure Searching: New Applications, Abstracts 186th
    National Meeting Am. Chem. Soc., Washington, D.C., March 29, 1983.

3. For a recent account, see Chem. Eng. News, May 9, 1983, p.22.

4. For selected reviews and articles, see, K.K. Agarwal, D.L. Larsen
    and H.L. Gelenter, Computers in Chemistry, 2, 75 (1978);
    M. Bersohn and A. Esack, Chem. Rev., 76, 269 (1976); E.J. Corey,
    W.J. Howe and D.A. Pensak, J. Am. Chem. Soc., 96, 7724 (1974);
    J. Gasteiger and C. Jochum, Top Curr. Chem., 93 (1978);
    J.B. Hendrickson and E. Braun–Keller, J. Comp. Chem., 1, 323
    (1980), T.D. Salatin, D. McLaughlin and W.L. Jorgensen, J. Org.
    Chem., 46, 5284 (1981); W.T. Wipke, W.J. Howe, eds., Computer–
    Assisted Organic Synthesis , ACS Symposium Series, No.61, American
    Chemical Society (1977); P. Gund, Ann. Rep. Med. Chem., 12, 288
    (1977), K. Hensler, Science, 189, 609 (1975).

5. E.J. Corey and W.T. Wipke, Science, 166, 178 (1969).

6. E.J. Corey, Quart. Rev., 25, 455 (1971).

7. W.T. Wipke and T.M. Dyott, J. Am. Chem. Soc., 96, 4825 (1974).

8. S. Hanessian and F. Major, unpublished results.

9. S. Hanessian and S. Léger, unpublished results.

10. See for example, G. Palmer, Wiswesser Line–Formula Notation ,
    Chem. Brit., 6, 422 (1970) and references cited therein.

11. See for example, E.F. Meyer, Jr., J. Chem. Ed., 55, 780 (1978).

INDEX

An asterisk* indicates a chiral starting material. Page numbers may refer to the text or the schemes. To avoid redundancy with the schemes, the index is organized according to types of reactions such as : oxidation, etc.